Early Mammals

Supplements to the Journals of the Linnean Society of London

Forthcoming titles:

Biology and chemistry of the Umbelliferae
Symposium held in September 1970 in conjunction with the Phytochemical Society and to be published as a supplement to the **Botanical Journal**

Behavioural aspects of parasite transmission
Symposium arranged with the British Society for Parasitology and the British Section of the Society of Protozoologists held in July 1971 and published as a supplement to the **Zoological Journal**

Fern taxonomy
Symposium to be held in Spring 1972

The relationships of fishes
Symposium to be held in Summer 1972

Already published:

New research in plant anatomy
The full-texts of papers read and submitted for a Symposium arranged by the Society's Plant Anatomy Group and held in London September 1970 and published as Supplement 1 to the Botanical Journal of the Linnean Society Vol. 63 1970

Published quarterly:

Biological Journal of the Linnean Society
Botanical Journal of the Linnean Society
Zoological Journal of the Linnean Society

Early Mammals

Edited by D. M. Kermack
Department of Zoology, Imperial College
of Science and Technology, London

and K. A. Kermack
Department of Zoology,
University College, London

Supplement 1 to the Zoological Journal of the Linnean Society Volume 50 1971

Published for the Linnean Society of London by Academic Press

ACADEMIC PRESS INC. (LONDON) LIMITED
Berkeley Square House
Berkeley Square
London W1X 6BA

U.S. Edition published by
ACADEMIC PRESS INC.
111 Fifth Avenue
New York
New York 10003

Library of Congress Catalog Card No. 70 170761
ISBN: 0 12 404750 5

Made and printed in Great Britain by
William Clowes & Sons Limited, London, Colchester and Beccles

Foreword

'Early Mammals' is the second symposium arranged by the Linnean Society, the proceedings of which appear as a supplement to one of its three Journals.

In arranging it the Society was grateful to the Society of Antiquaries for allowing the symposium to be held in its Rooms in Burlington House as the Linnean Society's Rooms were being modernized and redecorated at that time.

The travel and subsistence grants received from the Royal Society and the British Council enabled workers from many different countries to contribute to the symposium and informally discuss their current work.

Further symposia have been arranged for the coming sessions, this type of meeting now forming a regular part of the Linnean Society's programme.

Doris M. Kermack
Editorial Secretary

July 1971

List of contributors

CLEMENS, WILLIAM A., JR. *Department of Palaeontology, University of California, Berkeley, California 94720, U.S.A.* (pp. 117 & 165)

CROMPTON, A. W. *Museum of Comparative Zoology, Harvard University, Cambridge 02138, Massachusetts, U.S.A.* (p. 65)

EVERY, R. G. *Department of Zoology, University of Canterbury, Christchurch, New Zealand* (p. 23)

FOX, RICHARD C. *Departments of Geology and Zoology, The University of Alberta, Edmonton 7, Alberta, Canada* (p. 145)

HOPSON, JAMES A. *Department of Anatomy, The University of Chicago, Chicago, Illinois 60637, U.S.A.* (p. 1)

KERMACK, K. A. *Department of Zoology, University College, Gower Street, London, W.C.1* (p. 103)

KREBS, BERNARD. *Lehrstuhl für Paläontologie, Freie Universität Berlin, 1 Berlin 33, Schwendenerstrasse 8, Berlin* (p. 89)

KIELAN-JAWOROWSKA, Z. *Zakład Paleozoologii, Polska Akademia Nauk, Warszawa 22, Al. Zwirki i. Wigury 93, Poland* (p. 103)

KÜHNE, W. G. *Lehrstuhl für Paläontologie, Freie Universität Berlin, 1 Berlin 33, Schwendenerstrasse 8, Berlin* (p. 23)

LEES, PATRICIA M. *Department of Zoology, University College, Gower Street, London, W.C.1* (p. 117)

MILLS, J. R. E. *Institute of Dental Surgery, Eastman Dental Hospital, Gray's Inn Road, London, E.C.1* (p. 29)

SIMPSON, GEORGE GAYLORD. *5151 East Holmes, Tucson, Arizona 85711, U.S.A.* (p. 181)

SLAUGHTER, BOB H. *Shuler Museum of Palaeontology, Southern Methodist University, Dallas, Texas 75222, U.S.A.* (p. 131)

Preface

Prior to the end of the Second World War almost all the discoveries of Mesozoic mammals known had been made in the 19th century. Professor G. G. Simpson studied this material and published a series of papers between 1925 and 1930. So thorough and detailed was his work and so sound were his conclusions that it was clear that no major advance could be expected until further material was discovered.

The necessary discovery of new material and the consequent resumption of active research work on Mesozoic mammals were initiated by Professor W. G. Kühne, who commenced his field work just before the war. The other and quite independent impulse came from the work in Yunnan by Father Oehler under the direction of Father Harold Rigney.

In 1960 a most successful symposium on the evolution of lower and non-specialized mammals was organized in Brussels by Professor G. Vandebroek under the auspices of the Koninklike Vlaamse Academie voor Wetenschappen, Letteren en Schone Kunsten van Belgie. It gave an opportunity for many of the workers on the early evolution of mammals and of the mammalian dentition to meet and to discuss their ideas together. The proceedings of the symposium were published in 1961 and form a valuable permanent record.

Now in 1970, ten years after the Brussels Symposium, it seemed appropriate for the Linnean Society to hold a symposium upon early mammals and to publish the papers as a volume. We hope that this volume will indicate the great advance in our knowledge since 1960. Professor Simpson, whose work is the foundation of all subsequent studies of Mesozoic mammals, Professor Kühne and Father Rigney happily were able to attend the symposium. Father Oehler regrettably was unable to be present. To these four men this volume is respectfully dedicated.

Doris M. Kermack
Kenneth A. Kermack
Editors

June 1970

Contents

Symposium on Early Mammals

held in the Rooms of the Society of Antiquaries, Burlington House, Piccadilly, London, W.1, by kind permission of the President and Council.

PROGRAMME

Tuesday, 23 June 1970

Introduction by Prof. A. J. E. Cave, President, Linnean Society:

The origin and early evolution of the mammalian dentition

Morning session:

Chairman: Dr F. R. Parrington, F.R.S.

James A. Hopson (Chicago): Postcanine replacement in the gomphodont cynodont *Diademodon.*

R. G. Every and W. G. Kühne (Berlin): Bimodal wear of mammalian teeth.

Afternoon session:

Chairman: Prof. P. M. Butler

J. R. E. Mills (London): The dentition of *Morganucodon.*

A. W. Crompton (Harvard): The origin of the tribosphenic molar.

Evening session:

Chairman: Dr K. A. Kermack

Bernard Krebs (Berlin): Evolution of the mandible and lower dentition in dryolestids.

Panel discussion: Prof. G. G. Simpson, Prof. G. Vandebroek and Dr F. R. Parrington, followed by general discussion.

Wednesday, 24 June 1970

Patterns of mammalian evolution

Morning session:

Chairman: Father Harold Rigney, S.V.D.

K. A. Kermack (London) and Z. Kielan-Jaworowska (Warsaw): Therian and non-therian mammals.

Bob H. Slaughter (Dallas): Mid-Cretaceous (Albian) therians of the Butler Farm local fauna, Texas.

Richard C. Fox (Edmonton): Marsupial mammals from the early Campanian Milk River Formation, Alberta, Canada.

Afternoon session:

Chairman: Prof. G. G. Simpson, F.M.R.S.

William A. Clemens, Jr. (Berkeley): Mammalian evolution in the Cretaceous.

Panel discussion: Prof. A. W. Crompton, Dr J. A. Hopson and Dr K. A. Kermack, followed by general discussion.

Concluding remarks: Prof. G. G. Simpson, F.M.R.S.

Evening session

G. Vandebroek (Louvain): Demonstration of teeth of Mesozoic mammals—illustrated by coloured stereoscopic transparencies.

Z. Kielan-Jaworowska (Warsaw): Fossil collecting in Mongolia (film).

Friday, 26 June—Sunday, 28 June 1970

Excursion to the Mesozoic mammal localities in England and Wales.

Postcanine replacement in the gomphodont cynodont *Diademodon*

JAMES A. HOPSON

Department of Anatomy, University of Chicago, Illinois, U.S.A.

An analysis of postcanine replacement in a series of young individuals (skull lengths: 50 mm to 170 mm approximately) of the Lower Triassic cynodont *Diademodon* (adult skull length: about 300 mm) indicates that earlier interpretations suggesting a limited (and somewhat mammalian) degree of replacement are incorrect. A large amount of replacement occurs during ontogeny; however, because of the requirements for precise occlusion, the pattern of replacement has been greatly modified from the typical reptilian pattern seen in the more primitive *Thrinaxodon*. Crompton's (1963) model of replacement in *Diademodon* appears to be essentially correct except that the anterior conical teeth form a single replacement wave, with each tooth erupting consecutively.

The smallest specimen has a fully differentiated dentition although the total number of cheek teeth (7) is half the adult number (about 14). Brink's (1957) suggestion that very young individuals had a feeble dentition unsuited for dealing with resistant foodstuffs and that advanced cynodonts might have nursed their young on maternal milk is not supported. A mammalian type of limited tooth replacement, probably associated with intensive maternal care and nursing, is not found in gomphodont cynodonts and probably evolved only in the mammals or their immediate cynodont ancestors.

CONTENTS

INTRODUCTION

The dentition of reptiles, indeed, of most lower vertebrates, typically undergoes constant replacement throughout life and at any given time replacement activity may be observed in several regions of the tooth row. Edmund (1960) has shown that eruption of new teeth and shedding of old occurs in an ordered succession of waves, or *Zahnreihen*, which pass from front to back along the jaw. In the functional dentition this underlying pattern creates a superficial phenomenon of back to front waves of replacement in alternating teeth (see Edmund, 1960, 1969, for discussion). This apparent 'alternating' pattern of replacement is easily seen in fossils and is clear evidence of active continuous replacement.

The dentition of mammals consists of two sets of teeth with only a single replacement occurring in a portion of the dentition. Edmund (1960: 169) has suggested that the mammalian condition can be derived from that of reptiles by reducing the total dentition to two *Zahnreihen*—one comprising the deciduous teeth plus the true molars and a second partial one comprising the permanent incisors, canines, and premolars.

The transition from the reptilian to the mammalian pattern of replacement is not clearly documented in the fossil record. Despite a relative paucity of material, an apparently mammalian pattern of limited replacement has been demonstrated in several Late Jurassic orders of mammals (Triconodonta: Simpson, 1928; Docodonta: Kühne, pers. comm.; Eupantotheria: Butler, 1939; Kühne, 1968). Szalay (1965) has demonstrated diphyodonty in Late Cretaceous and Paleocene multituberculates. On the basis of this evidence, Hopson & Crompton (1969) have suggested that a diphyodont pattern of replacement, or a very close approach to this pattern, characterized the common ancestor of Jurassic and later mammals. This ancestor, they suggest, was a morganucodontid triconodont. Independent studies of tooth replacement in Late Triassic morganucodonts by Dr F. R. Parrington (pers. comm.) and Dr J. R. E. Mills (1971) indicate that a mammalian pattern was indeed already established in the earliest known mammals. The published results of these studies will undoubtedly shed welcome light into this critical corner of mammalian history.

Tooth replacement studies of advanced mammal-like reptiles are relatively few and have not yet demonstrated that the mammalian pattern of dental succession was established within the Therapsida, notwithstanding statements to the contrary by Tarlo (1964; quoted by Miles & Poole, 1967, and Edmund, 1969) and Ziegler (1969). Crompton (1955*a*, 1962) has shown that bauriamorphs were still basically reptilian in mode of replacement, as was the ictidosaur *Diarthrognathus*. Primitive cynodonts of the family Galesauridae also retain a pattern of multiple replacements (Parrington, 1936; Crompton, 1963; Hopson, 1964). Only in the later part of the Early Triassic (in the *Cynognathus* Zone of South Africa) do cynodonts begin to show a modification of the primitive 'alternate' pattern of replacement. The gomphodont cynodonts and their descendants the tritylodontids appear to show a slower rate and perhaps a reduced total amount of replacement compared with earlier cynodonts (Broom, 1913; Crompton, 1955*b*, 1963; Kühne, 1956; Fourie, 1963; Hopson, 1964; Ziegler, 1969). Though not studied in comparable detail, the advanced carnivorous cynodonts (Cynognathidae, Chiniquodontidae) also appear to have slowed the rate and also limited the amount of replacement. More work is required in order to determine the exact patterns of replacement in all groups of advanced cynodonts. Until this is done, it will not be possible to relate cynodont patterns of dental succession to that already established in the earliest known mammals.

The purpose of this paper is to describe certain aspects of post-canine tooth replacement in the Early Triassic gomphodont cynodont *Diademodon* based on new data derived mainly from small juvenile specimens. After presenting my interpretation of the replacement pattern in this genus, I shall speculate on: (1) the significance of this pattern in the life history of *Diademodon;* and (2) its contribution to an understanding of the origin of the mammalian mode of replacement. Finally, I shall offer some tentative suggestions about the usefulness of tooth replacement patterns in determining the position of the boundary between the classes Reptilia and Mammalia.

PREVIOUS STUDIES OF TOOTH REPLACEMENT IN *DIADEMODON*

The pattern of tooth replacement in *Diademodon* has long interested students of the therapsid-mammal transition because this animal has the most mammal-like dentition of any therapsid. Unlike other cynodonts, *Diademodon* possesses a postcanine dentition which is clearly divisible into distinct 'premolariform' and 'molariform' tooth types (see Fig. 1). In other cynodonts the simple anterior teeth grade gradually into the complex posterior teeth (as in carnivorous forms) or else only complex teeth are present (as in

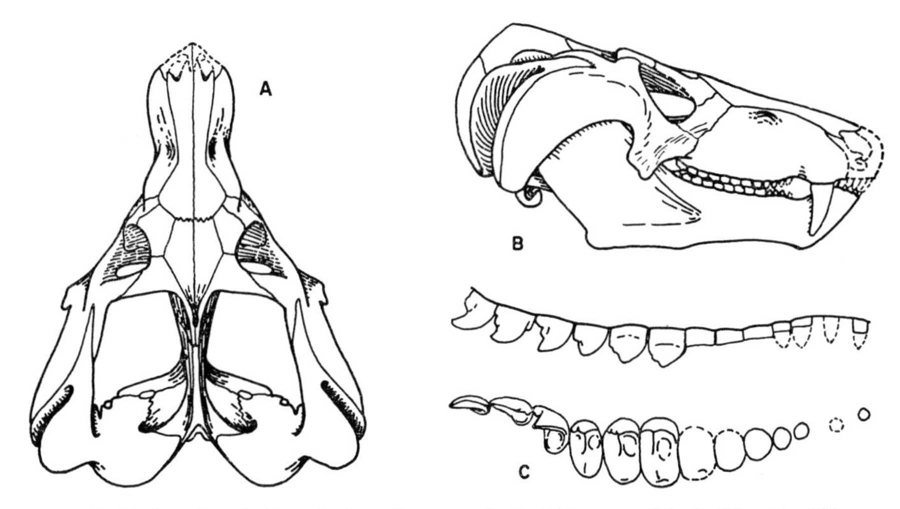

FIGURE 1. *Diademodon*. **A**. Dorsal view of mature skull of *D. grossarthi;* skull length: 313 mm. **B**. Lateral view of mature(?) skull of *D. rhodesianus;* skull length: 236 mm. **C**. Lateral and crown views of right upper-postcanine dentition of *Diademodon* sp. in Berlin Museum; length of tooth row: 70 mm. (**A** after Brink, 1963*b*; **B** after Brink, 1963*a*; **C** drawn from cast.)

all gomphodonts but *Diademodon*). Broom (1913) was the first to describe what appeared to be a mammalian replacement pattern in *Diademodon*. He noted replacement of incisors, canines, and simple conical 'premolars', but not of 'molars'. Brink (1955*a*, *b*, 1957, 1963*b*) has discussed tooth replacement in *Diademodon* and has consistently maintained the view that the amount of postcanine replacement was very limited, with no more than a single replacement anywhere in the cheek series. However, he believes it best not to distinguish premolars and molars in this genus.

Crompton (1955*b*) provided the first evidence of replacement of posterior (molariform) postcanine teeth. He described a maxillary fragment containing two longitudinally ovate foramina following the circular alveolus of a transversely expanded, or gomphodont, tooth (see Fig. 1). Above the first of these two foramina lay the unerupted crown of an apparently gomphodont-type tooth. Crompton interpreted the two ovate foramina as alveoli for lost sectorial teeth of the type which always terminate the tooth row in *Diademodon* (see Fig. 1). Brink (1955*a*, 1957), however, believed that these foramina might just as well be gubernacular canals as alveoli for sectorial teeth, and Ziegler (1969) has also expressed doubts that they represent alveoli. Crompton (1955*b*) also described a specimen showing replacement of the sixth postcanine, apparently of a gomphodont by a conical tooth although the nature of the replacing tooth has not been clearly established (see Crompton, 1955*b*: 661, fig. 15A, 1963: 516; Ziegler, 1969: 774, for different interpretations).

Fourie (1963) studied a large number of *Diademodon* mandibles and has stated that replacement and addition of teeth occurs at the posterior end of the postcanine row and loss occurs at the anterior end. Brink (1955*a*) had earlier noted that conical teeth are lost anteriorly; Fourie (1963) and Hopson (1964) independently suggested that this loss is related to canine replacement. Fourie's paper is a brief summary of his findings without a detailed presentation of his evidence, and the description and illustration of his model of the replacement process are not readily correlated with the observations made by Crompton (1955*b*). Crompton (1963) has combined the observations made by Fourie and earlier workers with observations of his own and has presented a model of the replacement sequence in the postcanine series of *Diademodon* which is the most complete and satisfactory interpretation of the observed data published so far. He was also the first to interpret the observed pattern of replacement in terms of the *Zahnreihe* theory of Edmund (1960). Though subject to a number of qualifications, to be discussed below, Crompton's model of the replacement process fits the available evidence better than any other yet proposed.

I (Hopson, 1964) discussed tooth replacement in *Diademodon* in a paper sent to press before the appearance in 1963 of Fourie's and Crompton's papers. Because I interpreted the sixth replacing tooth described by Crompton (1955*b*) as a gomphodont rather than a conical tooth, my hypothesis about postcanine succession in terms of successions of *Zahnreihen* was clearly incorrect, as I noted in an addendum written after I had seen Crompton's paper of 1963. Recently, Ziegler (1969) has published a theoretical analysis of tooth succession in *Diademodon* based solely on information derived from the literature. Ziegler's analysis is handicapped by the necessity of relying on conflicting and sometimes erroneous published descriptions for his data. His conclusion that *Diademodon* had an essentially mammalian replacement pattern is definitely incorrect, as I shall demonstrate.

MATERIALS AND METHODS

The following discussion of postcanine succession in *Diademodon* is based on a series of juvenile and subadult specimens ranging in skull length from about 50 to 170 mm. Presumed adult specimens range between 200 and 400 mm in skull length (Brink, 1963*b*). I have ranked the specimens in order of increasing size and, treating maxillary and mandibular dentitions separately, and have attempted to homologize the teeth from one individual to the next on the basis of similarity in tooth size. This is possible because the gomphodont teeth increase in size posteriad along the tooth row. Replacement teeth in the conical (premolariform) and sectorial series were exposed by means of careful dissection into the jaw. I have interpreted each dentition as a stage in an ontogenetic series, of which nine are illustrated diagrammatically in Fig. 4. Other specimens have been studied but are too poorly preserved or too incomplete to figure.

The shortcomings of this method and the tentative nature of some of the results must be emphasized. There is no certainty that only one species of *Diademodon* is represented in my material so the 'ontogenetic series' may include a mixture of several species. The taxonomy of this genus is extremely confused and a revision is long overdue. Furthermore, of the approximately 15 specimens available to me for study, only about half are sufficiently complete and well-preserved to allow accurate measurements to be made on the whole cheek dentition of either the upper or lower jaw. As a result, the total number of useful specimens is too small for a really detailed study. Of the homologies indicated in Fig. 4, I feel most confident of those made between the specimens in the 110–150 mm range and somewhat less so of those in the 50–75 mm range. Nevertheless, the available material has allowed me to demonstrate certain hitherto unobserved features of the replacement process in *Diademodon* and to settle several controversial points

The specimens used in this study are to be described in detail elsewhere (MS in prep.). In the following section, those specimens which contribute important new information are briefly described.

Institutional names referred to in this paper are abbreviated as follows:

A.M.N.H., American Museum of Natural History, New York.
B.M., British Museum (Natural History), London.
B.P.I., Bernard Price Institute, University of the Witwatersrand, Johannesburg.
Munich, Institut für Paläontologie und historische Geologie, Munich.
U.C.M.P., Museum of Paleontology, University of California, Berkeley.

DESCRIPTION OF SPECIMENS

Diademodon sp.
(Type specimen of *Sysphinctostoma smithi* Broili & Schröder)

The smallest specimen known to me which is undoubtedly referrable to the genus *Diademodon* is a tiny skull (53 ± 3 mm long) described as *Sysphinctostoma smithi* by Broili & Schröder (1936). This specimen is being redescribed elsewhere, but the main features of its dentition are summarized here (see Fig. 2).

This juvenile skull contains seven postcanine teeth. The most obvious differences between the cheek dentition of a mature individual of *Diademodon* and that of the tiny skull of '*S. smithi*' are the lower number and proportionally larger size of the teeth in the latter.

In the lower jaw there are two anterior conical teeth. The right upper jaw contains a single conical tooth which appears to be incompletely erupted. The second postcanine is damaged but is sufficiently preserved to indicate that it is part of the gomphodont series. On the left side, the first conical tooth is completely erupted. The second tooth position consists of an alveolus from which the functional tooth has been lost. Preparation into this alveolus exposed what appears to be the tip of an erupted tooth crown. I

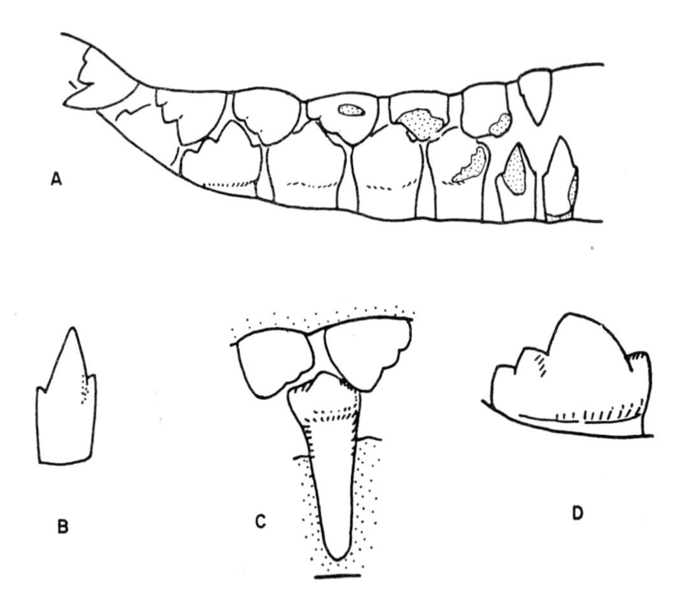

FIGURE 2. *Diademodon* sp. Postcanine teeth of the juvenile skull which was made the type specimen of '*Sysphinctostoma smithi*' Broili & Schröder. **A.** Lateral view of the right postcanine dentition, partially restored from the left side; fine stippling indicates damaged areas; *c.* × 4. **B.** Internal view of left lower first postcanine; *c.* × 7·5. **C.** Lateral view of left upper-third and fourth postcanines and left lower-fourth postcanine to show occlusal relations; *c.* × 3·75. **D.** Lateral view of right lower sixth postcanine; *c.* × 7·5.

interpret this as the conical replacement of a gomphodont predecessor such as is still in place on the right side. According to this intepretation, the left maxillary dentition is slightly in advance of the right in the replacement process.

The third to seventh postcanine teeth in both upper and lower jaws are multicusped. Except for the seventh tooth, which is sectorial, it is not possible to determine whether the teeth posterior to the fourth are fully gomphodont, semi-gomphodont, or sectorial because only their lateral surfaces are exposed. New preparation has revealed that the fourth lower postcanine is approximately oval in crown view with its transverse width being about half its length. (In increasingly larger individuals the crowns of lower gomphodont teeth become more circular in outline.)

As indicated in Fig. 2, the morphology of the postcanine teeth of '*S. smithi*' compares closely with that in larger specimens of *Diademodon* (see Fig. 1**C**).

As Broili & Schröder (1936) pointed out, the lower 'molariform' teeth of '*S. smithi*' show a relatively large amount of exposed root, with the amount of exposure decreasing in successively more posterior teeth (Fig. 2). This condition is also seen in adult specimens of *Diademodon* and in tritylodontids (Kuhne, 1956). In these forms the gomphodont teeth erupt in sequence from front to back (as further indicated by progressive increase in wear anteriorly), and this is also clearly the case in the extremely immature specimen of *Diademodon* described here.

The importance of this juvenile *Diademodon* specimen is that it possesses a fully differentiated dentition which shows sequential eruption in the gomphodont series and replacement of anterior gomphodont teeth by conical teeth. This indicates that the characteristic *Diademodon* replacement pattern commenced very early in the ontogeny of this cynodont.

Diademodon sp. (A.M.N.H. 5519)

Several larger specimens have contributed important details on the mode of replacement in the subgomphodont to sectorial series of teeth at the rear of the postcanine row. The most important of these is illustrated in Fig. 3. This specimen is a partial skull and two mandibular fragments from the same individual. The lower jaws were figured by Broom (1913) as belonging with the type skull (A.M.N.H. 5518) of *Diademodon platyrhinus* which was later made the type species of the genus *Cyclogomphodon*. Both specimens are from Winaarsbaken, Cape Province. Recent preparation and restudy of both skulls and the lower jaw fragments have shown that the latter belong with A.M.N.H. 5519. These mandibular fragments are particularly significant because it was mainly on them that Broom based his conclusion that *Diademodon* had a mammalian type of tooth replacement.

The right mandible contains alveoli for three conical teeth, the third of which contains an erupting conical tooth (Fig. 3 **A**). Broom interpreted the empty fourth alveolus as being for a conical tooth, but from its shape it is more likely to have contained a gomphodont tooth. This inference is supported by the discovery of a worn mandibular tooth of the proper size lying against the palate of the associated skull. Broom assumed that an erupting conical tooth in the third tooth position was in the process of replacing another conical tooth. Crompton (1963, fig. 17) and Ziegler (1959, fig. 2) have incorporated Broom's interpretation of conical tooth replacement into their respective models of *Diademodon* tooth replacement. However, the evidence of this and all other specimens available to me indicates that conical teeth replace only gomphodont teeth. This will be discussed further below.

Following the empty fourth alveolus, which almost certainly contained a gomphodont tooth in life, are four gomphodont teeth increasing in size posteriorly. The ninth postcanine is a subgomphodont tooth which is not fully erupted. Separated from it by a short gap is a longitudinally ovate alveolus containing the root of a small sectorial tooth. Broom (1913) believed the last alveolus contained an unerupted tooth, but preparation into the alveolus has revealed a damaged root from which the crown was lost prior to fossilization.

The most important feature of this specimen is the presence of two foramina on the medial surface of the jaw, a large circular foramen about 3 mm in diameter lying internal to the anterior half of the just-described root and a small slitlike foramen (or pit) lying internal to the posterior margin of the root. The larger opening leads into a crypt within the jaw containing the crown of an incompletely formed replacement tooth. The crown is very thin-walled and partially collapsed but sufficiently preserved to suggest that an internal shelf may have been present in the fully-formed tooth. (A definite internal shelf is present on the replacement tooth in the corresponding position in the associated upper jaw.) The small posterior opening (or pit) lies at the bottom of a shallow oval depression about 1 mm in diameter. The slit probably marks the place of entry of the dental lamina into the jaw. Though not excavated, I believe it marks the site

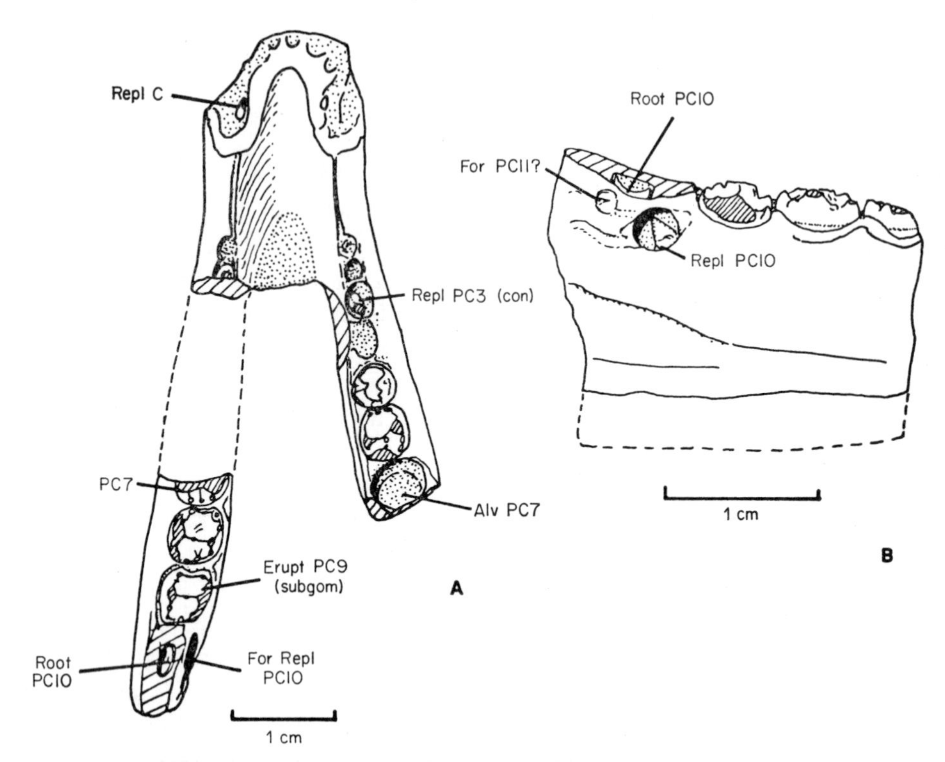

FIGURE 3. *Diademodon* sp. Partial lower jaw of immature specimen (A. M. N. H. 5519) showing replacement of both canines, right third postcanine and left tenth postcanine: **A**, dorsal view; **B**, internal view of posterior part of left horizontal ramus.

Broken or ground surfaces are represented by diagonal lines; matrix is represented by stipple; Alv, alveolus; C, canine; con, conical tooth; Erupt PC9, erupting ninth postcanine; For Repl PC10, foramen opening into crypt for replacing tenth postcanine; For PC11 ?, small foramen or pit interpreted as marking the site of a developing eleventh postcanine; PC, postcanine; Repl, replacement tooth; Root PC10, damaged root of tenth postcanine; subgom, subgomphodont tooth.

of a developing sectorial tooth. (An incompletely formed sectorial tooth was exposed in the corresponding position in the upper jaw.)

The interpretation of this dentition in terms of replacement waves is presented in the following section. For the present, I shall merely point out that it clearly demonstrates replacement in the posterior part of the tooth row and therefore disproves Ziegler's (1969) contention that replacement did not occur in this region.

INTERPRETATION OF THE REPLACEMENT SEQUENCE

In Fig. 4, the data from a series of nine *Diademodon* dentitions are summarized diagrammatically. The scale to the left represents the approximate length of the skull to which the dentition figured at that vertical level belongs. Upper and lower dentitions are distinguished. Where a tooth is illustrated with a broken outline its crown morphology is not known, although the presence of a tooth at that position is certain. Where a tooth has a broken outline and contains a question mark, its presence is not certainly known but it is assumed to have been present in the undamaged jaw.

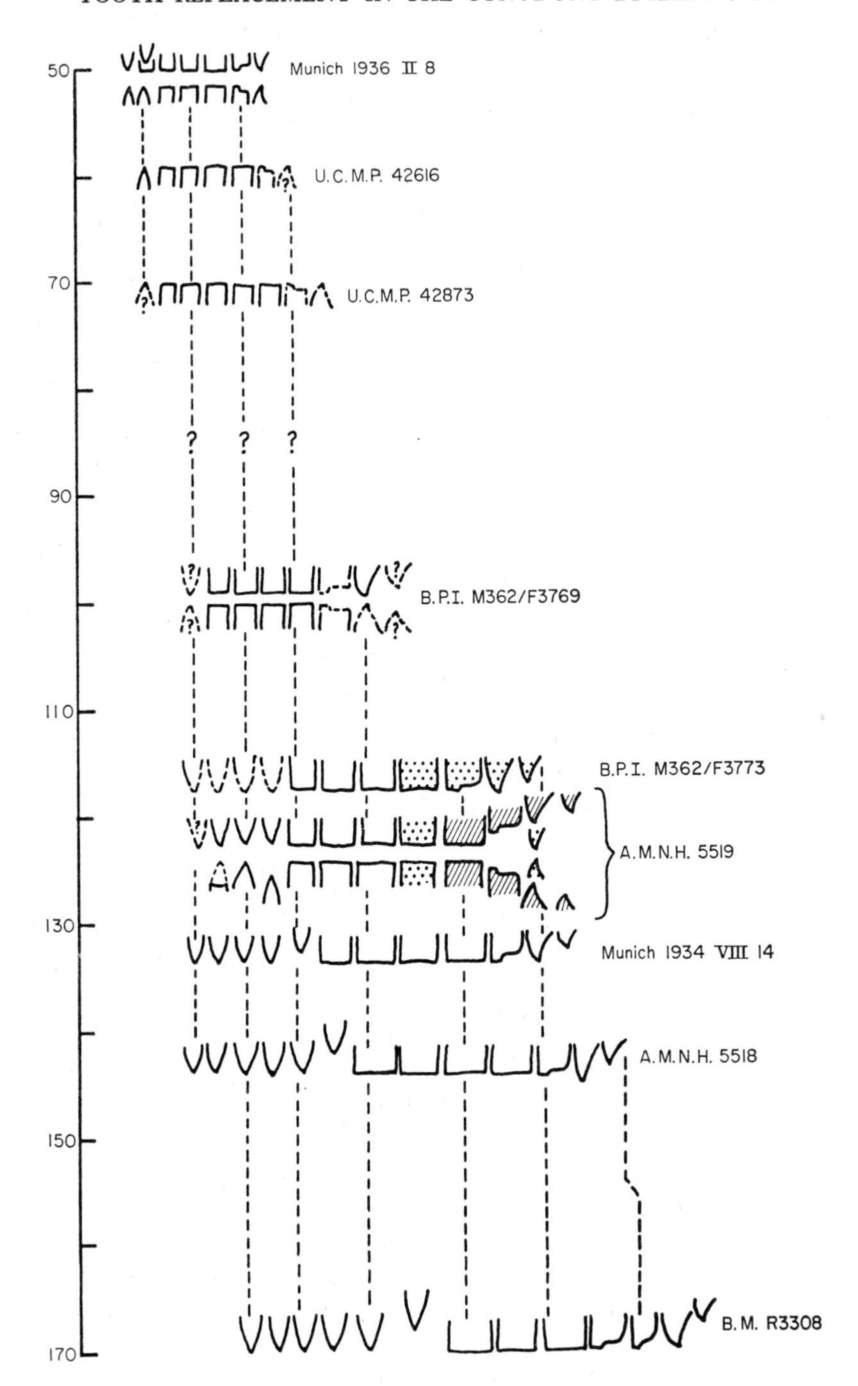

FIGURE 4. Diagrammatic representation of nine *Diademodon* dentitions. Institutions and catalogue numbers are indicated at the right of each specimen; scale at the left represents the approximate length of the skull containing the dentition at that vertical level. Broken lines connect teeth at inferred homologous positions. Members of one replacement wave in B.P.I. M362/F3773 are indicated by stipple, as are unreplaced members of the same wave in A.M.N.H. 5519. Members of subsequent replacement wave in A.M.N.H. 5519 are indicated by diagonal lines. See text for further discussion. Abbreviations of institutional names explained on p. 5.

As an example to illustrate the meaning of the symbols, the most differentiated dentition (B.M. R3308) consists of six conical teeth, the last of which is erupting, three gomphodont teeth, two subgomphodont teeth of which the first is more molariform than the second, a large sectorial tooth and an erupting small sectorial tooth. In all smaller specimens there is only one subgomphodont tooth which is uniformly represented by the symbol for the more molariform tooth in the largest specimen.

Broken lines connect tooth positions which on the basis of anteroposterior length measurements on individual teeth are considered to be homologous. Because the anterior gomphodont teeth of one specimen are older and more heavily worn than the more posterior and younger teeth in smaller specimens, a certain amount of reduction in tooth length has had to be taken into consideration when homologizing teeth. This method of identifying homologous teeth is admittedly subject to error in the absence of a large sample of specimens, so that I am not completely confident of the homologies I have figured in the specimens below 110 mm and above 150 mm.

The interpretation of replacement in A.M.N.H. 5519, as diagrammed in Fig. 4, will be compared with the information presented in Fig. 3 and the above description. Only the posterior teeth will be discussed. In the lower jaw, the last gomphodont tooth is followed by a subgomphodont tooth which is not fully erupted. The corresponding subgomphodont tooth in the maxilla is missing, but it was probably incompletely erupted at the time of death. Therefore, this tooth in both jaws was younger than the functional, or recently shed, sectorial tooth which lay behind it. The sectorial tooth, then, belonged to a replacement wave which preceded that to which the subgomphodont tooth belongs. If one looks at the next younger specimen (B.P.I. M362/F3773) in Fig. 4, one sees that the last four teeth are members of this earlier wave. In A.M.N.H. 5519, only the fully gomphodont and the small sectorial members of the earlier wave remain, and the latter tooth is shortly due to be lost. The two unerupted posterior teeth in this specimen are the last two members of the next wave, of which the last gomphodont tooth and the erupting subgomphodont tooth are the first two members. To clearly identify these two replacement waves, the members of the earlier wave are marked with a stipple pattern and those of the later wave with diagonal lines.

The interpretation of the pattern of replacement in the posterior teeth of *Diademodon* presented here is clearly the same as that presented by Crompton (1963), although in the younger individuals which I have studied the number of teeth in the replacement wave is less than that illustrated by Crompton (1963, fig. 17).

The main conclusions concerning tooth replacement in the growth series of *Diademodon* are summarized below. Additional observations from the literature are included where pertinent.

(1) There is a fully differentiated postcanine dentition containing the three main morphological types (i.e. conical, gomphodont, sectorial) in the smallest available juvenile specimen (type of '*Sysphinctostoma smithi*').

(2) The number of teeth of each morphological type is fewer in small juveniles than in adults and tends to increase with age. However, in the intermediate size range, the number of teeth of conical and gomphodont type may be greater than the number characteristic of adults because of variations in the time of: (a) replacement of one tooth type by another; and (b) shedding (and obliteration of roots and alveoli) of

anterior conical teeth. The number of conical teeth apparently becomes stabilized at four in adults (Brink, 1955*a*).

(3) Conical teeth erupt in sequence from front to back and replace anterior gomphodont teeth. There is no evidence of the replacement of conical teeth by other conical teeth (as thought by Broom, 1913, and Crompton, 1963).

(4) A replacement wave, or *Zahnreihe*, in the posterior part of the tooth row consists of a gomphodont tooth followed by a series of teeth grading from subgomphodont to sectorial. The number of teeth in the subgomphodont-to-sectorial series increases from three (perhaps fewer in very small individuals) to four in adults, but may be as many as five in exceptionally large individuals (Brink, 1955*a*, 1963*b*).

(5) Each replacement wave in the posterior series appears to begin one tooth position to the rear of the preceding wave; thus a new gomphodont tooth replaces a subgomphodont tooth and so on down the row, with a new sectorial tooth being added in a new tooth position at the posterior end of the series. This is exactly as described by Crompton (1963).

(6) It appears that all teeth in the posterior replacement wave become functional before the first tooth of the succeeding wave erupts. Because so few examples of an erupting tooth in the posterior series are known, it is probable that the time interval between the shedding of the crown of an old tooth and the eruption of its successor is relatively very short compared with the length of time a tooth spends developing within the jaw. This would partially explain why replacing teeth are rarely seen in the process of erupting, in spite of the large number of specimens which have been described.

DISCUSSION

Fourie (1963), Crompton (1963), and Ziegler (1969) have recently presented interpretations of the pattern of postcanine succession in *Diademodon*. Ziegler believed that replacement was limited to a single partial replacement comparable to that in mammals. The data presented here demonstrates this to be incorrect. Fourie described replacement at the posterior end of the tooth row and also noted that loss of teeth at the anterior end of the tooth row occurs. However, Fourie appears to have had only rather mature individuals at his disposal for his smallest specimen had fourteen cheek teeth. Crompton summarized earlier observations (including Fourie's) and, with additional observations of his own, attempted to account for the available information on postcanine succession in *Diademodon* in a theoretical model. As noted above, Crompton's interpretation of replacement in the posterior teeth is supported by my observations. Our only point of difference is in the replacement of the conical teeth. This, however, has an important bearing on our respective interpretations of replacement in terms of *Zahnreihen*.

In the figure which attempts to illustrate the main features of his model, Crompton (1963, fig. 17, reproduced here as Fig. 5 **A**) shows four conical teeth, seven gomphodont teeth, and a series of four posterior teeth which he terms 'intermediate gomphodont', 'intermediate sectorial', 'large sectorial', and 'small sectorial'. Replacement is indicated only in the anterior and posterior regions of the tooth row. In a single replacement

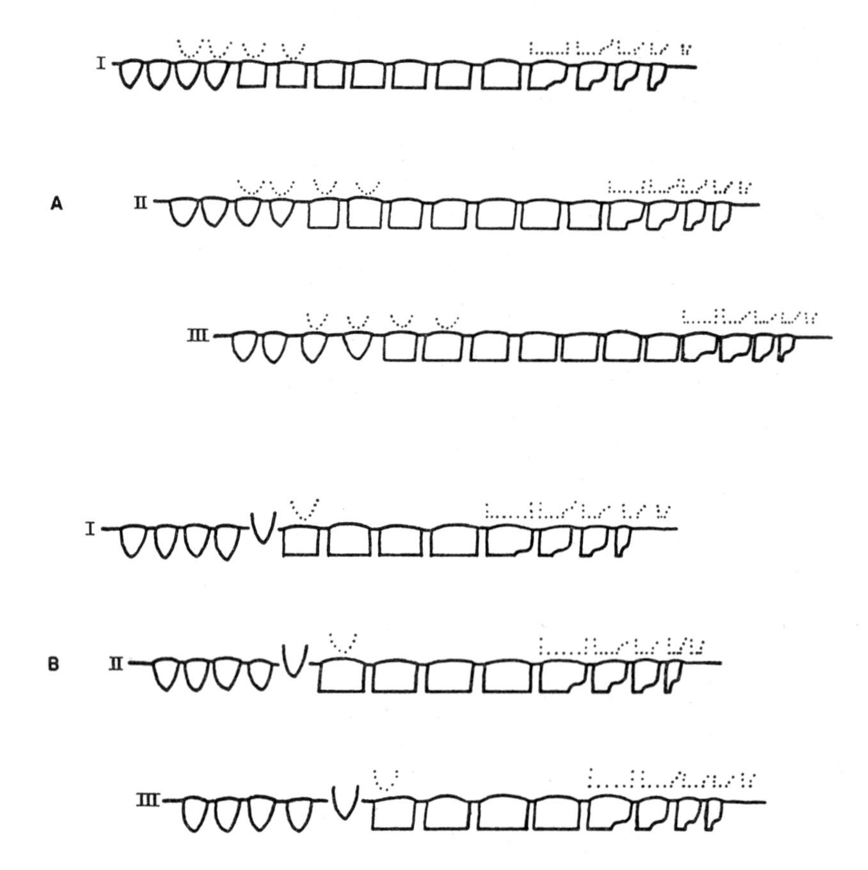

FIGURE 5. **A.** Interpretation of postcanine tooth replacement in *Diademodon* presented by Crompton (1963). **B.** Interpretation of postcanine tooth replacement in *Diademodon* presented in this paper. See text for explanation.

sequence at the anterior end, the first two conical teeth are shed without being replaced, the third and fourth conical teeth are replaced by new conical teeth, and the first two gomphodont teeth are replaced by conical teeth. At the posterior end of the tooth row, four teeth grading from 'intermediate gomphodont' to 'small sectorial' are replaced by a sequence consisting of a fully gomphodont tooth, an intermediate gomphodont tooth, an intermediate sectorial, and a large sectorial. A new small sectorial tooth is added at a new tooth position at the posterior end of the row. At a somewhat later time, indicated by stage II in his figure (see Fig. 5**A**), the same replacement sequence is repeated, and again at stage III, and so on through an unknown number of further stages. In this way, the whole tooth row migrates progressively further posteriorly in the jaw. At the same time, increasingly larger gomphodont teeth are added to the tooth row as the whole animal increases in size, and the amount of wear on the middle series of gomphodont teeth always decreases from front to back in an unbroken sequence.

The data obtained in the present study of young *Diademodon* specimens support Crompton's model in every respect but the manner of replacement of anterior conical teeth. In Fig. 5**B**, I have modified Crompton's diagram to show my conception of postcanine replacement in an adult individual which is not increasing the number of teeth in its dentition. The posterior part of the dentition is identical to that in Crompton's diagram. The anterior part, however, is quite different. The conical teeth are not replaced by new conical teeth in multiple waves as in Fig. 5**A**, but rather there is a *single* posteriorly moving wave of conical replacing teeth which replaces consecutive gomphodont teeth. Anterior conical teeth are shed consecutively so that the total number of conicals remains approximately constant.

Crompton's model attempts to account for the main features of postcanine succession only in mature individuals. Crompton speculates on the nature of the growth stages leading to the fully differentiated postcanine dentition of adults. He notes that in the youngest specimen (the size of which is not stated) described by Fourie (1963), the postcanine row is fully differentiated and shows no evidence that teeth were lost anteriorly. From this evidence he suggests that *Diademodon* probably 'commenced life with only a few teeth' and that 'the posterior region of the row, i.e. the subgomphodont to sectorial series was only added after a full complement of gomphodont teeth had erupted' (p. 517). Because Fourie's specimen does not appear to have lost anterior teeth, Crompton considers the possibility that 'replacement only commenced after a complete postcanine dentition had erupted'.

The evidence put forward in this study indicates that replacement was occurring in the youngest *Diademodon* specimens known, animals with skulls probably one-fifth or less the length of a full adult skull. As discussed earlier, the number of teeth in the postcanine series increased as the animal grew because the rate at which teeth were added to the posterior end of the row exceeded the rate at which they were lost at the anterior end. Because even the youngest known individuals (skull length: approximately 50 mm) replaced their teeth in the same manner as adults, the total amount of replacement which occurred in the lifetime of a single animal must have been very great, as is usually the case in reptiles. However, *Diademodon* differed from typical reptiles in the *pattern* of its replacement.

Most reptiles, including the ancestors of *Diademodon*, replace all teeth many times during their lifetime. At any one time newly-erupted teeth alternate with mature or old teeth and this pattern of apparent alternation characterizes dentitions undergoing active replacement in all parts of the tooth row. Edmund (1960) has developed a theoretical model in which the replacement pattern of reptiles is dependent on the spacing between a series of discrete impulses which move caudad along the dental lamina, initiating tooth development at each prospective tooth site. The row of teeth formed by a single impulse is designated a *Zahnreihe* by Edmund. The functional dentition of most reptiles consists at any one time of the products of many impulses, therefore it consists of portions of many *Zahnreihen*. In *Thrinaxodon*, for example, the seven teeth in the postcanine series of a mature individual are members of five *Zahnreihen*.

Crompton (1963: 518) has suggested that Edmund's *Zahnreihen* theory might be applied to postcanine replacement in *Diademodon* by supposing that the anterior and posterior replacing teeth (such as those at tooth positions 3 to 6 and 12 to 16 in Fig. 5**AI**) 'may constitute a . . . *Zahnreihe* of which the central members have been suppressed'. Thus, after the eruption of the initial sequence of postcanines, which Crompton suggested might have been formed by a continuous *Zahnreihe* I, subsequent waves of replacement occurred only in the anterior and posterior parts of the series as a result of the activity of discontinuous *Zahnreihen* II, III, IV, etc.

I believe the *Zahnreihen* theory can be used to interpret the pattern of postcanine replacement in *Diademodon* but my interpretation differs somewhat from that of Crompton. I consider each replacement sequence in the posterior teeth, i.e. one gomphodont tooth plus the subgomphodont to small sectorial series, to constitute a single complete *Zahnreihe* restricted entirely to the rear part of the jaw. The replacement of all conical teeth is, in my interpretation, the result of activity of one *Zahnreihe* and not of multiple *Zahnreihen* as suggested by Crompton. As indicated in Fig. 5**B**, all conical teeth develop in sequence down the jaw and are therefore most easily interpreted as products of a single *Zahnreihe* migrating slowly posteriad. This greatly simplifies the process of anterior postcanine replacement over what is suggested by Crompton. Furthermore, by assuming greater independence between the *Zahnreihe* activity in the anterior postcanines and that in the posterior postcanines one can more easily explain the rather large amount of variation in the number of teeth of different morphological types seen in the juvenile series (Fig. 4). Were the timing of replacement of anterior gomphodont teeth by conical teeth more closely correlated with replacement in the posterior series, the number of gomphodont teeth would not be expected to fluctuate as greatly as is indicated in the growth series described in this paper.

In Fig. 6 I have diagrammed my concept of the replacement process in *Diademodon* in terms of *Zahnreihen*. Each sequence at the posterior end of the tooth row, from the last gomphodont tooth back, constitutes a single complete *Zahnreihe*. Each new *Zahnreihe* begins one tooth position behind the preceding one so that the anterior-most tooth of the preceding *Zahnreihe* is not replaced and a new tooth is added to the posterior end of the tooth row. Each of the more anterior gomphodont teeth is the sole remnant of a *Zahnreihe* of which all of the other members have been replaced. The spacing of the impulses initiating development of successive *Zahnreihen* appears to be about equal to the length of the complete *Zahnreihe* because all members of a *Zahnreihe* erupt before replacement by the first member of the succeeding *Zahnreihe* takes place. Therefore spacing of *Zahnreihen* in Fig. 6 is at least four tooth lengths. In *Thrinaxodon* this spacing is 2·5 tooth lengths (Crompton, 1963).

At the anterior end of the tooth row in Fig. 6, a single *Zahnreihe* of conical teeth moves slowly posteriad, replacing gomphodont teeth. Since each gomphodont tooth represents a *Zahnreihe*, each tooth position occupied by a conical tooth can also be counted as representing a *Zahnreihe*. In addition, since conical teeth are shed at the anterior end of the tooth row, the evidence for an unknown number of *Zahnreihen* is lost as well.

The pattern of *Zahnreihe* activity in the posterior segment of the tooth row of *Diademodon* resembles the pattern described in *Thrinaxodon* by Crompton (1963: 511) in which 'each new impulse commences one tooth position behind the tooth position where the previous impulse commenced.' In *Thrinaxodon* the non-replacing anterior postcanines are shed relatively rapidly so the total number of cheek teeth does not continually increase with each wave of replacement. In *Diademodon* and other gomphodont cynodonts the non-replacing teeth become functionally the most important part of the dentition and they take up the major portion of the tooth row.

It is tempting to speculate on the origin of the gomphodont replacement pattern from that of a galesaurid such as *Thrinaxodon* as having occurred by gradually increasing the time that the non-replacing anterior cheek teeth were retained. As more anterior teeth were retained, the actively replacing part of the dentition became progressively restricted to the posterior part of the tooth row. This active portion would become progressively shorter, spanning fewer and fewer teeth, for those posterior teeth were becoming functionally less important than the more anterior non-replacing gomphodont teeth. In more advanced gomphodonts the replacing sectorial part of the tooth row becomes progressively shorter and in advanced traversodontids and tritylodontids it is

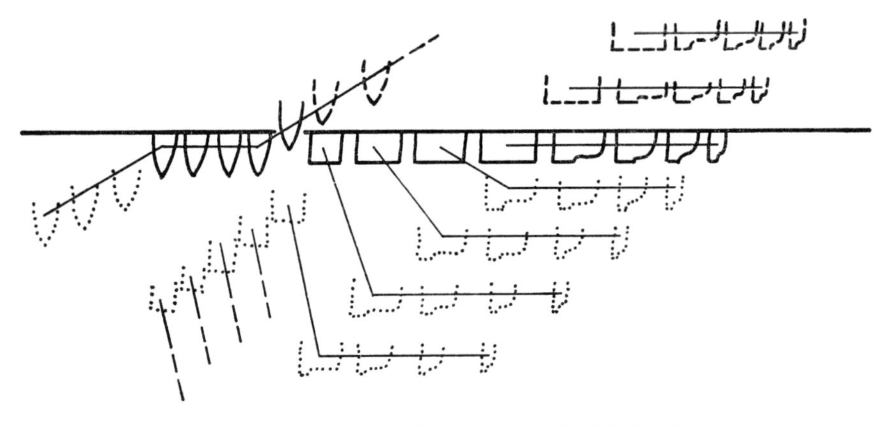

FIGURE 6. Interpretation of postcanine tooth replacement in *Diademodon* in terms of a large series of replacement waves or *Zahnreihen*. Above the functional dentition (in solid lines) are unerupted teeth (in broken lines); below the functional dentition are teeth which have been shed (in dotted lines). Lines connect teeth which are members of the same replacement wave. See text for further explanation.

eliminated completely. In these forms gomphodont teeth erupt directly and move into the tooth row without having had subgomphodont or sectorial predecessors. This picture is complicated in *Diademodon* by the presence of the conical series of teeth at the anterior end of the postcanine row. This series corresponds to a second *Zahnreihe* which during early ontogeny followed upon the first *Zahnreihe*, that which formed the first gomphodont teeth, in the normal fashion seen in *Thrinaxodon*. However, no third *Zahnreihe* succeeded it from the anterior end of the tooth row. Instead an entirely new series of multiple *Zahnreihen* began toward the posterior end of the tooth row and took over the function of producing new gomphodont teeth plus the subgomphodont to sectorial series. How this pattern of posterior *Zahnreihen* in *Diademodon* actually relates phylogenetically to the pattern of postcanine *Zahnreihen* in *Thrinaxodon* cannot be determined at present.

Origin of the mammalian pattern of limited tooth replacement

A shift from the 'alternating' pattern of continuous tooth replacement seen in *Thrinaxodon* and other primitive cynodonts to a pattern of front-to-back 'sequential' eruption occurred independently in the ancestry of gomphodont cynodonts and mammals. Tarlo (1964) has suggested that the lack of alternate succession in gomphodont cynodonts was due to a reduction of the total dentition to two *Zahnreihen*, as in mammals. In previous sections of this paper, I have shown that postcanine succession in *Diademodon* was not mammalian, either in the amount or in the pattern of replacement. *Diademodon* (and the more advanced gomphodonts) indicate that sequential eruption may evolve in animals with a relatively large amount of replacement (though probably less than occurred in their ancestor with 'alternating' replacement). The evidence provided by the gomphodonts indicates that the origin of sequential eruption was related primarily to the need for precise occlusion in complex cheek teeth. Though there is no direct evidence supporting the idea, I believe that this was also the most important factor in the origin of sequential eruption in mammals. Mammalian diphyodonty probably evolved at a later stage and was probably more directly related to the origin of mammalian patterns of growth and maternal care of the young. These factors were probably not at all related to the origin of sequential eruption in gomphodonts, as I shall discuss below.

As noted, but not elaborated on, by Crompton & Jenkins (1968: 445, footnote), complex postcanine occlusion would not have been possible in either gomphodonts or mammals if their teeth underwent frequent 'alternate' replacement. This is apparent for a number of reasons, only a few of which will be noted here. First, upper teeth typically lie opposite the contact between two lower teeth so that teeth in one jaw occlude with parts of two teeth in the opposite jaw. In order to maintain precise occlusal relations, adjacent teeth in the same jaw must maintain close contact with each other and gaps must not develop between them. Were new, larger replacement teeth to be continually taking the places of shed smaller teeth, constant shifting of teeth would be necessary in order to create room for the larger replacements. However, this would cause gaps to form between adjacent teeth and thereby prevent the maintenance of precise occlusion (for example, see *Bauria;* Crompton, 1962). Second, the strong masticatory pressures which are created between teeth with shearing and grinding occlusion require a stronger attachment to the jaw than is necessary in teeth which merely puncture and hold. Gomphodonts and mammals both possess a type of thecodont implantation which provides strong but resilient anchoring of the teeth. Thecodonty also confers additional advantages, to be discussed below, which made the evolution of precise occlusion possible in the first place. However, none of the advantages of mammalian thecodonty can be fully exploited unless replacement is reduced to a minimum. Thecodonty, or 'gomphosis', is the attachment of a rooted tooth to the wall of its alveolus by means of a periodontal ligament fixed at one end to the bone lining the socket and at the other end to the cement surrounding the root (Melcher & Eastoe, 1969). This type of tooth attachment is important in mammals for:

(1) transmitting strong masticatory pressures from the teeth to the bone of the jaw;

(2) gauging masticatory pressures by means of the proprioceptive innervation of the periodontium;

(3) development of proper occlusion by adjusting the erupting teeth to the proper height and position;

(4) maintaining occlusion and a closed tooth row despite coronal and interproximal wear by causing further eruption and longitudinal or transverse drift (see Noble, 1969).

It is clear from observation that points (1), (3), and (4) apply to gomphodont cynodonts as well as to mammals, and very likely that point (2) does also.

Implantation in the primitive cynodont *Thrinaxodon* was not truly thecodont for, though the rooted teeth are set in alveoli, they are anchored in the jaw by a ring of attachment bone around the neck of the tooth (Crompton, 1963). This mode of attachment was undoubtedly sufficient for *Thrinaxodon* because its cheek teeth did not occlude. They were adapted for holding and puncturing prey and therefore were not subjected to very great masticatory pressures. Furthermore, because the teeth did not occlude, post-eruptional movement of teeth (necessary for maintaining occlusion) was not necessary.

In *Diademodon* and other gomphodont cynodonts, and in the earliest mammals, the molariform teeth possessed long roots which provided large areas for anchorage by means of the periodontal ligament to the walls of the alveoli. Gomphodont teeth were used for grinding and shearing resistant (mainly vegetable) food by means of fore-aft movements of the jaws. In early mammals the teeth were used for cutting up animal food by means of primarily orthal but also slightly transverse movements of the jaws. In both cases, the pressures transmitted from the teeth to the walls of the alveoli were not only vertical (in the direction of the long axis of the root) but also horizontal (perpendicular to the long axis of the root). The periodontal ligament was an effective means for transmitting these stresses over the length of the socket rather than concentrating them in a restricted region of bony attachment. Furthermore, as the teeth wore, the periodontium allowed further eruption and horizontal migration ('drift') to occur, so that occlusion was maintained and at the same time support of the tooth remained strong. That this movement occurred in *Diademodon* is demonstrated by the large amount of exposed root and interproximal wear observed in the anterior gomphodont teeth.

Replacement of large-crowned thecodont teeth requires extensive resorption of their roots by the developing replacement teeth. Later, the newly-erupted replacements come into use before their roots are fully formed. In both cases, attachment of the tooth to the jaw is not as strong as it is in a completely-rooted tooth. In an animal with numerous replacements, there would always be weakly attached teeth in the jaw, some occupying the region where the greatest masticatory pressures are developed, i.e. close to the posterior end of the molariform series. It is clearly advantageous to reduce the amount of replacement in teeth which are subjected to strong masticatory forces. In mammals, these are primarily the molars and deciduous molars; in gomphodont cynodonts they are the gomphodont (molariform) teeth (the posterior sectorial teeth do not appear to occlude in *Diademodon*). Mammals do not replace their molars and the molariform deciduous teeth are replaced only when the region where greatest masticatory force is developed has shifted to a more posterior part of the jaw. The gomphodont teeth of cynodonts are also shed only when they lie in front of the functionally most important part of the tooth row. In both cases, the integrity of an uninterrupted molariform series is never disturbed. This, then, is probably the primary functional reason for the shift from an 'alternating' to a 'sequential' type of replacement.

Although similar functional reasons underlay the shift from 'alternating' to 'sequential' replacement in gomphodont cynodonts and mammals, there is no evidence that the total amount of replacement in gomphodonts (or any other known cynodont) significantly approached the very limited amount characteristic of mammals. In the transition to the mammalian diphyodont replacement pattern, additional selective factors appear to have operated. One of these was probably an increased rate of growth early in life so that the adult size was reached in a relatively shorter period of time than occurred in the reptilian ancestors. More important, however, was the evolution of maternal care of the young and particularly the origin of lactation.

As has been frequently pointed out (e.g. Ewer, 1963; Edmund, 1969), a reptile must from the time of hatching be capable of feeding and defending itself, and so must come from the egg with a fully functional dentition. Because it grows relatively slowly (as compared with a mammal) and must increase in size many times before reaching maturity, its teeth must go through many replacements, each of gradually increasing size, in order to keep pace with the lengthening jaws. Mammals, on the other hand, grow more rapidly and, because of the availability of maternal milk, can delay eruption until they are well on the way to adult size; thus they are able to get through the growing period with but a single replacement of teeth. As Ewer (1963) notes, 'Suckling is a prerequisite for the development of a mammalian type of tooth replacement'.

Brink (1957), in an attempt to assess the degree of approach to a mammalian level of organization which was achieved within the Therapsida, cited the dentition of *Diademodon* as giving 'some encouraging indirect support to the view that milk glands could actually have been present' (p. 89). He speculated that a newly 'born' (i.e. either live-born or hatched) *Diademodon* could not have had a postcanine dentition much more elaborate than the four conical teeth of the adult and that the more complex posterior teeth probably erupted only after birth. He further speculated that the 'delicate' dentition of the newly born was unsuitable for dealing with the type of diet characteristic of the adult and that the young may therefore have been nursed by the mother on milk-like glandular secretions (1955*a*: 30). Tatarinov (1967) has supported this hypothesis but appears to have interpreted Brink's unproven inference regarding the 'delicate' nature of the neonate dentition as an observed fact.

The evidence provided by the present study of tooth succession in *Diademodon* indicates that in the youngest individuals for which material is available the dentition is fully differentiated. The animals, therefore, appear to have been fully capable of obtaining and processing food which may not have been greatly different from that taken by adults. In this regard, several tiny skulls from the *Cynognathus* Zone offer corroborating evidence that very young cynodonts possessed functional dentitions. *Gomphodontoides megalops* (Brink & Kitching, 1951) is based on a tiny skull only 40 mm long which possesses four cheek teeth, one apparently conical tooth and three gomphodont teeth. This skull is clearly that of an extremely young gomphodont, either *Trirachodon* or *Diademodon*. A small carnivorous form named *Cistecynodon parvus* by Brink & Kitching (1953) on the basis of a skull 55 mm long, possesses six postcanine teeth of which the last four have three cusps. The dentary is very large and

I suspect that the specimen is a young cynognathid rather than a galesaurid. These specimens provide additional evidence that *Cynognathus* Zone cynodonts were as capable of fending for themselves at birth as are the hatchlings of living reptiles. This does not disprove Brink's suggestion that the young of advanced cynodonts might have been nursed on maternal milk, but it strongly suggests that they were not. Taken with the evidence for a large amount of tooth replacement in *Diademodon*, I believe it is more reasonable to assume that known cynodonts were more reptilian than mammalian in reproductive biology and ontogenetic development.

Tooth replacement and the reptile-mammal class boundary

In recent years, the presence of a dentary-squamosal contact has been used as the practical diagnostic character for classifying an animal as a mammal (Kermack & Mussett, 1958; Simpson, 1959). It is now universally recognized that the class Mammalia as thus defined is a polyphyletic taxon. In addition to being possessed by the traditionally-recognized groups, a dentary-squamosal contact has also been described in a ictidosaur (Crompton, 1958) and a tritylodontid (Fourie, 1968) from the Late Triassic and in a chiniquodontid cynodont from the Middle Triassic (Romer, 1969, 1970).

Recent studies of early mammals indicate that they probably had a monophyletic origin (Hopson & Crompton, 1969), or, at the least, a diphyletic origin (Kermack, 1967). Also, the ancestry of mammals is almost certainly to be found among the cynodonts (Barghusen, 1968; Crompton & Jenkins, 1968, Romer, 1970; but also see Kermack, 1967; Kermack & Kielan-Jaworowska, 1971). Because of this new information, several workers (Hopson & Crompton, 1969; Romer, 1970; Barghusen & Hopson, 1970) have urged the adoption of additional criteria for restricting the definition of a mammal so that it includes only the traditional groups of mammals and excludes ictidosaurs, tritylodontids, and other cynodonts and cynodont-descendants which have not achieved the mammalian level of organization in most other respects or which are phylogenetically distant from the true mammals.

Hopson & Crompton (1969) and Romer (1970) have suggested that among other criteria which might be used for separating reptiles and mammals the mode of tooth succession is probably significant. The evidence provided by the present study supports this; it indicates that mammalian diphyodonty is not only a useful diagnostic criterion but also that a great deal of biological significance can reasonably be attached to it. Diphyodonty at the present time is known only in mammals; all cynodonts (broadly including ictidosaurs and tritylodontids) show a much greater amount of replacement than do mammals even though deviations from the primitive 'alternate' pattern occur. I believe it can be reasonably argued that diphyodonty is correlated with the possession of a high degree of dependence of the young on the maternal parent for care and nourishment (milk), whereas the multiple replacement patterns of cynodonts are correlated with independence of the young from time of hatching and the presence of little or no parental care. The mammalian mode of development has probably allowed mammals to perfect their homeothermy and to evolve a much more complex central nervous system than is seen in reptiles. Therefore, the origin of the mammalian mode of tooth replacement appears to be correlated with the origin of other features which are basic to the success of the mammals. For this reason, I would strongly advocate the adoption of the criterion of diphyodonty to join the criterion of the jaw joint as a diagnostic feature for defining mammals. Furthermore, because monotremes possess what I consider to be the basic physiological features of mammals, and because the other nontherian orders are diphyodont, I would retain the class boundary where it has traditionally been drawn (i.e. to include both the therians and the nontherians). Despite certain objections (MacIntyre, 1967), this seems to me to be the most biologically meaningful point of separation which has yet been proposed.

ACKNOWLEDGEMENTS

For allowing me to study specimens under their care I wish to thank Drs A. J. Charig, R. Dehm, J. T. Gregory, and B. Schaeffer. Thanks are also due Dr A. W. Crompton for the loan of stereophotographs of specimens in the Bernard Price Institute. I have profited from discussions with Drs H. R. Barghusen, A. W. Crompton and F. R. Parrington F.R.S.

The research upon which this paper is based was supported by National Science Foundation Research grants GB-03877 and GB-7836.

REFERENCES

BARGHUSEN, H. R., 1968. The lower jaw of cynodonts (Reptilia, Therapsida) and the evolutionary origin of mammal-like adductor jaw musculature. *Postilla*, **116**: 1–49.

BARGHUSEN, H. R. & HOPSON, J. A., 1970. Dentary-squamosal joint and the origin of mammals. *Science, N.Y.*, **168**: 573–575.

BRINK, A. S., 1955*a*. A study on the skeleton of *Diademodon*. *Palaeont. afr.*, **3**: 3–39.

BRINK, A. S., 1955*b*. Note on a very tiny specimen of *Thrinaxodon liorhinus*. *Palaeont. afr.*, **3**: 73–76.

BRINK, A. S., 1957. Speculations on some advanced mammalian characteristics in the higher mammal-like reptiles. *Palaeont. afr.*, **4**: 77–96.

BRINK, A. S., 1963*a*. Two cynodonts from the Ntawere Formation in the Luangwa Valley of Northern Rhodesia. *Palaeont. afr.*, **8**: 77–96.

BRINK, A. S., 1963*b*. Notes on some new *Diademodon* specimens in the collection of the Bernard Price Institute. *Palaeont. afr.*, **8**: 97–111.

BRINK, A. S. & KITCHING, J. W., 1951. Some theriodonts in the collection of the Bernard Price Institute. *Ann. Mag. nat. Hist.*, **4** (12): 1218–1236.

BRINK, A. S. & KITCHING, J. W., 1953. On some new *Cynognathus* Zone specimens. *Palaeont. afr.*, **1**: 29–48.

BROILI, F. & SCHRÖDER, J., 1936. Ein neuer Galesauride aus der Cynognathus-Zone. *Sitzungsber. Bayer, Akad. Wiss., Math.-naturw. Abt.*, **1936**: 269–282.

BROOM, R., 1913. On evidence of mammal-like dental succession in cynodont reptiles. *Bull. Am. Mus. nat. Hist.*, **32**: 465–468.

BUTLER, P. M., 1939. The teeth of the Jurassic mammals. *Proc. zool. Soc. Lond.*, **109**: 329–356.

CROMPTON, A. W., 1955*a*. A revision of the Scaloposauridae with special reference to kinetism in this family. *Navors. nas. Mus., Bloemfontein*, **1**: 149–183.

CROMPTON, A. W., 1955*b*. On some Triassic cynodonts from Tanganyika. *Proc. zool. Soc. Lond.*, **125**: 617–669.

CROMPTON, A. W., 1958. The cranial morphology of a new genus and species of ictidosaurian. *Proc. zool. Soc. Lond.*, **130**: 183–216.

CROMPTON, A. W., 1962. On the dentition and tooth replacement in two bauriamorph reptiles. *Ann. S. Afr. Mus.*, **46**: 231–255.

CROMPTON, A. W., 1963. Tooth replacement in the cynodont *Thrinaxodon liorhinus* Seeley. *Ann. S. Afr. Mus.*, **46**: 479–521.

CROMPTON, A. W. & JENKINS, F. A., JR., 1968. Molar occlusion in Late Triassic mammals. *Biol. Rev.*, **43**: 427–458.

EDMUND, A. G., 1960. Tooth replacement phenomena in the lower vertebrates. *Contr. R. Ont. Mus., Life Sci. Div.*, No. **52**: 1–190.

EDMUND, A. G., 1969. Dentition. In C. Gans, A.d'A. Bellairs & T. S. Parsons (Eds), *Biology of the Reptilia*, **1**, *Morphology A*, xv + 373 pp. London: Academic Press.

EWER, R. F., 1963. Reptilian tooth replacement. *News Bull. zool. Soc. sth. Afr.*, **4** (2): 4–9.

FOURIE, S., 1963. Tooth replacement in the gomphodont cynodont *Diademodon*. *S. Afr. J. Sci.*, **59**: 211–213.

FOURIE, S., 1968. The jaw articulation of *Tritylodontoideus maximus*. *S. Afr. J. Sci.*, **64**: 255–265.

HOPSON, J. A., 1964. Tooth replacement in cynodont, dicynodont and therocephalian reptiles. *Proc. zool. Soc. Lond.*, **142**: 625–654.

HOPSON, J. A. & CROMPTON, A. W., 1969. Origin of mammals. In T. Dobzhansky, M. K. Hecht & W. C. Steere (Eds), *Evolutionary biology*, **3**: viii + 309. New York: Appleton-Century-Crofts.

KERMACK, K. A., 1967. The interrelations of early mammals. *J. Linn. Soc. (Zool.)*, **47**: 241–249.

KERMACK, K. A. & MUSSETT, F., 1958. The jaw articulation of the Docodonta and the classification of Mesozoic mammals. *Proc. R. Soc. (B)*, **149**: 204–215.

KERMACK, K. A. & KIELAN-JAWOROWSKA, Z., 1971. Therian and non-therian mammals. In D. M. Kermack & K. A. Kermack (Eds), *Early Mammals. Zool. J. Linn. Soc.*, **50**, Suppl. 1: 103–115.

KÜHNE, W. G., 1956. *The Liassic therapsid* Oligokyphus, x + 145 pp. London: Br. Mus. (Nat. Hist.).

KÜHNE, W. G., 1968. Origin and history of the Mammalia. In E. J. Drake (Ed.), *Evolution and environment*. New Haven: Yale University Press.

MACINTYRE, G. T., 1967. Foramen pseudovale and quasi-mammals. *Evolution, Lancaster, Pa.*, **21**: 834–841.

MELCHER, A. H. & EASTOE, J. E., 1969. The connective tissues of the periodontium. In A. H. Melcher & W. H. Bowen (Eds), *Biology of the periodontium*, xii + 563 pp. London: Academic Press.

MILES, A. E. W. & POOLE, D. F. G., 1967. The history and general organization of dentitions. In A. E. W. Miles (Ed.), *Structural and chemical organization of teeth*, **1**: xv + 525 pp. New York: Academic Press.

MILLS, J. R. E., 1971. The dentition of *Morganucodon*. In D. M. Kermack & K. A. Kermack (Eds), *Early mammals. Zool. J. Linn. Soc.*, **50**, Suppl. 1: 29–63.

NOBLE, H. W., 1969. The evolution of the mammalian periodontium. In A. H. Melcher & W. H. Bowen (Eds), *Biology of the periodontium*, xii + 563 pp. London: Academic Press.

PARRINGTON, F. R., 1936. On the tooth-replacement in theriodont reptiles. *Phil. Trans. R. Soc. (B)*, **226**: 121–142.

ROMER, A. S., 1969. Cynodont reptile with incipient mammalian jaw articulation. *Science, N.Y.*, **166**: 881–882.

ROMER, A. S., 1970. The Chañares (Argentina) Triassic reptile fauna: VI. A chiniquodontid cynodont with an incipient squamosal-dentary jaw articulation. *Breviora*, **344**: 1–18.

SIMPSON, G. G., 1928. *A catalogue of the Mesozoic Mammalia in the Geological Department of the British Museum*. x + 215 pp. London: Br. Mus. (Nat. Hist.).

SIMPSON, G. G., 1959. Mesozoic mammals and the polyphyletic origin of mammals. *Evolution, Lancaster, Pa.*, **13**: 405–414.

SZALAY, F. S., 1965. First evidence of tooth replacement in the subclass Allotheria (Mammalia). *Am. Mus. Novit.*, **2226**: 1–12.

TARLO, L. B. H., 1964. Tooth replacement in the mammal-like reptiles. *Nature, Lond.*, **201**: 1081–1082.

TATARINOV, L. P., 1967. Development of the system of labial (vibrissal) vessels and nerves in theriodonts. *Paleont. Zh.*, **1967** (1): 3–17. [In Russian.]

ZIEGLER, A. C., 1969. A theoretical determination of tooth succession in the therapsid *Diademodon*. *J. Paleont.*, **43**: 771–778.

Bimodal wear of mammalian teeth

R. G. EVERY

Department of Zoology, University of Canterbury, Christchurch, New Zealand

and

W. G. KÜHNE

Geologisch-Paläontologisches Institut, Freie Universität, West Berlin, Germany

The attempt is made to prove the thesis of bimodal wear on mammalian teeth. The modus of abrasion by contact of tooth and food is a process of blunting and polishing, producing finally excavations in the dentine surrounded by high and smooth enamel ridges. Abrasion is counteracted by *thegosis*, a contact of tooth on tooth, producing cutting edges and striate and evenly curved surfaces, flush when passing from enamel into dentine.

The results of abrasion and thegosis and their interrelationship are described.

Among mesozoic mammals thegosis is not found in multituberculates; it is found in dryolestids.

The subject matter of this paper is based on Dr Every's lectures which he gave while guest professor at the Freie Universität Berlin during the winter term 1967–68.

Wear on the crown of mammalian teeth lends itself to analysis because two wear surfaces of very different kind can be observed.

In order to do so we choose a sizeable molar for instance of *Cervus* or the fossil *Lophiodon*. In the case of the fossil, the specimen has to be clean, that is, not covered by any protective laquer and it has to come from a nonclastic deposit, where features of the tooth surface are not obliterated by water transport between sand and stones.

Obviously we neither choose a young unworn tooth nor an outworn one; a specimen having some dentine already exposed is best fitted for our purposes.

The relevant observations are only possible with direct light throwing shadows; an illumination for instance with neon lighting is unsuitable. The features of the two different wear surfaces we put down below:

(1) *Delimitation of the wear surface*

From the apex of a cusp, or the edge of a loph, the surface-features decrease in intensity and fade out towards the valleys or the side walls of the crown.

The surface is circumscribed by distinct boundaries, some of which are characteristically sharp.

(2) *Surface-features seen with ×100 magnification* (Plate **1 A, B**)

The surface consists of deep and shallow scratches crossing each other in many directions.

The surface consists of deep scratches only, strictly parallel in direction.

(3) *Surface-features observable with the naked eye*

Polish.

Striations on a shiny surface.

(4) *Curvature of the surfaces*

The surface is irregularly curved in many planes and, except where the dentine is exposed, more or less follows the form of the tooth. It covers cone-shaped cusps as well as bowl-shaped depressions.

The surface is generally curved in only one plane.

(5) *Features of the surface when passing from enamel to dentine and vice versa*

When both enamel and dentine are cut, the dentine is excavated leaving the enamel raised above as a rounded wall; the differential hardness of the two tissues under the impact of the frictional force producing the surface is responsible for this kind of relief.

A surface passing from enamel into dentine is flush. The frictional force producing this kind of surface deletes any macro-relief previously present.

(6) *Correspondence in upper and lower teeth*

There are no fitting surfaces in upper and lower teeth.

There are only fitting surfaces in upper and lower teeth.

(7) *Interrelation of both kinds of surface*

The surface features are also found superimposed on surfaces of the other kind (right-hand column), gradually deleting them.

In the fresh condition the surface features are seen for only a short while; surface features of the other kind (left-hand column) are soon (in hours) superimposed and gradually obliterate them.

(8) *Function*

The surfaces are blunting points and edges and obliterate the unworn form of the tooth.

The surfaces form bevels to sharp cutting edges previously blunted by abrasion.

(9)

The direction of the surface and of the striae does not always coincide.

The enumeration of the observed features on the two kinds of wear surfaces leads to the conclusion that two different kinds of forces are responsible for the formation of the two surfaces; in other words wear is bimodal.

The features in the left-hand column are caused by a contact of tooth and food. None of the enumerated features conflicts with this cause. Features (1), (2), (4), (5) and (6) are only to be explained by the contact of the tooth with a loose abrasive medium, acting non-directional; this action could be compared with sand blast. The features are a consequence of mastication; they are passive wear or abrasion. Abrasion features are clearly dominant on the surface of the worn tooth but they are not governing this surface.

The features in the right-hand column can only be caused by contact of tooth on tooth, last but not least, because there is no other frictional wearing force available in the mouth. None of the enumerated features in the right-hand column can be caused by another action. The described surfaces form shear edges; whenever these shear edges serve to comminute food, the features under (7) develop; that is, while shear edges are used in mastication, they lose their feature which they develop during a powerful and short action of sharpening. This action we call active wear or *thegosis* (from the Greek, to sharpen).

The direction of the thegosis striae is the direction of the lower jaw in respect to the upper jaw during the thegosis stroke. Mills (1966), without referring to the function of this tooth on tooth contact, has mentioned this fact. Butler illustrated thegosis striae in 1952 and later.

The fact that abrasion marks are soon superimposed on thegosis facets is evidence that the latter is not a consequence of mastication. During mastication no thegosis striae are formed.

The manifestation of thegosis and abrasion on mammalian teeth is very different: abrasion is shown almost always, thegosis less. The latter cases are those where thegosis occurs in short intervals maintaining a continuously functional cutting edge. The carnassial apparatus of P^4 and M_1 of Felidae, the canines of *Sus* and *Hippopotamus* are of this kind. The respective edges are long and distinct: the flattened thegosis surfaces are easily observed and the function as a scissor and slashing weapon well known or self evident. That the shattering noise before the attack of pig and peccary are by-products of thegosis is obvious, the respective canines are ever-growing, are specialized almost exclusively as weapons which are freshly ground for each attack (Brehm, 1877: 564; Frädrich, 1967: 27–28). A functional analysis of a carnassial apparatus is given for the miacid *Ictidopappus* by MacIntyre (MacIntyre, 1966: 143).

When thegosis occurs in long intervals, superposition of abrasion marks on thegosis marks is the rule. The M^3 of *Lophiodon*, Plate 1**C**, Fig. 1, is a pertinent case. The individual striae of thegosis are already deleted, but the gross relief of the thegosis surfaces still allows one to observe the direction of the thegosis stroke. The posterior cutting edges of the thegosis surfaces though blunted are still the sharpest part of the tooth.

To the authors it seems plausible that here the function of thegosis is also the planing of excessively high enamel ridges, produced by abrasion. Such seems to apply to *Equus:* the region of P_4M_1 where the maximum of masticating work is performed does not show thegosis while P_2 and M_3 show it distinctly.

Further evidence for thegosis is derived from its absolute absence, in cases where opposing teeth can not touch—for instance, anterior premolars of mustelids—or, in the absence of antagonistic teeth, for instance, the anterior lower dentition of cavicornia (bovoidea).

The authors hope to have made their point: thegosis and abrasion are corollaries. Thegosis as well as abrasion are general features of wear of mammalian dentitions since the Jurassic. The rhythm of thegosis and abrasion, of sharpening and blunting, seems to them an intelligent means of interpreting the function of mammalian dentitions.

Readers of this paper on the lookout for thegosis may be disappointed not to find it. In consequence we may be allowed to state: thegosis is not found in all mammalian

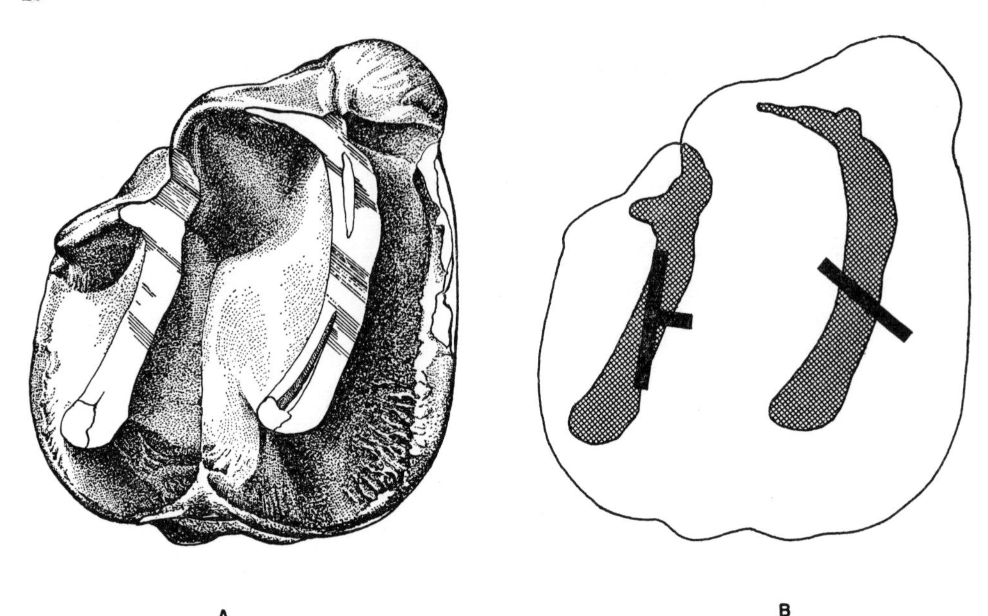

Figure 1**A.** *Lophiodon* M³dex. In order to illustrate thegosis, the tooth has been drawn in dot-manner.

B. The same. Left: long bar = horizontal on thegosis surface (strike); crossbar = maximal inclination of the thegosis surface (dip). Right: direction of the thegosis striae in projection on thegosis surface.

taxa; thegosis is not found in all members of a population of one taxon at the same time; thegosis is not always found in all teeth of a dentition. If this were not so, the paper would not have been written.

It is no accident that only now the worn tooth gains importance. The mainstay in mammalian systematics are still the teeth albeit in the unworn or little worn condition. The enormous success of the mammalian teeth as indicators of systematic relationship barred the way to investigations of the worn dentition and to the functional aspects of it.

Since 1952, however, Butler and Mills have been active in this field. They figure thegosis striae though the non-masticating nature of thegosis has not been realized by these authors.

REFERENCES

Brehm, A., 1877. *Thierleben*, 2nd ed., **3**: 564.

Butler, P. M., 1952. The milk molars of Perissodactyla with remarks on molar occlusion. *Proc. zool. Soc. Lond.*, **121**: 777–817.

Every, R. G., 1965. The teeth as weapons. *Lancet*, **1965**: 685–688

Every, R. G., 1970. Sharpness of teeth in man and other primates. *Postilla*, **143**: 1–30.

Frädrich, H., 1967. Das Verhalten der Schweine. *Handbuch der Zoologie*, pp. 27–28.

MacIntyre, G. T., 1966. The Miacidae, Part 1. *Bull. Am. Mus. nat. Hist.*, **131**: 117–209.

Mills, J. R. E., 1966. The functional occlusion of the teeth of Insectivora. *J. Linn. Soc.* (*Zool.*), **47**: 1–24.

R. G. EVERY AND W. G. KÜHNE

(*Facing p.* 27)

EXPLANATION OF PLATE

Plate 1

A and **B.** Lower Molar of *Palaeotherium* from the Upper Eocene of Sosis Catalonia, ×100. Photographs by R. Stiemen, Institut für Metallkunde Technische Universität Berlin.

A. Thegosis. Scratches are unidirectional and of equal depth.

B. Abrasion. Scratches are multidirectional and of different depth.

C. *Lophiodon* M^3dex. The direction of thegosis striae is brought out by horizontal illumination from the right.

The dentition of *Morganucodon*

J. R. E. MILLS

Institute of Dental Surgery and University College, University of London

This paper consists of a description of the very large collection of *Morganucodon* teeth at University College, London, including information on tooth replacement which has not previously been published.

The affinities of *Morganucodon* are discussed. It is suggested that in the Upper Triassic three groups of mammals have evolved independently from mammal-like reptiles: the Morganucodontidae (including *Erythrotherium*), the Sinoconodontidae (a provisional family including *Sinoconodon* and *Megazostrodon*), and the Kuhneotheriidae, of one genus. The two former are more closely related to each other than either is to the last. There is strong evidence that *Morganucodon* is related to the Triconodontinae (but not the Amphilestinae), and probably to the Docodonta. Possibly also to the monotremes, while evidence linking them to the multituberculates is inconclusive.

CONTENTS

INTRODUCTION

Morganucodon watsoni was first described by Kühne (1949) from a fissure filling in Glamorganshire. Subsequently, very large numbers of this species were found in three other quarries in the same area (Kermack, Kermack & Mussett, 1956). Through the kindness of Dr K. A. Kermack and his team at University College, London, the description in the present paper is based on this collection, and also on a small collection of teeth in my own possession, which have been extracted from matrix given to me by Dr Kermack. Other descriptions of teeth from the same sources have been published by Parrington (1967) and Crompton & Jenkins (1968) under the name of *Eozostrodon parvus*. This assumes synonomy with the premolar tooth described by Parrington (1941). The provenance of these specimens has been described by their discoverers, and their age is believed to be Lower Rhaetic.

Morganucodon oehleri is at present confined to a single skull, in a quite exceptional state of preservation, which was collected by Father Oehler in Yunnan, China, and has been described by Rigney (1963). This is of a similar age to the other species.

THE TEETH OF *MORGANUCODON WATSONI*

Incisors

The number of lower incisors would seem to vary between individuals. In a well-preserved specimen from the quarry at Ewenny there appear to be only four teeth anterior to the lower canine, while in specimen M24563 there were six alveoli mesial to that of the canine. This variation in the number of teeth is not uncommon, even between specimens from the same site, and would seem to represent individual variation rather than variation between different species. The lower incisors are spatulate, having convex labial and concave lingual surfaces, and are bluntly pointed. The first incisor is the largest, and is so procumbent as to be almost horizontal. The remaining incisors decrease in size and become more upright as one passes backwards, so that the final tooth is almost vertical and situated immediately in front of the canine. The first incisors of the two dentaries were probably in contact on their mesial surfaces, but the two dentaries were not fused together and there may well have been movement in life.

Not surprisingly, the dentary bones and lower dentitions are represented in the University College collection by many more specimens than are those from the upper jaws, and one cannot therefore be equally certain in describing the latter. There are three intact premaxillae, without teeth (M23204, M23206 and M24629) and each of these contains alveoli for three incisors. One broken premaxilla (M24661) has two incisors, probably the second and third, *in situ*. It would appear that three was the usual number of incisors, and that they were spaced from each other, there being a particularly large diastema in the mid-line. The teeth were quite small, with a single rounded cusp. Blunt crests run mesially and distally from this cusp. There is a tiny cuspule at the distal end of the distal crest, and on the presumed second incisor only, an even smaller one at the mesial end of the tooth. The second incisor is rather larger than the third. The roots are short and stout, dilating slightly towards their ends.

In the maxilla there is an alveolus for a small tooth in front of the canine (e.g. in M26160) which may be regarded as a precanine but was probably functionally a fourth incisor.

Canines

The upper and lower canines are both large, typically caniniform teeth. The lower canine is, with the possible exception of the final lower premolar, the highest tooth in the lower jaw, and its length compares with that of the molars. Its crown is slightly recurved, so that its mesial surface is convex and its distal concave, and there are rather rounded crests on these mesial and distal surfaces. There are no additional cuspules. The root of the lower canine is dilated and slightly recurved, although this is less marked than in the upper jaw. It is grooved on mesial and distal aspects and may divide into two roots at its extreme end—it usually has two pulp canals.

The upper canine (Plate 4**A**) is essentially similar. Again oval in cross-section, with blunt mesial and distal crests. There is no suggestion of the type of wear produced by shearing between upper and lower teeth on any of the canines examined, and as a result they are more reminiscent of reptilian than mammalian canines, lacking the sharp distal crest of the latter. The root of the upper tooth is quite striking. It is about twice as long as the crown, and dilates continuously towards its end so that, on cursory examination, it is difficult to see where the crown ends and root begins. The end of the root is the longest (mesiodistally) part of the whole tooth. It has a single, oval root canal. It would seem that the canines were used for seizing a struggling prey, the excessively large roots preventing their dislocation.

Premolars

Throughout this paper I have used the words 'premolars' and 'molars' rather than 'premolariform teeth' and 'molariform teeth' for simplicity, although the latter terms would be more strictly correct.

The lower premolars apparently vary in number between four and five. The first is a very small tooth, with a single rather rounded cusp, slightly flattened bucco-lingually. A crest runs distally from its tip, to end in a very faint heel. The lingual surface slopes towards the gum margin rather less steeply than does the buccal side. The mesial surface is rounded and even bulbous, and this rounded crest runs mesially to a point just above the gum margin, where it turns distally to run along the lingual surface, forming a very weak lingual cingulum.

The remaining anterior premolars are essentially similar to the first, but increase in size as one passes backwards, with such details as the heel and lingual cingulum becoming slightly more marked. I have not found any evidence of wear on these more anterior premolars, and it seems probable that they did not come into occlusion with their opponents, as is the case in many recent carnivores.

The penultimate premolar is essentially similar to the more anterior ones but shows some differences. The mesial and distal crests of the principal cusp are more developed, and the distal one shows wear, apparently produced by shearing against its upper antagonist. The minor features are better developed than on the anterior premolars, and the lingual cingulum increases in height in the position where one might expect to find a kuhneocone (*vide infra*), but there is no actual cusp.

The final premolar, although related to the more anterior ones, is sufficiently different to be worthy of note. The principal cusp rivals that of the canine in height and may even be slightly higher. It lies mesial to the centre of the tooth, and is less bulbous and more flattened, bucco-lingually, than the more anterior teeth. Both mesial and distal crests end in small cuspules, the more posterior one, or heel, being the larger. There is quite marked wear on the buccal side of this distal crest. There is again no buccal cingulum, but the lingual cingulum is much better developed than on the more anterior teeth, and more reminiscent of those of the molars. The mesial cuspule is duplicated on its lingual side, as on the molars, and this cingulum has a variable number of tiny cuspules on its mesio-lingual corner, where it forms the margin of a small basin. It is almost obsolete opposite the widest part of the principal cusp, but, behind this, opposite the groove between main cusp and distal cuspule, it rises to form its largest cuspule. This would seem to correspond to the kuhneocone of the molars. The distal end of the cingulum is quite well developed and incipiently cuspidate.

The roots of the premolars are similar to those of the molars (Plate 3 **A**, **B**) and quite striking. They are normally two in number, arranged mesio-distally, but on the first premolar they are apparently imperfectly divided, while on the second there is sometimes only a single, flattened root. The roots themselves are unexpectedly long, reaching almost to the lower border of the dentary; in fact the roots of both premolars and molars occupy almost all the cancellous part of the bone. They are flattened buccolingually, and therefore oval in outline. Unlike recent mammals, they do not taper towards their ends. Their sides are parallel until near to their end, when they dilate, giving a shape reminiscent of a blunderbuss. The end of the root shows in many, but not all specimens an open pulp canal, and it would seem that this foramen finally reduces to microscopic dimensions rather late in life.

I am grateful to Mrs F. Mussett for drawing my attention to specimen M24562, a dentary bone from which most of the teeth have been lost. This specimen shows quite clearly an unerupted final lower premolar, lying in its crypt on the point of eruption, with, adjacent to it, the almost resorbed root of its predecessor. There is also in the University College collection a specimen from Ewenny, which shows a lingual exposure of an immature dentary (Plate 4**B**). On it are three molariform teeth. The most distal, which is in the later stages of eruption, is the largest, and one would therefore assume this to be the second molar. If this is so, the two more anterior teeth would be the first molar and final milk molar. This possible milk molar is slightly different in appearance from the true molars. The cusps are more squat and rounded. The lingual cingulum is rather better developed than on the adjacent true molars and has a continuous row of cuspules, giving it a beaded appearance. Apart from these differences, and its smaller size, it has the appearance of a typical *Morganucodon* molar. Being exposed only from its lingual side, it is not possible to assess the relative amounts of occlusal wear on these three teeth, but the presumed true molars have the appearance of teeth recently erupted, whereas the 'milk molar' appears more worn.

From the first of these specimens alone, there can be little doubt that the final premolar was replaced. In the many hundreds of dentaries in the collection there is no evidence of replacement of the more anterior premolars. Such a situation is, of course, found in other groups of unrelated mammals, notably the marsupials, and a possible explanation presents itself. The final premolar is a very large, and especially a very high tooth. If it were present from an early age, on eruption it would be situated near the back of the mouth, and would be an embarrassment to an immature animal. A lower milk tooth, functioning as a molar near the back of the mouth, and later replaced by a functional premolar, would seem more satisfactory.

The anterior upper premolars are shed at an early age, so that for much of its life the animal has a diastema between the canine and penultimate premolars. Later in life this procedure continues, the posterior premolars being shed, followed by the molars, from before backwards. Indeed one dentary in the collection shows only a single alveolus, apparently for the fifth molar. This process, which has been described, not very happily, as replacement from the distal, is a feature of cynodont dentitions (Crompton, 1964).

Also of interest is the manner of this resorption (Plate 3 **C**, **D**). It would seem that the first stage consists of cutting through the root, just below the alveolar margin, presumably by osteoclastic action, so that the crown is lost. Many dentaries in the collection have roots of premolars from which the crowns have apparently been so removed, but not surprisingly, there are none showing this actually taking place. Bone then grows in from the margins of the socket, gradually covering the root, until in the final stage its former position is marked only by the pulp canal, which is visible after the root itself has disappeared. At the same time, the root is invaded and replaced by bone as can be seen in several dentaries where fracture has occurred through a tooth socket. Presumably the tooth is resorbed by osteoclasts, and the bone laid down by normal osteoblastic activity. It will be recognized that this mechanism is apparently identical with that described by Kermack (1956) as occurring in canine teeth of the Gorgonopsia and Therocephalia as part of normal tooth replacement. In *Morganucodon* it would seem to be a retained reptilian feature, but here, of course, the teeth are not replaced, and it is not possible to tell whether they are milk or permanent teeth.

The more anterior premolars are also lost early in life, and this has caused Crompton & Jenkins (1968) to suggest that there were only two upper premolars. In fact there were certainly more, and there is no reason to think that the number was different from that in the lower jaw. Certainly the same method of resorption occurs.

The crowns of the more anterior premolars consist of a principal cusp situated rather mesially to its centre (Plate 4 **A**). This cusp is flattened bucco-lingually, and has anterior and posterior crests which, unlike the lower premolars, are both straight. They meet at about 70° at the blunt tip of the main cusp, this being a much higher angle than on the lower premolars. The anterior crest runs down to the gum margin, where there may be a diminutive cuspule; it seems probable that this is absent on the more anterior premolars. The posterior crest runs down to a shallow groove which separates it from the distal cuspule. The lingual surface is flat or slightly concave. There is a suggestion of a lingual cingulum, which is always very narrow and may be confined to the mesial end of the lingual surface, or to its mesial and distal ends, but it seldom exists opposite the widest part of the principal cusp. Again it seems probable that this is better developed on the more posterior premolars. The buccal surface is rounded and is featureless apart from the groove between the principal and posterior cusps, which runs about half-way to the gum margin. There are two roots arranged antero-posteriorly. These are of morganucodontid form, although the dilation is less extreme in some specimens than in the lower teeth, and they may taper slightly.

Table 1. Lengths and widths of last two premolars and M1–4 of *Morganucodon*

	PM X-1		PM X		M1		M2		M3		M4	
Mandibular teeth	L	W	L	W	L	W	L	W	L	W	L	W
M. watsoni												
M24560	0·84	0·36	1·12	0·52	1·28	0·54	1·44	0·70	1·20	0·62	0·82	0·56
M24590			1·06	0·50	1·16	0·46	1·40	0·70	1·30	0·64		
M22679					1·10	0·44	1·50	0·70				
M22768					1·10	0·46	1·38	0.66				
Ewenny	0·86		1·00									
Ewenny									1·32	0·80	0·70	0·52
Ewenny							1·50		1·36		0·84	
Ewenny*					0·74	0·34	0·88	0·44				
Ewenny	0·96		1·04						1·18		0·60	
M. oehleri	1·18	0·50	1·74	0·70	1·90	0·72	2·36	1·14	1·78	0·88		
Maxillary teeth												
M. watsoni												
M23408			1·24	0·58	1·18	0·52	1·26	0·60	1·18	0·54		
M24626			1·30	0·62	1·26	0·60	1·50	0·74	1·20	0·70		
M24633			1·06	0·66	1·08	0·66	1·22	0·80	1·06	0·74	0·78	0·48
Ewenny U4*					0·80	0·42	1·00	0·54				
Ewenny U1	0·50	0·42	1·40	0·80	1·10	0·62	1·46	0·84	1·28	0·80	0·92	0·56
Ewenny U3			1·00		0·92		0·96					
M. oehleri	0·84		1·68		1·80		2·24		1·52			

* Possibly these should be last milk molar and M1.

The final premolar is essentially similar but considerably larger. It is as long as the longest molar, and rather higher. The lingual cingulum may be continuous along the lingual aspect, although it is very narrow opposite the principal cusp. Unlike the more anterior premolars there is also a cingulum, albeit very narrow, at the mesial and distal ends of the buccal surface.

There is evidence of shearing wear on at least the final upper premolar. There is no evidence of tooth replacement, but in view of the smaller number of specimens, this is not surprising, and it seems likely that the final premolar would, like its antagonist, be replaced.

The lower molars

(Plates 1 **A, B** and 2 **A, B**.) The molar teeth have been previously described by Kühne (1949), Kermack (1965), Parrington (1967) and Crompton & Jenkins (1968). It is proposed to repeat this here for the sake of completeness, but more especially because of the additional evidence which becomes available with a large collection. So far as possible, I propose to describe cusps and other structures topographically, because this is easier for the reader to follow, but where questions of possible homology arise, I shall use the same terminology as Crompton & Jenkins.

There are four and sometimes five lower molars. The most striking feature of these is the marked variation in crown pattern which exists between individuals. This is essentially a difference in the degree of molarization and, to some extent also, of size. It may be that we shall ultimately find that we have representatives of more than one species, but the teeth do not obviously fall into groups; the variation is continuous except that the Ewenny specimens are perhaps rather different from those from the other two sites.

The principal cusp is situated rather mesial to the mid-point of the tooth, and usually closer to its buccal margin. Behind it is the distal cusp, which is usually the second largest. The discrepancy in size between the two cusps is largest on M_2, least on M_4, and is about equal on the first and third molars. The buccal surface of these cusps is smooth and continuous down to the gum margin; there is no external cingulum. There is invariably a lingual cingulum, which commences anterior to the principal cusp where there is a mesial cusp of variable size. It may be little more than a vestige, or may be as large as the distal cusp; the variation is partly individual, but largely a difference between the different molars of the series. The mesial cusp is usually smallest on the first molar and increases in size as one passes backwards in the mouth, so that on the fourth molar, it is equal in size to the distal cusp, giving an apparent triconodont appearance. The lingual cingulum continues around the mesio-lingual corner of the tooth as far as a point opposite the widest part of the principal cusp, where the cingulum is correspondingly narrow and may be obsolete. The width of this mesio-lingual part of the cingulum varies, being narrrowest on M_1, where it is not usually cuspidate, and widest on M_4 where, indeed, it is wide all the way along its length. On the second and third molars this part of the cingulum is divided into about three cuspules, of which the most anterior lies directly lingual to the mesial cusp which it duplicates.

The cingulum again widens behind the greatest width of the principal cusp and immediately lingual to the groove between principal and distal cusps is a very constant cuspule which Parrington (1967) has called the kuhneocone. It is usually second only to the mesial cusp in size, among the cingulum cusps. Around the disto-lingual corner of the tooth the cingulum is divided into a small number of rather inconstant cuspules, and finally ends immediately behind the distal cusp, in a small distal cuspule. This last is easily seen on an isolated tooth, but in a tooth series, especially on more anterior molars, it is somewhat hidden, lying vertically below the mesial cusp of the succeeding molar to form an interlocking mechanism of the teeth.

Most previous authors have regarded the lower molars as 'triconodont', but this would seem to be an over-simplification. Both the principal and distal cusps might be classed as true cusps, but the mesial cusp is a cingulum cusp. On the more posterior teeth it may be as large as the distal cusp, but on the other teeth, especially those towards the front of the molar series, it is small and may even be indistinguishable from the anterior end of the cingulum. It would seem quite illogical, except as wishful thinking, to 'count' this as a cusp and not other cingulum cusps such as the kuhneocone which may be as large or even larger.

It is, nevertheless, quite common for major cusps to develop from cingula. It may be that this is actually occurring here, and we probably see a later stage of the process in the Tricondontinae.

The fifth molar

In the above description, I have concentrated on the first four molars. In a proportion of dentaries a socket is present behind the fourth molar. None of the specimens in the collection has the tooth *in situ;* it has a single rather tapering root, and it is not surprising that they are lost after death. A number of lower molars are known which seem to correspond to the size and shape to this socket, and which seem undoubtedly to be fifth molars.

Most dentaries do not show evidence of this fifth molar, although it might be that it only erupts late in life. I have very thoroughly dissected a dentary in my own collection without finding any evidence of a fifth molar tooth germ, and the most probable explanation seems to be that this tooth, erupting late in life, is not always present, so that the molars, like premolars and lower incisors, vary in number.

The teeth in question are clearly lower molars, although only about two-thirds the size of the fourth molars. They have a single root, flattened to a varying extent bucco-lingually and sometimes grooved on their buccal and lingual sides. In a few examples the roots actually separate at their extreme ends. The crown is essentially a modification of the more anterior molars. The mesial cusp is at least as high as the distal cusp, and is less obviously a cingulum cusp. The cingulum cusp which is normally lingual to the mesial cusp (cusp E of Crompton & Jenkins) has a tendency to be displaced buccally, so that in some (but not all) specimens it is almost anterior to the mesial cusp. The lingual side of the main cusps slopes gently downwards, with the lingual cingulum and its cuspules poorly developed, although the kuhneocone can usually be recognized.

The upper molars

(Plates 1 **C, D** and 2 **C, D.**) There are certainly four upper molars, and while we have no maxillae with sockets for a fifth upper molar, we have teeth which would seem to fall into this category. It seems therefore rather likely that the situation is similar to that in the dentary; a fifth upper molar is present in a proportion of cases.

The outline of a typical molar tooth is oval, and the completely unworn tooth is surrounded by a low cingulum. There are three 'true' cusps, arranged in a straight line mesio-distally. The principal cusp is always the largest and highest, while the mesial cusp is nearly always lower than the distal. The relative size of the mesial cusp increases as one passes backwards in the mouth, being almost equal to the distal cusp on M^3, and quite equal on M^4 and the presumed M^5.

Although the cingulum is continuous around the unworn tooth, it rapidly wears away on the lingual side. It is split up into a very variable number of cuspules, looking rather like a string of beads. At the distal end, in line with the 'true' cusps is a rather larger cuspule which forms part of the rather weak interlocking mechanism between adjacent teeth. At the mesial end of the tooth the cingulum often becomes narrow, but it invariably passes mesially to the mesial cusp, which, unlike the lower cusp, never arises from the cingulum. In the mesio-lingual corner of the cingulum there is often a larger cuspule, which may be of such a size as to appear to duplicate the mesial cusp. This cusp is usually at its largest on the third molar. On the buccal side there is a tendency for the largest cuspules to lie in line with the grooves between the main cusps, corresponding to the situation on the lingual side of the lower molars.

The difference between the molars of a series is even less marked than in the lower jaw. The fourth molar is fairly distinctive. It is smaller than the first three, has mesial and distal cusps equal in size, and the buccal cingulum is proportionately much wider than on the more anterior molars.

The second molar is always the largest, although this does not help to identify an isolated tooth. The first and third are about equal in size. The mesial cusp is almost as large as the distal one on the third molar, but definitely smaller on the first. Moreover, the buccal cingulum becomes wider towards the distal end of the tooth on the first molar, wider at the mesial end of M^3, while the two ends are of equal width on the second molar. The teeth also tend to become wider relative to their length as one passes backwards in the mouth, largely due to the increased width of the buccal cingulum. These differences are not sufficiently marked to enable one to identify an isolated molar with certainty, but there should be no difficulty in identifying the molars on a maxillary fragment with at least two such teeth.

When set in the maxilla, the long axis of the upper molars leans noticeably lingually, while the mesio-distal axis of individual teeth is set at an angle to the line of the dentition. The mesial end of each molar has a tendency to overlap the more anterior tooth on its buccal side, although to a variable extent.

The presumed fifth molar shows the general characteristics of an upper molar, with modifications corresponding to those seen on the lower fifth molar. It has a single stout root. The three main cusps are present, but the mesial cusp is larger and higher than the distal. The cingula are present on both buccal and lingual sides of the tooth, but are reduced in both height and width when compared with more anterior molars, especially on the buccal side. The tooth is, again, only two-thirds the size of the fourth molar.

THE TEETH OF *MORGANUCODON OEHLERI*

(Plate 5)

Morganucodon oehleri is known from a single, beautifully preserved skull which has been described by Rigney (1963). It comes from Yunnan and is probably broadly contemporaneous with *M. watsoni*. It has been suggested (K. A. Kermack, 1967) that some of the fragments referred to *Sinoconodon* may in fact belong to *M. oehleri*, but this suggestion does not include any specimens of the former which carry teeth, so that this description must be based on the one excellent specimen.

In the lower jaw, there were at least three incisors, essentially similar in shape to those of *M. watsoni*, while the canine is also similarly large, bluntly pointed and with a large root. There are two premolars present in the specimen, and apparently roots of two more, of which the more anterior is undergoing the *Morganucodon* type of resorption. The posterior two premolars, while recognizably morganucodontid, differ in details from those of the other species. On the penultimate premolar the lingual cingulum is narrower than in *M. watsoni*, while on the last premolar the lingual cingulum is vestigial and has no suggestion of cusps. The latter tooth, but not the former, has a tiny mesial cuspule, and both have a small distal cuspule, which fits beneath the mesial cuspule of the succeeding tooth; a condition only seen on the molar teeth of the other species. Surprisingly, the penultimate premolar has a narrow buccal cingulum which runs only on the distal side of the groove between principal and distal cusps. The corresponding region of the last premolar is broken, and cannot be examined.

There are four lower molars present, of which the fourth is only just erupting from the bone. Their general form is similar to that of *M. watsoni*, but with minor differences. These mostly arise from the fact that the lingual cingulum is less well developed. On the first two molars it becomes obsolete lingually to the principal cusp, although it continues as a very narrow ledge in this region on M_3. It is not possible to make out any detail of the lingual cingulum on M_4, although it is certainly present. Because the cingulum is reduced the mesial cusp is less obviously derived from it, although, as in *M. watsoni*, it increases in size as one travels backwards in the mouth. There is a very much smaller cuspule directly lingually to the mesial cusp on $M_{2,3}$, and on M_1 this cuspule has taken up a more buccal position, so that it lies mesio-lingually to the mesial cusp. As in *M. watsoni*, the distal cuspule lies beneath the mesial cusp of the succeeding molar in the case of M_1, but lies anterior to it in the case of M_2, parallelling the relationship of M_3 to its successor in the case of the other species.

All the lower molars have small kuhneocones, and the distal end of the cingulum is reduced and, apart from the two cuspules mentioned, is not cuspidate. On M_3 there is a suggestion of a cingulum on the buccal surface of the tooth, just above the alveolar margin; an occurrence never seen on *M. watsoni*.

The roots can be clearly seen on P_4 and less clearly on P_3 and the first two lower molars. They are similar to those of *M. watsoni* stout, parallel, with dilated ends, and they extend to the lower border of the dentary.

One half of the dentary has been removed from the skull by Mrs Mussett, and it has therefore been possible to describe both buccal and lingual aspects of the lower teeth. In doing so the upper teeth were necessarily sacrificed, although this has given us access to their roots, which would not otherwise be visible, and they seem generally similar to those of *M. watsoni*, but possibly slightly more slender. The molar roots are dilated, at least in some cases, although this is apparently not true of the premolars. Their crowns must be described from the other side of the skull, and only their buccal aspect is visible.

There were three, and probably four upper incisors, of which the most posterior was rooted in the maxilla, and therefore strictly a precanine. The canine is large and recurved, and has the dilated root which is typical in this genus. There are four premolars, all *in situ*, and these again differ from *M. watsoni* in details. The principal cusps are slightly recurved distally, unlike the other species, but the difference is slight. The mesial cuspule is present only on the last two premolars, and even here is smaller than on *M. watsoni*. As with the lower premolars, there is a small distal cuspule behind the distal cusp. There is a buccal cingulum on both the two posterior premolars, which is mildly cuspidate on the final premolars. The last three premolars have two roots each, but the most anterior one may be single-rooted.

There are three molars in place on the side of thc skull from which the mandible has been removed, but on the other side the base of a broken fourth molar is identifiable, so that there were certainly four molars present, as in the lower jaw. They are essentially similar to those of *M. watsoni*, so far as one can see from the external view, except that they are again rather larger than the average size of the other species. It is noticeable that the external cingulum is wider and better developed on each molar than is usual on the corresponding molars of *M. watsoni*. That on M^3 forms a buccal 'basin' corresponding to that seen on M^4 of the other species. The fourth molar of *M. oehleri* is represented only by the tooth on the side from which the dentary has been removed, and this is broken off at the base of the crown, so that no details are visible.

It would seem, then, that the teeth of *M. oehleri* are generally very much like those of *M. watsoni*, and fully justify their inclusion in a single genus. There are minor differences, notably in size, and in the fact that the buccal cingula of the upper cheek teeth are better developed in *M. oehleri*, while on the lower molars the lingual cingulum is less well developed, although there is a suggestion of a buccal cingulum on two teeth of *M. oehleri* which is never seen on *M. watsoni*. These differences would accord with the separation into two species.

FUNCTION OF THE CHEEK TEETH

The function of the cheek teeth of *Morganucodon* has been deduced from a study of the teeth of *M. watsoni*. It would be difficult to confirm it in the case of *M. oehleri* because of the inaccessibility of the lingual side of the upper teeth. The single specimen is of a fairly young individual, and the teeth are not heavily worn, but from an examination of the lower molars, there can be little doubt that they functioned exactly as do those of *M. watsoni*. Crompton & Jenkins (1968) have described the pattern of wear facets on the molar teeth of this species and I do not propose to do more than summarize my findings as these are essentially similar to theirs.

Wear on a typical molar tooth commences by the principal cusp of the upper tooth shearing down in the groove between principal and distal cusps of the corresponding lower molar, producing facets A in Fig. 1**A**. Almost simultaneously, but possibly a little later, wear is produced by the principal cusp of the lower molar shearing between the principal and mesial cusps of the upper tooth, producing facets B in the figure. It is noticeable that at this stage the wear on the upper molar is largely localized to the tip of the principal cusps and to the lingual cingulum, with a large unworn area in between. The lingual cingulum of the upper molar holds the main part of the approximated surfaces of the teeth slightly apart at this stage.

Figure 1**B** shows the situation at a slightly later stage, when further wear has taken place. The main wear still occurs in the same way, but the areas of wear, A and B, are tending to become confluent. On the upper molars it is, however, still noticeable that

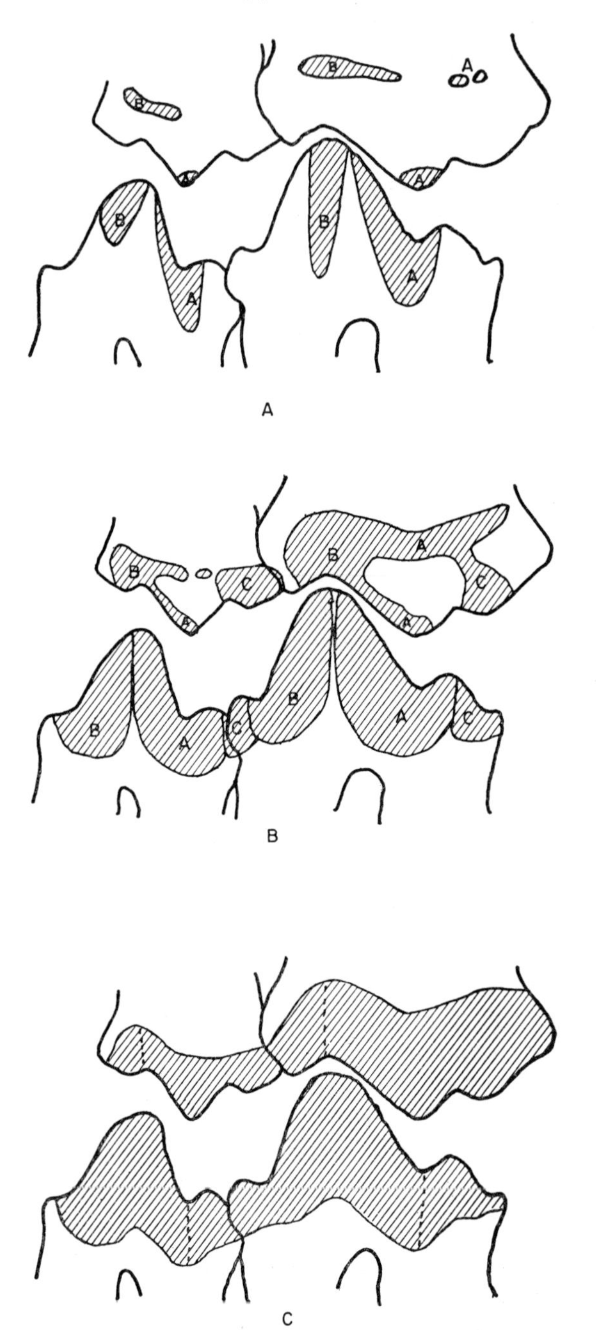

Figure 1. Lower left and upper right first and second molars to show stages in the development of wear facets.

wear is confined to the cingulum and the region near the tips of crests and ridges, with an unworn area in between. At this stage, also, a further facet system has appeared due to the action of the distal cusp of the upper molar. This shears in the shallow groove which exists between the distal cusp of a lower molar and the mesial cusp of the succeeding lower molar. This latter cusp is closely applied above the distal cuspule of the more anterior molar. This is the area marked C in Fig. 1**B**.

The Fig. 1**C** shows the situation in a worn tooth. The whole lingual surface of the upper molar and buccal surface of the lower is worn. I have suggested elsewhere that the vulnerable point in a shearing dentition is the point of contact between the teeth (Mills, 1964), and in the therian mammals the principal cusps shear not on this point of contact but down a groove slightly removed from it. The same is true in *Morganucodon*,

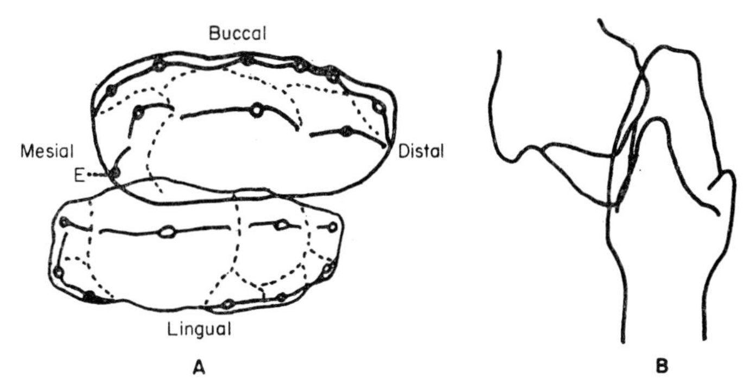

FIGURE 2. **A**. Occlusal view of upper and lower right second molars of *Morganucodon* to show occlusal relations. **B**. The corresponding left molars seen in bucco-lingual 'cross-section'. Note the lingual slope of the upper tooth.

so that even when wear covers all the opposed surfaces of the teeth, it is not a flat shearing surface, but rather a zig-zag one. There is a re-entrant between the principal and mesial cusps of the upper teeth into which the lower principal cusp shears. The depth of this re-entrant varies, and if the cuspule on the lingual side of the upper mesial cusp (Crompton's cusp E) is well developed (as in Fig. 2**A** and Plate 4**D**), the distal side of this cusp will be taken into the shear, and wear will produce a hook-shaped appearance reminiscent of the parastyle of a tribosphenic dentition. In other individuals the groove is quite shallow, as in the example illustrated by Crompton & Jenkins (1968) and Plate 2**D**. There will also be a second re-entrant, between the lower principal and distal cusps, where the upper principal cusp shears, which may appear analogous to that between the protoconid and hypoconid of therian mammals.

In the above description there are certain minor differences from that given by Crompton & Jenkins in the area labelled C in Fig. 1**B**. These are not due to differences of opinion, but due to the fact that they have described the contact point between the third and fourth lower molars, which is different from that between the more anterior molars.

This describes the wear as seen from the buccal side. The most unexpected aspect of this wear appears in a bucco-lingual 'cross-section' (Fig. 2**B**). Wear on the buccal side of an isolated lower molar is at a steep angle relative to the long axis of the tooth, whereas on the upper molar the angle between the wear facet and the long axis of the tooth is greater, giving the appearance of a more transverse type of wear as may be seen by comparing Plate 2**A** and **D**. Indeed the first point at which wear occurs on the upper molars is on the lingual cingulum, then the tips of the cusps, with for some time an unworn area between them. In fact the facets on upper and lower teeth must be parallel; they could not otherwise have been produced. The cause of this apparent inconsistency is shown in Fig. 2**B**; the upper cheek teeth lean sharply lingually in the maxilla, as can be seen in several specimens in the collection.

Crompton & Jenkins (1968) drew attention to the fact that 'the corresponding upper and lower teeth do not have accurately matching occluding surfaces as is the case in molars of primitive therians.' In therian mammals generally, the part of a molar tooth on which a wear facet forms already exhibits the form and shape of the ultimate wear facet, even before wear takes place. It lies in contact with its antagonist in occlusion, so that, as wear occurs, it does so equally over the whole area. Crompton & Jenkins suggest that this is not the case in *Morganucodon*, and that this dentition represents an early mammalian stage, through which the therian mammals passed earlier still.

Certainly, they would appear to be right about the upper molars of *Morganucodon*, because of the presence of the lingual cingulum which protrudes beyond the rest of the shearing face. On the lower molar, however, the wear facet closely follows the shape of the tooth (Plates 2**A** and 4**C**). While the dentition of *Morganucodon* is undoubtedly that of a very primitive animal, I would suggest that, in this context at least, it is not more primitive than *Kuhneotherium*, but merely different. The occlusal relations of *Morganucodon* are quite constant; as constant as those of any therian mammal. The difference lies in this apparently perverse persistence of the lingual cingulum of the upper molars. A similar situation exists in the Triconodontinae of the Upper Jurassic, and it seems unlikely that this cingulum would persist so long if it merely interfered with efficient shearing. It may be that, in holding the two surfaces slightly out of contact, and therefore preventing the surface of the upper tooth from becoming highly polished, it improves the efficiency of the shearing and grinding action between the two opposed surfaces. Jenkins (1969) has suggested that the wrinkling of the enamel of *Docodon* serves the same purpose.

The cuspules seen on the lingual cingulum of the lower molars and buccal cingulum of the uppers have no very obvious function. However, it will be noted that the largest of these are usually situated immediately opposite a groove between two main cusps. This is, inevitably, the widest part of the cingulum, and this fact alone may explain the presence of a large cuspule at this point. It seems rather more probable that they have a function in diverting the food, which spills over the groove between cusps as the opposing cusp shears down this groove, away from the gum margin; this would certainly seem to be the function of the kuhneocone. Gregory (1922) pointed out that the mesostyle in a tribosphenic dentition served this purpose.

THE AFFINITIES OF *MORGANUCODON*

Contemporaneous mammals

Broadly contemporaneous with *Morganucodon*—that is from the Upper Triassic—there are five animals which might be considered to be mammals and of which adequate specimens are known. Of these *Diarthrognathus*, although deserving of further study, is almost certainly unrelated to any of the others. This leaves us with *Erythrotherium* from Lesotho, *Kuhneotherium* from Wales, *Megazostrodon* also from Lesotho, and *Sinoconodon* from Yunnan. The dentary articulation of the last genus is not preserved, so that its mammalian status is inferred. K. A. Kermack (1967) has suggested that all the specimens referred to that genus may not belong there, but all the useful teeth except one certainly come from the same individual and the exception, a lower fourth molar exposed in its crypt, may safely be referred.

Erythrotherium parringtoni

This animal has been the subject of a preliminary report by Crompton (1964) made before the preparation of the specimen was complete. It apparently had four or five lower incisors. Crompton states that the crown of the lower canine is not exposed 'but it was clearly a small tooth'. Such a term is, of course, comparative, but judging from the drawing it may be that this canine was somewhat, but not grossly, smaller than that of *Morganucodon*. There was then a small diastema with a pit which probably represents the pulp chamber of a resorbed first premolar. There were seven post-canine teeth, and these were almost certainly four premolars and three molars, of which the last was in process of eruption. Crompton suggests that the second premolar present was in the process of eruption, and might be a replacement tooth. If so, this would be very interesting, but the evidence is inconclusive. There is only one upper molar present, of which only the buccal surface is exposed. The lower post-canines are very broken.

It would seem that the lower molar teeth have an internal cingulum which is typically morganucodontid, although less well developed than the average. This is also true of the external cingulum of the upper molar, but, since *M. watsoni* is so variable, either could easily be matched from the University College collection. Crompton, while considering the two genera to be closely related, lists a number of features by which *Erythrotherium* differs from *Morganucodon*, but, as a result of our greater knowledge of the latter genus, none of these is now valid; if *Erythrotherium* had been discovered in South Wales it would certainly have been hailed as a specimen of *Morganucodon*. On the knowledge at present available any taxonomic change would not be desirable, but there can be no doubt of their very close relationship; they are certainly members of the same family.

Megazostrodon and *Sinoconodon*

These two animals came respectively from Lesotho and the Yunnan province of China. *Megazostrodon* has been the subject of a preliminary description by Crompton & Jenkins (1968), and Professor Crompton has also kindly allowed me to examine some excellent stereoscopic photographs of the molar teeth of this specimen. The information is nevertheless limited, since at the time the photographs were taken the lingual surfaces of the teeth had not been freed from matrix. *Sinoconodon* has been completely freed from matrix, and I have been fortunate in having access to the original specimen. It has been described by Olson & Patterson (1961). Unfortunately most of the teeth were damaged and severely weathered before recovery.

Despite the somewhat unsatisfactory nature of the material, there are a number of features by which the two genera seem to be distinguished from *Morganucodon* and, less certainly, in which they resemble each other. The following points of comparison and contrast are relevant:

(1) *Lower molars.* Both *Sinoconodon* and *Megazostrodon* have at least four lower molars, as has *Morganucodon*. In *Sinoconodon* at least the posterior molars erupt late, after the premolars have been lost; a feature shared with *Morganucodon* and certain cynodonts. On the lower molars of *Morganucodon* there are only two main cusps, although the mesial cusp, arising from the cingulum, may attain a size similar to that of the distal cusp. In *Sinocodon* there is a mesial cuspule anterior to the mesial cusp, which latter does not apparently arise from a cingulum. The corresponding situation in *Megazostrodon* is not clear; the mesial cusp is larger in size than is usual in *Morganucodon*, but is still smaller than the distal cusp. In *Morganucodon* the division between the principal and distal cusps is at a low level, giving the appearance that they have arisen separately from the base of the tooth. In the other two genera they divide less completely, the distal cusp giving the appearance of having split off from the principal cusp. The lingual surface of the lower molars of *Sinoconodon* is severely weathered, and there is no evidence of a lingual cingulum. There is a very narrow featureless cingulum on the lingual aspect of the unworn M_3, and a similar one could have been present originally on the more anterior molars. Professor Crompton tells me (Crompton, 1970 pers. comm.) that the lingual aspects of the lower molars of *Megazostrodon* have a lingual cingulum not unlike that of *Morganucodon*. There is a difference in the way the teeth contact their neighbours. In *Morganucodon* the distal cuspule of the more anterior tooth fits below the mesial cusp of the more posterior one, whereas in both *Sinoconodon* and *Megazostrodon* it fits lingually to the mesial cusp.

(2) *Upper molars.* The crowns of the upper molars of *Sinoconodon* are known only by a single, very damaged tooth, probably an M^2. There are four upper molars of *Megazostrodon*, known only from their buccal sides. Both genera apparently have three cusps, of which the central is the highest and the mesial the lowest. This is similar to *Morganucodon*, although the cusps themselves are more slender than in that genus. In *Megazostrodon* the buccal cingula are wider and more cuspidate than in *Morganucodon*, increasing in width as one passes backwards in the mouth. On the sole molar of *Sinoconodon* the buccal surface is intact at the mesial end. There is a small cuspule directly in line, buccally, with the mesial cusp, but the cingulum, posterior to this, if originally present, is broken away. The posterior margin of the tooth has a bulbous appearance, not unlike that of *Megazostrodon*. It seems unlikely that the upper molar of *Sinocondon* has a buccal cingulum as wide as that on M^{2-4} of *Megazostrodon*, but the tooth could well have resembled M^1 of that genus. *Sinoconodon* certainly had no lingual cingulum on the distal half of its one known upper molar, but this region is still hidden in *Megazostrodon*.

(3) *Dental occlusion.* It is in this feature that the molar teeth of *Megazostrodon* and *Sinoconodon* most resemble each other and differ from the morganucodontids. As Crompton & Jenkins (1968) have pointed out, the main cusp of the upper molar of *Morganucodon* shears between the main and distal cusps of the lower molar, whereas in *Megazostrodon* it shears behind the distal cusp and against the distal cuspule.

It is not possible to interpret the wear facets in *Sinoconodon*, because functional wear is slight and the teeth are heavily weathered, but there can be little doubt that it was similar to that in *Megazostrodon*, from the general shape of the teeth and especially because the groove between the principal and distal cusps is simply not deep enough to accommodate the upper cusp. Crompton & Jenkins have suggested that at the 'eozostrodontid' (=here, *Megazostrodon*, and my morganucodontids) level of development 'accurate dental occlusion was only incipiently developed, and slight variations in the relative position of upper and lower molars would result in different occlusal patterns'. In the very large number of *Morganucodon* teeth in the University College collection, and in my own, there is no more variation in occlusal pattern than one would find in a comparable group of recent therians. I have discussed this point in detail above. For the morganucodontid type of occlusion to develop into the *Megazostrodon* type it would be necessary to envisage an intermediate stage in which the principal cusp of the upper tooth occluded directly with the distal cusp of the lower, while the lower principal cusp similarly occluded with the mesial cusp of the upper tooth. It is very difficult to believe that such an arrangement would be advantageous, or even viable, in a shearing dentition. Although less so, it seems unlikely in a non-shearing reptilian dentition. Such a dentition, used for seizing a struggling prey, requires upper and lower teeth to be quite close together, and this can be better achieved if the larger cusps alternate. If cusp stood opposite cusp, there would be a greater danger of food jamming the jaw.

It would seem, from a number of features, that the two groups; the Morganucodontidae and the group including *Sinocondon* and *Megazostrodon*, which for convenience I am including together as the **Sinoconodontidae (fam. nov.** see Appendix) are closely related. A possible hypothesis for the difference of molar occlusion would assume that their common ancestor had upper and lower molars similar to those shown diagrammatically in Fig. 3**A**; not unlike the posterior premolars of *Morganucodon*. The upper teeth were basically single cusped, with a small mesial and distal cuspule. The lowers also had a single main cusp, with a small distal cuspule and a lingual cingulum ending anteriorly to the principal cusp. There could well have been other basal cingula—they are irrelevant to the present point. With the commencement of shearing, the principal cusps alternated, and all was well except that the mesial cuspule on the upper molar sheared directly against the tip of the lower molar, as in Fig. 3**A**. This arrangement was unsatisfactory, and was solved differently in the two lines. In one the mesial cuspule, as it increased in size, grew mesially so as to occlude anteriorly to the lower principal cusp, as in Fig. 3**B**, giving rise to *Morganucodon* shown in Fig. 3**C**. It is worth noting that as wear becomes advanced, the upper mesial cusp of this genus does, in fact, tend to wear away before the other cusps, as in Plate 2**C** and **D**. The other group required rather high, slender cusps with narrow grooves between them, so that the upper mesial cuspule could not migrate forwards. Instead, a groove appeared in the tip of the lower principal cusp to accommodate it, splitting off a fresh distal cusp, as in Fig. 3**C** and **D**. If this is correct, the cusps of the two genera are homologous except that the distal cusp of *Morganucodon* is homologous with the distal cuspule of *Megazostrodon*, while the distal cusp of the latter genus has no homologue in the former. This is, at best, only a hypothesis, but is one which does not involve a shearing cusp migrating 'across' its antagonist in occlusion. The two lines would diverge at the very development of a shearing action; the question of whether this was before or after the animal crossed the hazy and wholly artificial line between reptiles and mammals is pointless; I prefer to see it as a part of a normal radiation.

Crompton & Jenkins (1968) have suggested *Thrinaxodon* as a possible ancestor for these groups. While quite inconclusive, there is nothing in my hypothesis to disagree with this.

(4) *Tooth replacement.* The more anterior teeth of all Mesozoic mammals are rather similar, and in the present context not very helpful. *Megazostrodon* had five premolars

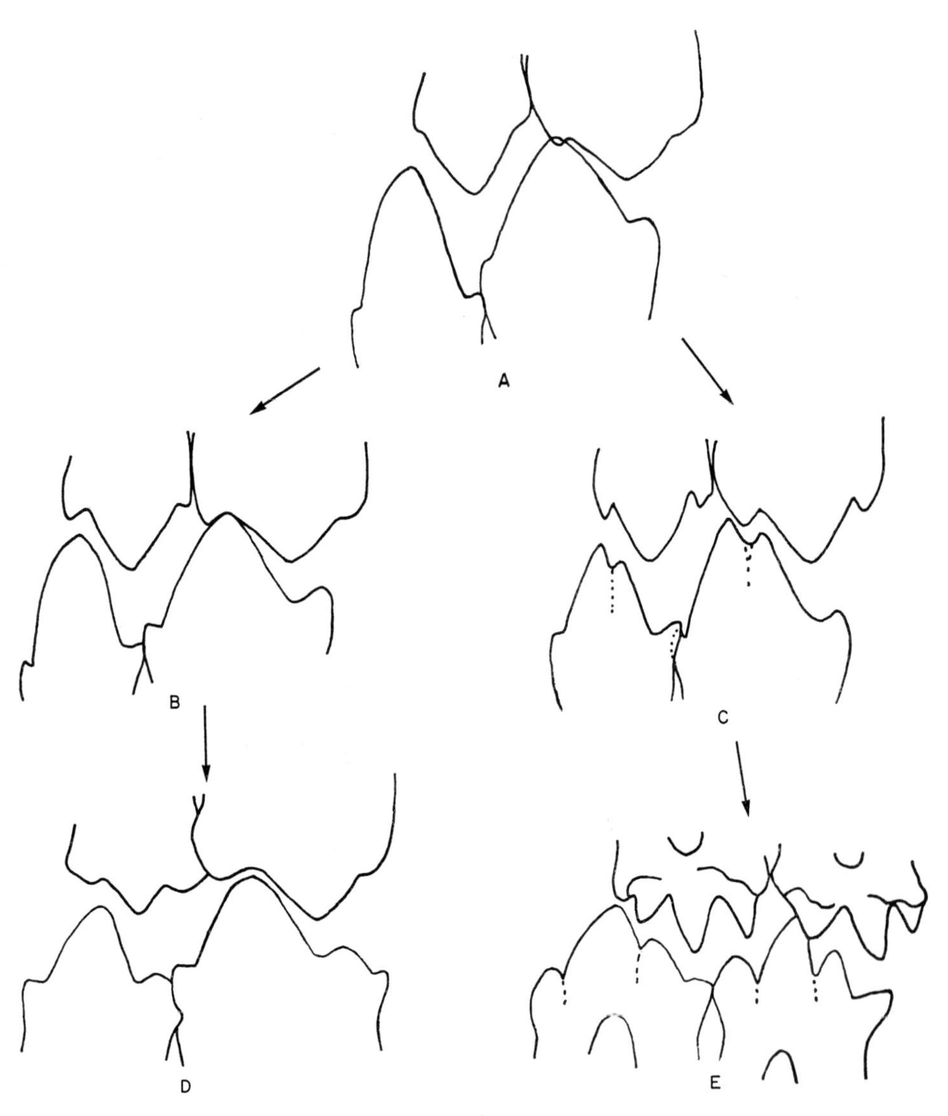

FIGURE 3. Stages in possible development of *Morganucodon* and *Megazostrodon*: **A**, hypothetical common ancestor, at point where a shearing dentition is first commencing; **B**, **C**, hypothetical intermediate forms. **D**, *Morganucodon watsoni*; **E**, *Megazostrodon* (drawn from photographs of molar teeth lent by Prof. A. W. Crompton).

and *Sinoconodon* none. However, there can be little doubt that those of *Megazostrodon* are in the early stage of being lost, without replacement, from the front. In *Sinoconodon* on the left dentary of specimen CUP1, a partly resorbed root can be distinguished, so that it seems probable that here again the anterior premolars were resorbed early in life.

Kuhneotherium

This animal was first described, in the form of an isolated upper molar tooth ('Duchy 33') by Kühne (1950), although he did not name the genus. Since then a very large collection of isolated teeth has been made by Dr K. A. Kermack and his team at University College, London (Kermack, Lees & Mussett, 1965; Kermack, Kermack & Mussett, 1968). In addition to the published accounts, I have had access to the original material, although time has prevented a complete study. There are, in addition to teeth, a number of dentary and maxillary fragments, but unfortunately none of these has teeth in place. It is generally accepted that *Kuhneotherium* is a late Triassic representative of a group which D. M. Kermack *et al.* (1968) define as the Infraclass Pantotheria, and which includes not only this animal but Simpson's (1945) Pantotheria and Symmetrodonta.

The incisors of *Kuhneotherium* are not certainly known, while canines and premolars of Mesozoic mammals are so similar as to be of little use taxonomically. There are 14 dentary and six maxillary fragments with alveoli including tooth sockets present, and in none of these is there any evidence of the early resorption of the anterior premolars which is a marked feature of *Morganucodon* and the Sinoconodontidae. If the premolars are shed in *Kuhneotherium* in the same way as in these two groups and the cynodonts, then it must have occurred much later in the animal's life, and it seems rather likely that it did not occur. Also if, as I have suggested, only the last premolar was replaced in *Morganucodon*, it would seem unlikely that this was also the case in *Kuhneotherium* and the Pantotheria generally. Among this Infraclass certainly lies the ancestor of the therian mammals which undoubtedly replace their premolar teeth. Such a reversal of trend, if not impossible, seems most unlikely.

The main evidence of affinity or otherwise between this and the other two Triassic groups must, however, lie with the molar teeth.

(1) *Lower molars*. The lower molars of *Kuhneotherium* have three main cusps together with mesial and distal cuspules. This is similar to the sinoconodontids, but *Morganucodon* has only two main cusps, although the mesial, cingulum cusp may be quite large on some teeth. Again, the main cusps in *Kuhneotherium* are arranged as an obtuse-angled triangle, with the posterior face of the triangle more transverse than the anterior, as in the Pantotheria generally, but unlike both other Triassic groups. I have suggested elsewhere (Mills, 1967) that this is associated with a jaw movement which is not only medial but also has an anterior component; both I (Mills, 1967), and Crompton & Hiiemae (1970) believe this to be the primitive jaw movement in the theria. In both *Morganucodon* and the Sinoconodontidae the cusps are arranged antero-posteriorly and movement is apparently at right-angles to the antero-posterior long axis of the tooth. Since the two halves of the dentary converge anteriorly, this will in fact mean that movement is medially and slightly posteriorly relative to the skull.

Morganucodon and *Kuhneotherium* both have lingual cingula, but they differ very much in structure. In *Morganucodon* it is cuspidate and often basined near its anterior and posterior ends. In *Kuhneotherium* it is always narrow, of uniform width, and although there is usually a slight peak near its centre, nothing which could reasonably be called a cusp. *Sinoconodon* has probably a very narrow lingual cingulum on its lower molars; *Morganucodon watsoni* never has a buccal cingulum (although *M. oehleri* has a very slight suggestion on one molar only) whereas *Kuhneotherium* has a buccal cingulum on the anterior and posterior ends of the buccal surface.

The mesial and distal cusps of *Kuhneotherium* separate from the principal cusp incompletely and quite high on the tooth; this is similar to the sinoconodontids but even more marked. It differs very much from the situation in *Morganucodon*, as has been discussed in connection with the sinoconodontids.

The contract point between molar teeth in *Kuhneotherium* is formed by the distal cuspule (hypoconulid) fitting in a groove between the two mesial cuspules on the molar behind it. This differs from both the other groups.

(2) *Upper molars.* The difference between the upper molars of *Kuhneotherium* and *Morganucodon* is, superficially at least, less marked. Both have three main cusps, and in *Kuhneotherium* the division between them is more marked than on the lower teeth. Both have mesial and distal cuspules, and the contact point arrangement probably does not differ greatly between the two. Both have basal cingula on buccal and lingual sides, although these differ very considerably. The buccal cingulum is cuspidate on *Morganucodon* and even more markedly so on *Megazostrodon*, but not so on *Kuhneotherium.* The function of the lingual cingulum is different in the two groups (*Kuhneotherium* and *Morganucodon*), being worn by the lower teeth in the latter but not in the former where parts of it act as a 'stop' against which the lower teeth occlude. As in the lower jaw, the cusps on *Kuhneotherium* are arranged in an asymmetrical triangular form, with the antero-lingual face more transverse than the postero-lingual, presumably for the same reason.

(3) *Wear and function of the teeth.* As Crompton & Jenkins (1968) have pointed out, the main cusp of the upper molar of *Morganucodon* shears between the main and distal cusps of the lower molar, whereas in *Kuhneotherium* it shears between the distal cusp and distal cuspule, and in the sinoconodontids between the distal cusp of one tooth and mesial cusp of its successor, against the small distal cuspule. Similarly, in *Morganucodon*, the principal lower cusp occludes between the principal and mesial upper ones, perhaps with a re-entrant if the cuspule E (Fig. 2) is well developed. In *Kuhneotherium* the principal lower cuspule shears anterior to the upper mesial cusp, between this and the mesial cuspule. On *Kuhneotherium* this occlusal relationship is well developed, with the shape of the wear facets fully foreshadowed by the shape of the tooth. For reasons which I have advanced above, I cannot believe that this difference between *Morganucodon* and *Kuhneotherium* is less than a very basic one, although I would be prepared to accept affinity between *Kuhneotherium* and the sinoconodontids if there were no other contrary evidence (which there is).

Moreover, apart from the different relationship of the cusps in occlusion, the actual method of chewing differs in the two genera. The cusps of the molar teeth of *Kuhneotherium* have sharp ridges running mesially and distally from their tips. The buccal surfaces of the lower molars are pressed very closely against the lingual surfaces of the upper teeth, so as to prevent food intervening, and the ridges act exactly like a pair of scissors to cut up food which is held on the morsel surfaces of the teeth. The wear on opposed surfaces of the teeth is rather slow, and produces highly polished facets, its purpose being, as Every (1967 pers. comm.) has pointed out, to keep the cutting edges sharp. The cusps serve only as the highest points of the cutting edges, which are thus sloping, like the blade of a guillotine, for great efficiency. In *Morganucodon*, and probably also in the Sinoconodontidae, the chewing action consists of comparatively blunt cusps moving down grooves. Wear of the teeth is more rapid, and it would seem that the food is milled between the teeth, the wear tending to blunt the cusps. It would not be desirable for the wear surfaces to become mirror-smooth; a 'glazed' grind-stone is a useless object. For this reason, the upper cingulum holds the surfaces slightly apart, and retards such 'glazing' wear. A similar contrast is seen in recent mammals between, for example, the Tenrecidae, with a *Kuhneotherium* type of dentition, and the Erinaceidae which resemble *Morganucodon.*

It may be that the former type of dentition is adapted for cutting through the chitinous covering of some insects and crustaceans, while the latter is more suitable for chewing up muscles of invertebrates and small vertebrates, and even the bones of the latter.

(4) *The roots of the teeth.* As previously mentioned, the roots of the teeth of *Morganucodon* are quite unlike those of therian mammals. The roots of *Kuhneotherium* teeth, on the other hand, are more like those of therian mammals; they taper towards their ends, they diverge and are usually slightly curved, with the concavity of the curvature of the two roots facing each other.

All three groups of Upper Triassic mammals are very primitive, with comparatively simple dentitions, and at this level a three-cusped (more or less) molar is common in both mammals and carnivorous mammal-like reptiles. Apart from this resemblance their teeth seem almost as unlike as one could imagine. They function differently, and the relationship of upper and lower cusps is different. Teeth are notoriously ready to modify their shape to a new function, but it is difficult to believe that the common ancestor of *Morganucodon* and *Kuhneotherium* was a mammal; that it had a shearing dentition. The difference in root structure of the two genera, and the absence of the early resorption of the premolars in *Kuhneotherium* would seem to place this common ancestor some way below the mammalian level; they may well have evolved from different groups of mammal-like reptiles. The poorly-known Sinoconodontidae, although having features of both groups, and marked differences from both, would seem to lie much closer to the Morganucodontidae and, perhaps, to the cynodonts.

Eozostrodon

Eozostrodon was the first mammal to be described from this zone, although its provenance is a little doubtful, and probably not strictly contemporaneous with the animals so far described. Parrington (1941) described two teeth from Holwell in Somerset, which he named *E. parvus* and *E. problematicus.* Since then, Parrington (1967) and Crompton & Jenkins (1968) have suggested that *M. watsoni* is synonymous with *E. parvus*, although other authors, notably D. M. Kermack *et al.* (1968) do not necessarily agree.

E. parvus consists of a single tooth with an intact crown and broken roots. It would appear to be a premolar, probably from thc upper jaw. It could well belong to

Morganucodon. The intact part of the roots would tend to confirm this, but there are minor points of difference from the typical pattern of a final upper *Morganucodon* premolar; the mesial cuspule is rather larger and the lingual cingulum better developed lingually to the main cusp. These could be individual variations. On the other hand, D. M. Kermack *et al.* (1968) have shown that it could equally well belong to *Kuhneotherium*, and, indeed, in the present state of our knowledge, it seems very difficult, if not impossible to identify an isolated premolar tooth with any certainty.

E. problematicus consists of rather more than half of a molariform tooth, with one of apparently two roots intact. This root is curved and tapers slightly, in a manner quite unlike that seen in *Morganucodon.* The crown carries two of probably three major cusps, together with a terminal cuspule lying on a cingulum. The two major cusps are divided well above the gum margin, and are rather slender and recurved backwards—all features which differ from *Morganucodon.* Of the known Upper Triassic mammals *E. problematicus* would seem to bear the closest resemblance to the lower molars of *Sinoconodon*, but both specimens are too damaged to be certain. *Sinoconodon* appears to lack the basal cingulum on the erupted molars which is seen on *E. problematicus*, although a lingual cingulum is present on the partly erupted third molar, and may have been lost by weathering on the more anterior ones. The trivial name of this species was well chosen.

Haramiyidae

This family is known only from tiny isolated teeth. They are therefore doubtfully mammalian, and our understanding of them must await more satisfactory specimens. The teeth are quite different from those of *Morganucodon*, and no close relationship has ever been suggested. If they are, indeed, mammals, their closest known relatives may be the multituberculates.

Later Mesozoic mammals

Many of the mammals from the Jurassic and Cretaceous would fall into Kermack's Infra-class Pantotheria or into the Theria proper, and for these the remarks I have made about *Kuhneotherium* would equally apply. This leaves us, therefore, with three groups of non-therian Mesozoic mammals; Triconodonta, Docodonta and Multituberculata.

Triconodonta

This order, comprising a single family, is divided by Simpson (1945) into two subfamilies. I propose to consider these separately.

Triconodontinae

Most authors are agreed that there is a relationship between *Morganucodon* and the Triconodontinae (Kühne, 1949; D. M. Kermack *et al.*, 1956; Parrington, 1967; Crompton & Jenkins, 1968).

The incidence of parallel evolution in teeth is so high that any relationship based on them may be fallacious, but in this case the evidence is very strong. The following points would seem to be the important ones:

(1) The general form of the dentition is similar in the two groups. The incisors are similar in shape, while the premolars are generally similar, although somewhat more developed; the mesial cuspule and lingual cingulum are rather larger. The four premolars and three or four molars of the Triconodontinae could well have been derived from the four or five of each in *Morganucodon*. On the other hand, in none of the later Mesozoic mammals are the premolars apparently resorbed early in life. In both groups the final premolar is enlarged, as is the canine. These are all rather general points, and their value negative rather than positive.

(2) There is evidence in *Morganucodon* that the final premolar is replaced, and no evidence for the replacement of any other premolar. The same is true for this group.

(3) The upper molars in both groups have three main cusps, arranged in line. The lower molars of the triconodonts have three cusps in line, followed by a small cuspule which forms part of the interlocking mechanism. In *Morganucodon* the mesial cusp is variable in size and arises from the cingulum, whereas in the triconodonts it is a large cusp, subequal to the other cusps, as on the posterior molars of *Morganucodon*. It will be noted in the triconodonts that the lingual cingulum passes behind the distal cusp to the distal cuspule, while at its mesial end it runs into the mesial cusp, which might indicate its origin from a cingulum (Fig. 5**B**).

Both groups have buccal and lingual cingula on the upper molars, but only a lingual cingulum on the lowers. On the upper molars these are cuspidate in both groups, with a tendency, on the buccal side, for the cuspules to be enlarged buccally to the mesial and distal cusps.

The lingual cingulum of the lower molars is cuspidate in *Morganucodon* but not in the triconodonts. Parrington (1967) has drawn attention to a small cuspule on the lingual cingulum of *Trioracodon*, BM 47775, which he believes to be homologous with the kuhneocone. This is confined to the three molars of this specimen and also to M_3 of the only other specimen of this genus in the British Museum (Natural History) which is exposed from this side, M21891; it is not seen in *Triconodon*. It is a minor point of resemblance, but may well be a valid one.

(4) The pattern of wear is apparently similar in the two groups. Seen in cross-section, wear on the lower molars is almost parallel to their long axis, whereas it is more transverse on the upper molars, because of their different angles in their alveoli. As in *Morganucodon*, the first point on the upper molars which wears is the lingual cingulum.

When we consider wear as seen from the buccal view, there is a possibility of controversy. There can be little doubt, from the pattern of wear on the teeth, that each cusp occludes in the groove between two opposing cusps. Since the primitive single-cusped reptilian teeth alternated, so that each tooth occluded opposite the space between opposing teeth, it may seem logical to suppose that the principal cusp of each tooth will occlude against the point of contact of opposing teeth, as suggested by Simpson (1928). The resulting occlusion in *Trioracodon* would be as shown in Fig. 4**B**. The final molar of *Trioracodon* has only two cusps, and in this case the more distal of these would occlude distally to the distal cusp of M_3, although none of the known specimens in the British Museum shows wear in the appropriate place. The occlusion of the final upper premolar behind the mesial cusp of M_1, however, seems highly unlikely. Moreover, when we consider *Triconodon*, its final upper molar (actually M^4) has three cusps, so that the final cusp would take up a position, as I have shown by the dotted outline, well behind the lower molar.

It seems altogether more probable that the occlusion was as I have indicated in Fig. 4**A**, with a more probable premolar occlusion, and with the more distal cusp of the last upper molar of *Trioracodon* occluding mesially to the distal cusp of the lower molar, while in *Triconodon* the upper distal cusp lies behind the distal cusp of the lower tooth. If this is so, the occlusion was essentially similar to that of *Morganucodon*.

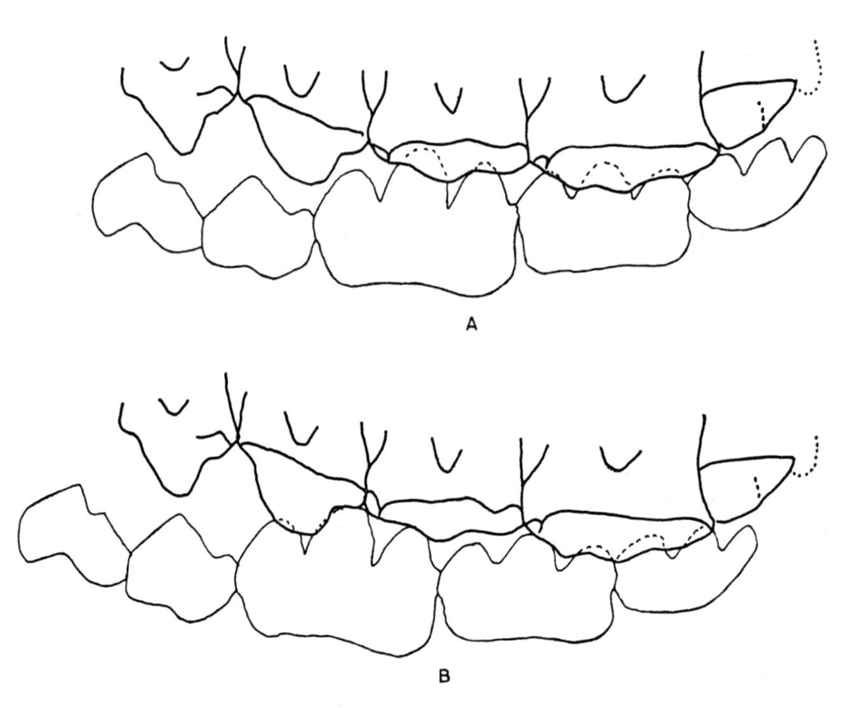

FIGURE 4. Buccal view of last two premolar and all molar teeth of *Trioracodon* showing **A**, occlusion suggested in this paper; **B**, alternative occlusion suggested by Simpson (1928).

(5) The roots of the teeth of the Triconodontinae are essentially similar to those of *Morganucodon*. In several specimens they can be seen to be long, with parallel sides, while in the specimen M21890 the ends of the roots of an upper molar are visible, and apparently dilated as in *Morganucodon*. Moreover, the root of the upper canine shares the dilated shape seen in *Morganucodon*.

(6) The contact point between adjacent molars of the Triconodontinae resembles *Morganucodon* in that the distal cuspule or cingulum lies beneath the mesial cusp of the succeeding molar. It differs in that the distal cingulum lies in a vertical groove in the triconodonts, but not in *Morganucodon*. This may be a development made possible by the greater thickness, bucco-lingually, of the younger genus.

(7) It is well known that in *Morganucodon* the posterior teeth pass on the lingual side of the vertical ramus of the dentary, whereas in the triconodonts they are in line with it, the final molar lying just anterior to it. This is an undoubted difference between the

genera, but presumably the arrangement could have changed and this one factor cannot be held to invalidate the probability of a relationship between *Morganucodon* and the Triconodontinae.

Amphilestinae

The family of Triconodontidae is traditionally divided into two subfamilies, of which this is the second (Simpson, 1928, 1945). Recently a number of authors (e.g. Simpson, 1961; Olson & Patterson, 1961; K. A. Kermack, 1967) have cast doubt on this relationship, and, after examination of the specimens of *Amphilestes* and *Phascolotherium* in the British and Oxford Museums, I would suggest that we should now abandon this arrangement. The English Amphilestinae would seem to belong to the Infra-Class Pantotheria, and their closest relations, so far as tooth morphology is concerned, would seem to be *Kuhneotherium* and the 'obtuse-angled' symmetrodonts such as *Tinodon* from the uppermost Jurassic. Since the known *Amphilestinae* come from the Middle Jurassic, they would lie between these two groups in time. My reasons for this suggestion are as follows:

(1) The dental formula of *Phascolotherium* is $I_4C_1P_2M_5$ and that of *Amphilestes* is $I_?C_?P_3M_5$. While either of these could have been derived from *Morganucodon*, it seems unlikely that they could have given rise to the Triconodontinae. The antemolar teeth of the amphilestines could have been derived either from *Morganucodon* or *Kuhneotherium* so that, as usual, distinctions must be made from the molars, and in this case the upper molars are unknown.

(2) The lower molars have three cusps, whose tips are more or less in line mesio-distally. The central one is the highest and also bulges buccally more than the others, so that below their tips the crowns are definitely angular, with the posterior face more transverse than the anterior. In fact, if this subfamily were known only from a specimen of *Amphilestes* which was exposed from its buccal surface, its resemblance to *Kuhneotherium* would be so striking that a relationship would not be doubted, and this would probably also be true of *Phascolotherium*, although the cusps are rather more substantial than the gracile cusps of *Amphilestes*. The difference from *Kuhneotherium* is apparent on the lingual side, which, instead of being flat or concave as in the pantotheres, is here convex.

(3) The mesial cusp of the Amphilestinae is not a cingulum cusp. There is, anterior to it, a mesial cuspule which is continuous with the lingual cingulum, and neither this nor the distal cuspule take part in the interlocking mechanism.

(4) The lingual cingulum of the Amphilestinae is, as in *Kuhneotherium*, narrow and non-cuspidate. In *Amphilestes* it rises to its highest point lingually to the tip of the principal cusp, as in *Kuhneotherium*, although in *Phascolotherium* there are usually a mesial and distal peak.

(5) *Phascolotherium* has a narrow cingulum on the distal end of the buccal surface of the molars, which is also seen as a vestige in *Amphilestes*. The latter, but not the former, also has a narrow buccal cingulum at the mesial end of the buccal face. Both these features are seen in *Kuhneotherium*, but not *Morganucodon*.

(6) The dental occlusion in *Phascolotherium*, as deduced from the pattern of wear, is essentially of the *Kuhneotherium* type. That is the principal cusp of the upper molar

apparently sheared behind the distal cusp and against the distal cuspule, an arrangement very similar to that in *Kuhneotherium* but very different from that in *Morganucodon*, where the principal upper cusp shears anterior to the lower distal cusp. The groove between the principal and distal cusps of the lower molars of *Phascolotherium* would not have been deep enough to accommodate the principal upper cusp, apart from the evidence of wear. The same is essentially true of *Amphilestes* except that here the distal cuspule of the lower molar is reduced, and the principal upper cusp occludes against the mesial cuspule of the more posterior lower molar. Such a development from *Kuhneotherium* requires only a minor change, and Crompton & Jenkins (1967) have pointed out that this is the occlusal relationship seen in *Tinodon*. I have explained above why I believe the type of occlusion seen in *Kuhneotherium* and, now, the Amphilestinae could not have been derived from *Morganucodon*, and the same arguments would prevent the Triconodontidae from being derived from the Amphilestinae or their close relatives.

Docodonta

This order consists of a single family of three known genera. *Docodon* is represented only in America, and I have not had the opportunity to examine the original specimens. Dr K. A. Kermack has kindly lent me models of upper and lower molars, given to him by Dr F. A. Jenkins, while Professor P. M. Butler has lent me *camera lucida* drawings of upper and lower teeth, made by him at the United States National Museum while in America. On these drawings he has indicated the position of wear facets.

The other two genera are European. *Peraiocynodon* is known by a single specimen of a fragment of dentary with four teeth and a developing fifth; probably the last two premolars and first three molars. It is Purbeckian in age, and is in the British Museum (Natural History), where I have examined it. *Haldanodon*, unlike the other two genera, is Kimmeridgian, and for details I have relied on the published account by Kühne (1968). The two latter genera are very similar; indeed, from comparison of a single broken specimen with the illustration in Kühne's article, I have been unable to find any difference between them, although these may emerge when both are better known.

Kühne (1950) was the first to suggest a relationship between *Morganucodon* and *Docodon*, apparently after discussion with P. M. Butler. Since then Patterson (1956), K. A. Kermack (1965) and Crompton & Jenkins (1968) have supported his view, although Vandebroek (1961) and Parrington (1967) cast doubt on any close relationship. The teeth of the docodontids are apparently specialized for a somewhat different function from those of *Morganucodon*, and this has involved a marked increase on the number of molars in *Docodon* (but not, apparently, in the more primitive European genera). Nevertheless, they must have been derived from a simpler form and of the known upper Triassic groups, it would seem that the morganucodontids are the only possibility. The dental evidence is not overwhelming, but there is certainly nothing to prevent such a derivation, and some points in favour of it. The dental evidence is supported by other evidence in the skull structure (K. A. Kermack, 1967), and in all there seems a fair, if not an overwhelming probability that the docodontids were derived from a member of the Morganucodontidae. The main items of dental interest are as follows:

lamina here quickly disappears. It will be recalled that the upper incisors of *Morganucodon* are small teeth.

(2) Two of the molars in each jaw are large and well-developed teeth. Both have two main cusps, and since the three-cusped condition is so widespread in the earliest mammals, at least in the upper jaw, it seems probable that one has been lost. In both jaws the more mesial cusp develops consistently before the more distal and it is therefore tempting to assume that this represents the original reptilian cone. If so, then the mesial cusp has been lost, as in the docodontids, and the remaining cusps may be designated principal and distal.

(3) The lower molars have a wide lingual cingulum but no buccal cingulum, as in *Morganucodon*. *Sinoconodon* has only a narrow lingual cingulum in this region, although in *Megazostrodon* it is apparently better developed. It would be all too easy to homologize the cuspules of this cingulum with those of *Morganucodon*.

(4) The upper molars have a wide buccal cingulum. This is cuspidate, although the arrangement of the cuspules is not constant. It is noticeably divided by a groove into mesial and distal parts, a feature also seen in *Megazostrodon*, and there is no lingual cingulum, as was probably the case in *Sinoconodon*.

(5) Both Green and Simpson agree that the more mesial cusp of the upper molar occludes between the two lower cusps; if my identification of the cusps is correct, the principal upper cusp occludes between principal and distal lower cusps, as in *Morganucodon*. The more mesial ('principal') lower cusp occludes on the contact area between the two upper teeth. In *Morganucodon* it would occlude between the principal upper cusp and the mesial cusp which has here disappeared. The occlusal evidence would therefore fit in with my identification of the cusps. In order to derive the molars of *Ornithorhynchus* from those of the Sinoconodontidae it would be necessary to assume that in the upper jaw, the principal cusp was represented by the smaller, more distal cusp, the original distal cusp having been lost, while in the lower jaw the situation was as I have already described it. The former arrangement seems more likely.

DISCUSSION

The dentition of *Morganucodon* shows features of both its mammalian status and its reptilian ancestry. The molar occlusion, associated with a considerable degree of lateral jaw movement and a dentary-squamosal contact, has a well-developed shearing mechanism between upper and lower teeth, and a precise occlusal relationship. These features are typically mammalian. On the other hand, the variation in the number of teeth, and the presence of an incisiform tooth in the maxilla, anterior to the canine, are reptilian features as is the form of the canine. The 'replacement' of cheek teeth from the rear, and late eruption of the posterior molars, is a feature of the cynodonts.

Among recent vertebrates, the mammals form a well-defined class, differing from other classes in a number of ways, one of which is the possession of a dentary-squamosal jaw contact. This feature has been regarded as diagnostic because it involves parts which are often preserved as fossils, and it is therefore convenient. When we turn to the period discussed here, in the later Triassic, its importance is very much less. Parrington (1967) has pointed out that the important feature of theriodonts from the Late Triassic was their increased activity, which alone made possible their survival in a world dominated by dinosaurs. This was made possible by the ability to maintain their temperature above that of their surroundings, which in turn would require a vastly increased food supply; especially true in very small animals with a large surface area relative to volume, as is seen in the recent shrews. Food would be more efficiently used if it could be cut into smaller pieces—chewed—before ingestion. This would involve the ability to shear the lower teeth against the upper, which reptiles do not normally possess.

Mussett (1967) has pointed out that with a cutting dentition of the type seen in all the Upper Triassic mammals, pressure exerted on food held between the cheek teeth would place great strain on the jaw joint, tending to dislocate it. For this reason it would need to be strengthened. I have suggested elsewhere (Mills, 1966) that strengthening of the jaw joint is usual where chewing consists of a unilateral, near-vertical shear. It may not be chance that the shrew has also evolved a double jaw-joint. In the immediate ancestors of the mammals not only might chewing dislocate the jaw joint, it could also dislocate the post-dentary bones from the dentary. This may well be the reason for the development of the dentary-squamosal joint. Parrington (1967) and Crompton & Jenkins (1968) have pointed out that chewing, with precisely related upper and lower dentitions, was present in the gomphodont cynodonts. These animals were, however, herbivores, chewing their food between near-horizontal surfaces, and this would place very much less strain on the joint. The jaw joint of recent herbivores is comparatively weak, with a small rounded condyle functioning on a flat fossa. As usual it was the carnivorous group which survived.

The question inevitably arises as to whether mammals are monophyletic, diphyletic or multiphyletic. More correctly, as to where the base of the phylum lies, and whether this coincides with, from among all the characteristics of recent mammals, the development of a dentary-squamosal joint. In the present state of our knowledge only an inspired guess is possible. There is increasing evidence that the therian mammals were derived from, at least, a kuhneotherioid mammal, and that many of the other groups were related to the morganucodontids. If I am correct in thinking that *Morganucodon* and *Megazostrodon* diverged as the shearing contact of the teeth began, then this was probably also the point at which a dentary-squamosal contact first became functional. In this case *Kuhneotherium* certainly diverged at an earlier stage, quite possibly at a much earlier stage. Parallel evolution is a well-known phenomenon, and the approach of the dentary bone to the squamosal was a wide-spread feature of the later mammal-like reptiles, as illustrated by the surely unrelated *Diathrognathus*.

ACKNOWLEDGEMENTS

I wish to express my thanks to the many people who have helped in the preparation of this paper, especially the Keeper of Palaeontology and Dr A. J. Sutcliffe of the British Museum (Natural History) and the Keeper of Geology of the Oxford Museum for their hospitality and for allowing me to work on their specimens. To Father Harold

Rigney for allowing me access to *Morganucodon oehleri* and to *Sinoconodon*. To Professor P. M. Butler for lending me drawings which he made of *Docodon* and to Professor A. W. Crompton for photographs of *Megazostrodon*.

I would especially thank Dr K. A. Kermack for inviting me to join his team at University College, London, in order to work on *Morganucodon*, and him and Mrs F. Mussett for the assistance which they offered me, and for many hours of useful discussion.

ADDENDUM

Immediately after reading this paper at the Symposium on Early Mammals, I received a copy of an important paper by Hopson & Crompton (1969). This covers some of the ground covered in the present communication, and on points of fact the two papers corroborate each other. The main points of difference concern the closeness of relationship between *Morganucodon* and *Kuhneotherium*. The evidence is so inadequate as to make any statements little more than conjecture, and I have discussed the matter fully in connection with the paper by Crompton & Jenkins (1968) see p. 47. The more recent paper, in addition, discusses the marked differences in the brain case.

REFERENCES

Crompton, A. W., 1964. A preliminary description of a new mammal from the Upper Triassic of South Africa. *Proc. zool. Soc. Lond.*, **142**: 441–452.

Crompton, A. W. & Hiiemae, K., 1970. Molar occlusion and mandibular movements during occlusion in the American opossum *Didelphis marsupialis* L. *Zool. J. Linn. Soc.*, **49**: 21–48.

Crompton, A. W. & Jenkins, F. A., 1967. American Jurassic Symmetrodonts and Rhaetic 'Pantotheres' *Science, N.Y.*, **155**: 1006–1009.

Crompton, A. W. & Jenkins, F. A., 1968. Molar occlusion in late Triassic mammals. *Biol. Rev.*, **43**: 427–458.

Green, H. L. H. H., 1938. The development and morphology of the teeth of *Ornithorhynchus*. *Phil. Trans. R. Soc.* (Ser. B), **228**: 367–420.

Gregory, W. K., 1922. *The original and evolution of the human dentition.* Baltimore: Williams & Wilkins Co.

Hopson, J. A. & Crompton, A. W., 1969. Origin of mammals. *Evolutionary biology*, **3**: 15–72.

Jenkins, F. A., 1969. Occlusion in *Docodon* (Mammalia, Docodonta). *Postilla*, **139**: 1–24.

Kermack, D. M., Kermack, K. A. & Mussett, F., 1956. Specimens of new Mesozoic mammals from South Wales. *Proc. geol. Soc.*, No. 1533: 31.

Kermack, D. M., Kermack, K. A. & Mussett, F., 1968. The Welsh pantothere *Kuehneotherium praecursoris*. *J. Linn. Soc.* (*Zool.*), **47**: 407–423.

Kermack, K. A., 1956. Tooth replacement in mammal-like reptiles of the suborders Gorgonopsia and Therocephalia. *Phil. Trans. R. Soc.* (Ser. B), **240**: 95–133.

Kermack, K. A., 1963. The cranial structure of the Triconodonts. *Phil. Trans. R. Soc.* (Ser. B), **246**: 83–103.

Kermack, K. A., 1965. The origin of mammals. *Science Journal*, **1** (9): 66.

Kermack, K. A., 1967. The interrelations of early mammals. *J. Linn. Soc.* (*Zool.*), **47**: 241–249.

Kermack, K. A., Lees, P. M. & Mussett, F., 1965. *Aegialodon dawsoni*, a new trituberculosectorial tooth from the Lower Wealden. *Proc. R. Soc.* (Ser. B), **162**: 535–554.

Kühne, W. G., 1949. On a triconodont tooth of a new pattern from a fissure-filling in South Glamorgan. *Proc. zool. Soc. Lond.*, **119**: 345–350.

Kühne, W. G., 1950. A symmetrodont tooth from the Rhaeto-Lias. *Nature, Lond.*, **116**: 696–697.

Kühne, W. G., 1968. Origin and history of the Mammalia. In E. J. Drake (Ed.), *Evolution and environment*. New Haven: Yale University Press.

Mills, J. R. E., 1955. Ideal dental occlusion in the Primates. *Dent. Practnr. dent. Rec.*, **6**: 47–61.

MILLS, J. R. E., 1964. The dentitions of *Peramus* and *Amphitherium*. *Proc. Linn Soc. Lond.*, **175**: 117–133.
MILLS, J. R. E., 1966. The functional occlusion of the teeth of Insectivora. *J. Linn. Soc.* (*Zool.*), **47**: 1–25.
MILLS, J. R. E., 1967. A comparison of lateral jaw movements in some mammals from wear facets on the teeth. *Archs. otal. Biol.*, **12**: 645–661.
MUSSETT, F., 1967. The phylogeny of the mammalian jaw joint. *Bull. Mammal Soc. Br. Isl.*, pp. 2–3.
OLSON, E. C. & PATTERSON, B., 1961. A triconodont mammal from the Trias of Yunnan. International colloquium on the evolution of lower and non-specialized mammals. *Kon. Vlaamse Acad. Wetensch. Lett. Sch. Kunsten België*, pp. 129–191. Brussels.
PARRINGTON, F. R., 1941. On two mammalian teeth from the Lower Rhaetic of Somerset. *Ann. Mag. nat. Hist.*, **8** (11): 140–144.
PARRINGTON, F. R., 1967. The origin of mammals. *Advmt Sci., Lond.*, **24**: 165–173.
PATTERSON, B., 1956. Early Cretaceous Mammals. *Fieldiana, Geol.*, **13** (1): 1–100.
PEYER, B., 1956. Über Zähne von Haramiyden, von Triconodonten und von wahrscheinlich synapsiden Reptilien aus dem Rhät von Hallau Kt. Schaffhausen, Schweiz. *Schweiz. palaeont. Abh.*, **72**: 1.
RIGNEY, H. W., 1963. A specimen of *Morganucodon* from Yunnan. *Nature, Lond.*, **197**: 1122–1123.
SIMPSON, G. G., 1928. *A catalogue of the Mesozoic Mammalia in the Geological Department of the British Museum* (*Natural History*). London: Br. Mus. (Nat. Hist.).
SIMPSON, G. G., 1929. The dentition of *Ornithorhynchus* as evidence of its affinities. *Am. Mus. Novit.*, No. 390: 1–15.
SIMPSON, G. G., 1945. The principles of classification and a classification of mammals. *Bull. Am. Mus. nat. Hist.*, **85**: 1–350.
SIMPSON, G. G., 1961. Evolution of Mesozoic mammals. *Kon. Vlaamse Acad. Wetensch. Lett. Sch. Kunsten België*, pp. 57–95. Brussels.
VANDEBROEK, G., 1961. The comparative anatomy of the teeth of lower and non specialised mammals. *Kon. Vlaamse Acad. Wetensch. Lett. Sch. Kunsten België*, pp. 219–320. Brussels.

APPENDIX

Family Sinoconodontidae nov.

Diagnosis

Differs from the Morganucodontidae Kühne 1958, in the following particulars:

(1) Lower molars have three cusps, none of which arises from the cingulum. The principal and distal cusps divide well above the alveolar margin, giving the appearance that the distal cusp has split off the main cusp, unlike the Morganucodontidae, where the division is more complete.

(2) At the point of contact between adjacent lower molars, the distal cuspule of the more mesial tooth lies on the lingual side of the mesial cuspule of the more distal, whereas in the Morganucodontidae the former fits below the latter.

(3) In shearing occlusion, the principal cusp of the upper molar shears between the distal cusp and distal cuspule of the lower molars, whereas in the Morganucodontidae it shears between principal and distal cusps of the lower molar.

Megazostrodon Crompton & Jenkins 1966; Hallau I, LVI, LXIV and LXV; Peyer 1956; and '*Eozostrodon*' *problematicus* Parrington 1941 may tentatively be placed in this family.

EXPLANATION OF PLATES

PLATE 1

Last two lower premolars and first four molars of *Morganucodon* seen **A**, from the lingual and **B**, from the buccal. Final upper premolar and first three molars seen **C**, from the buccal and **D**, from the lingual. × 9.

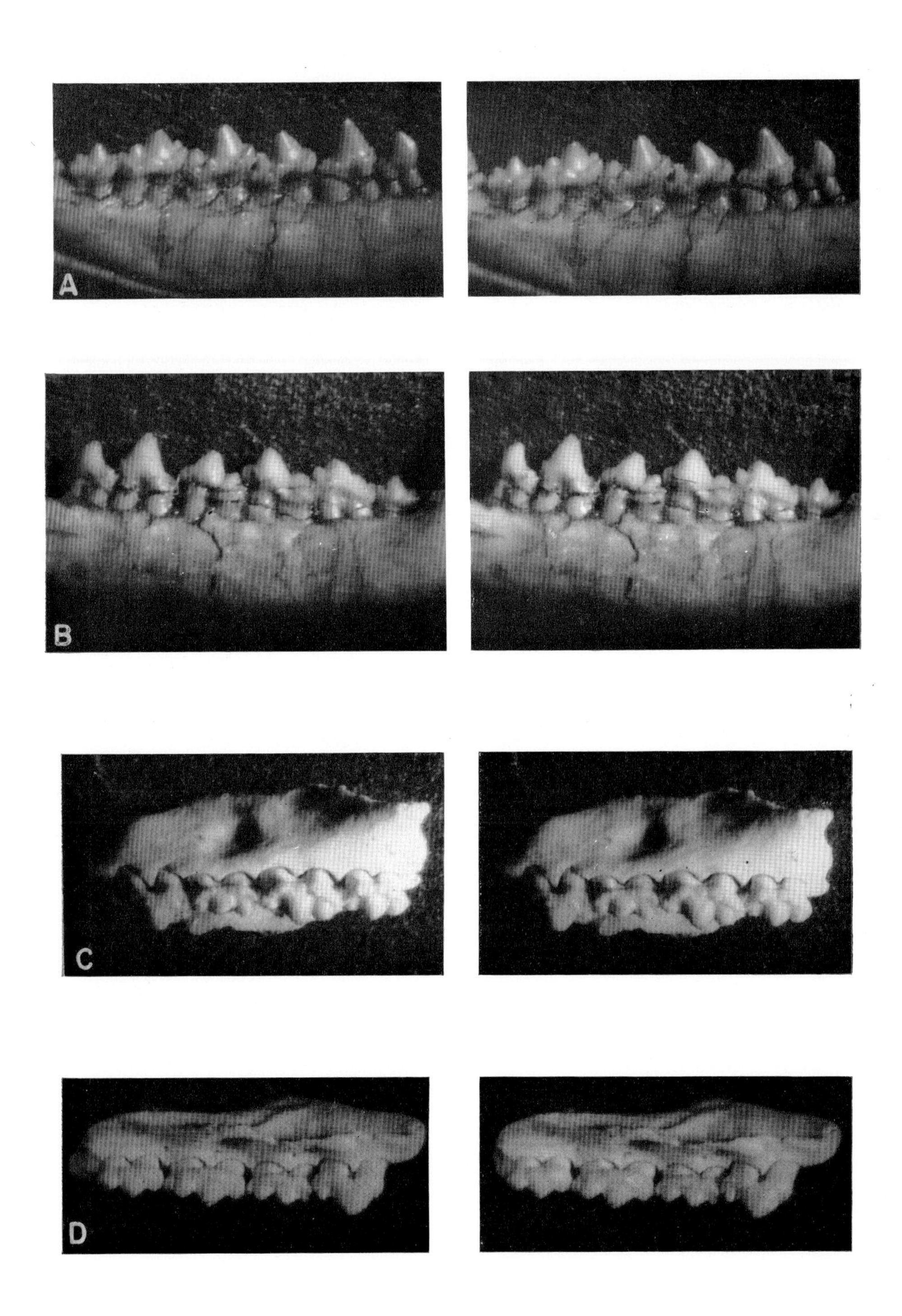
A
B
C
D

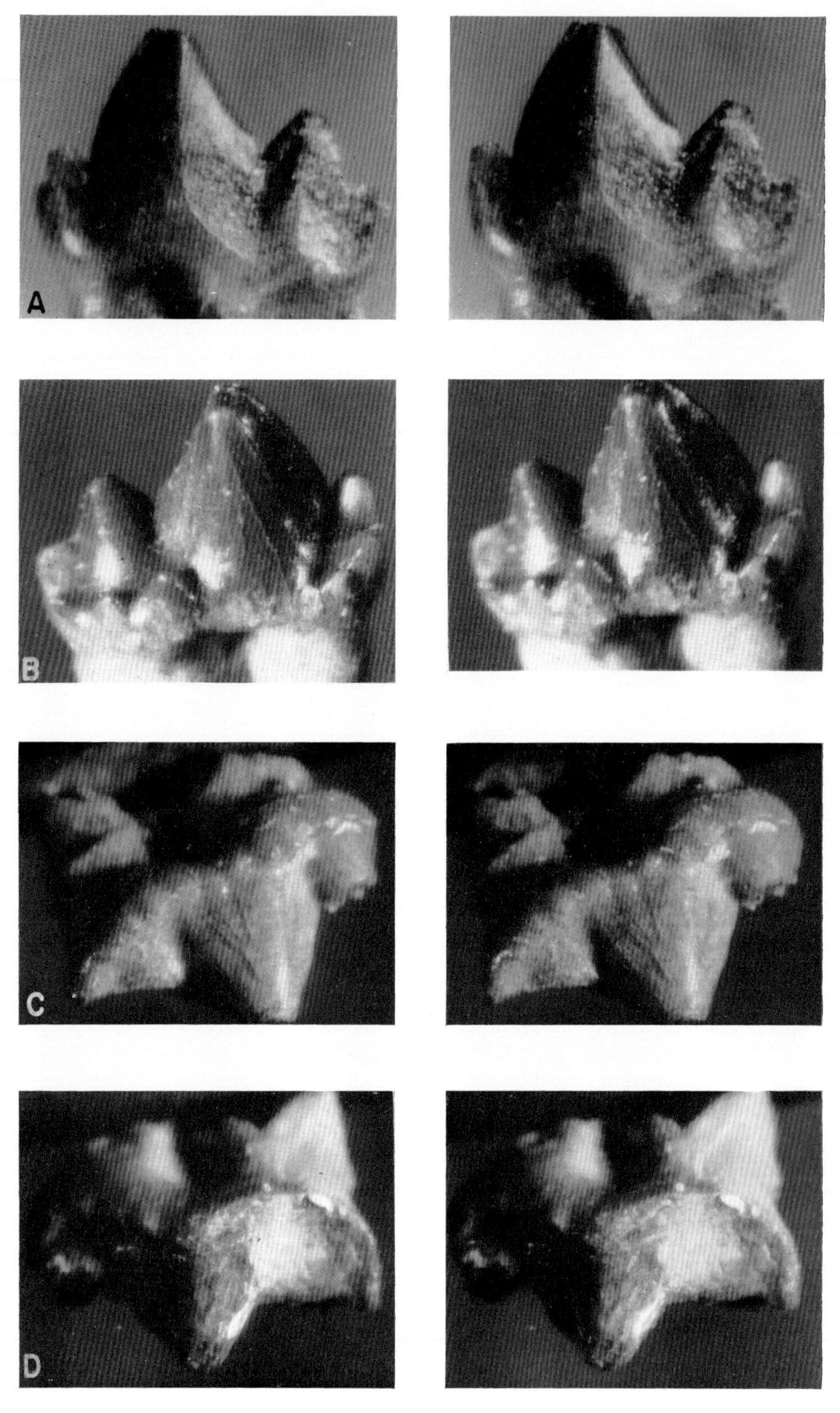

J R. E. MILLS

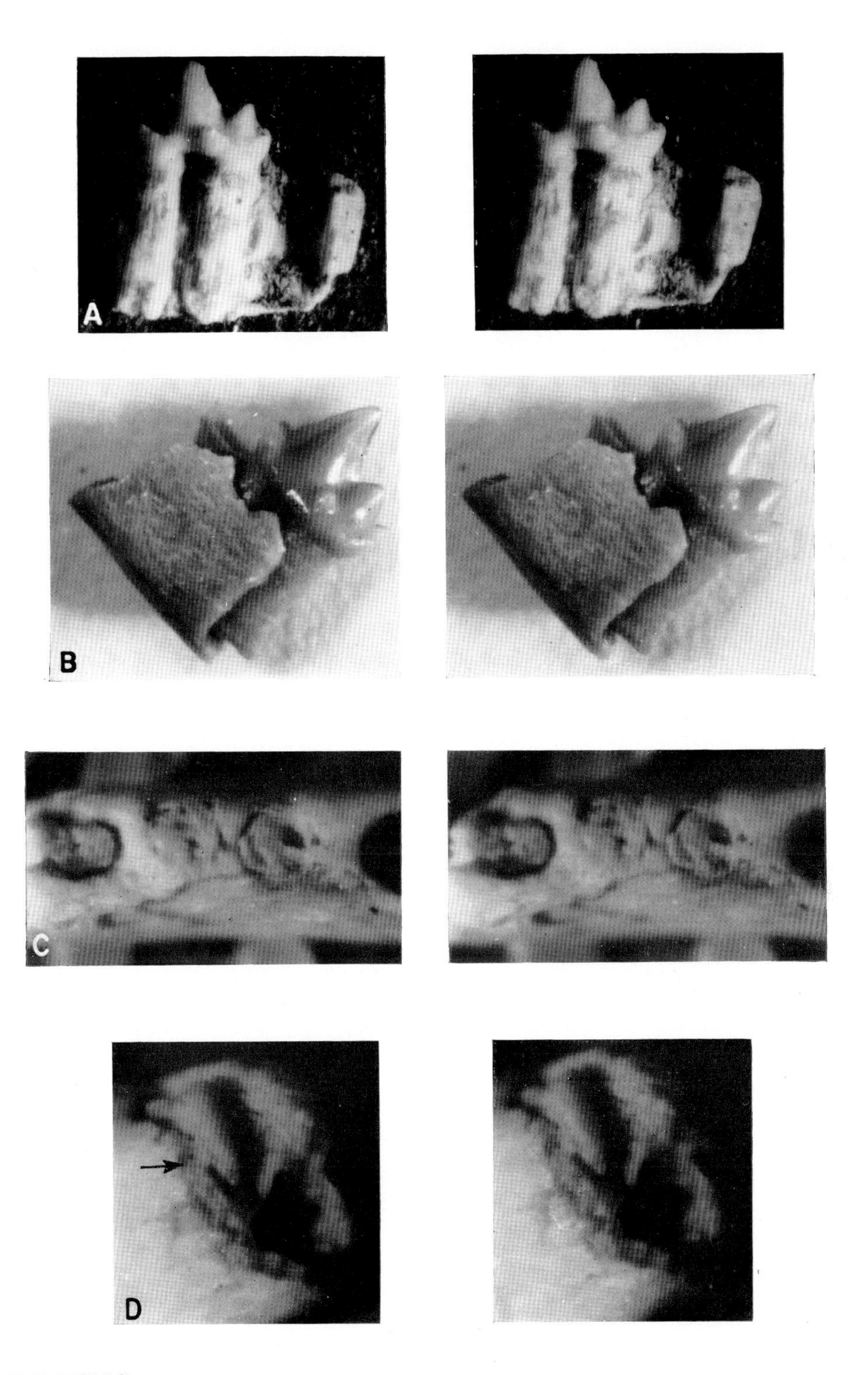

J. R. E. MILLS

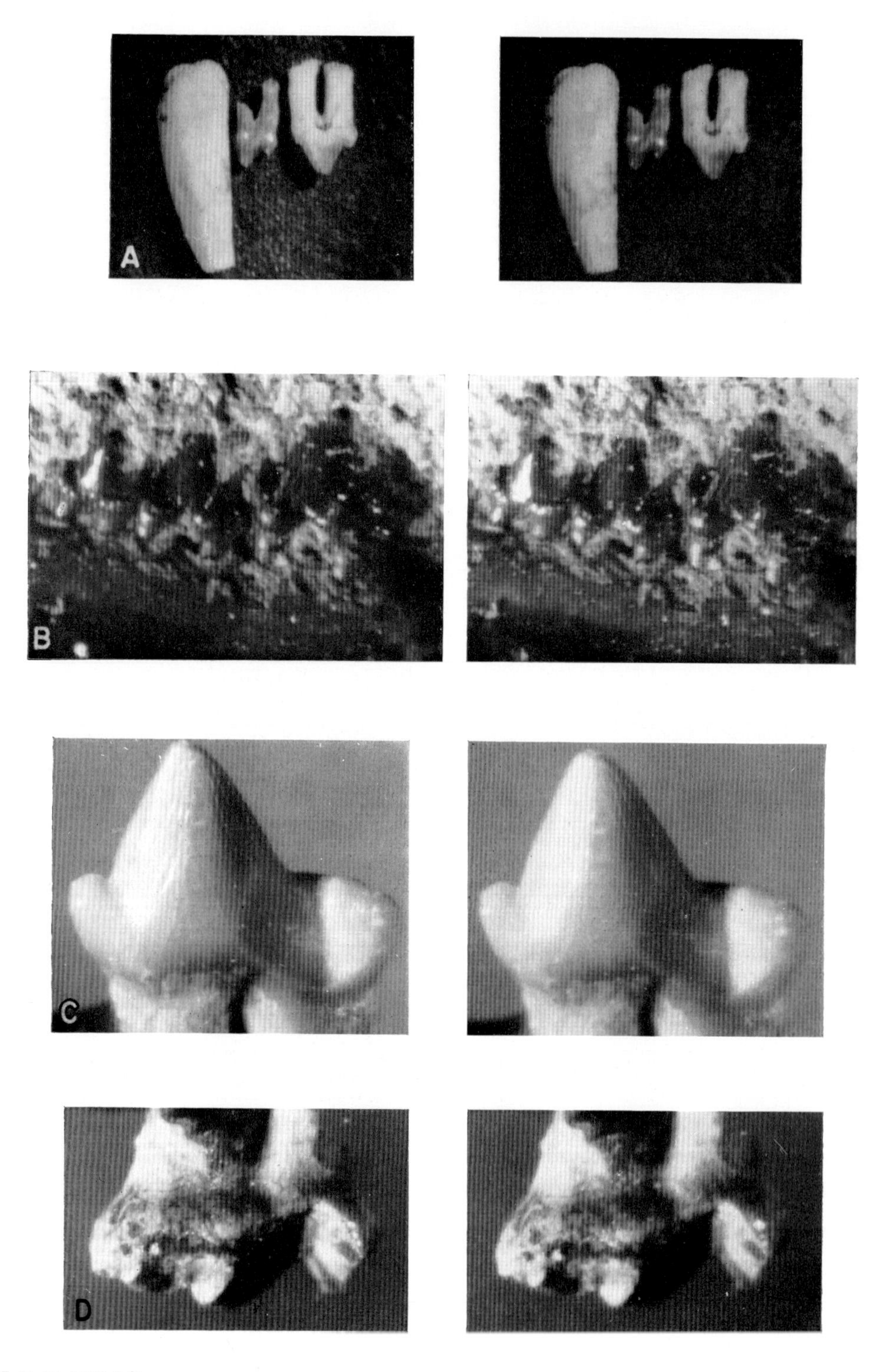

J. R. E. MILLS

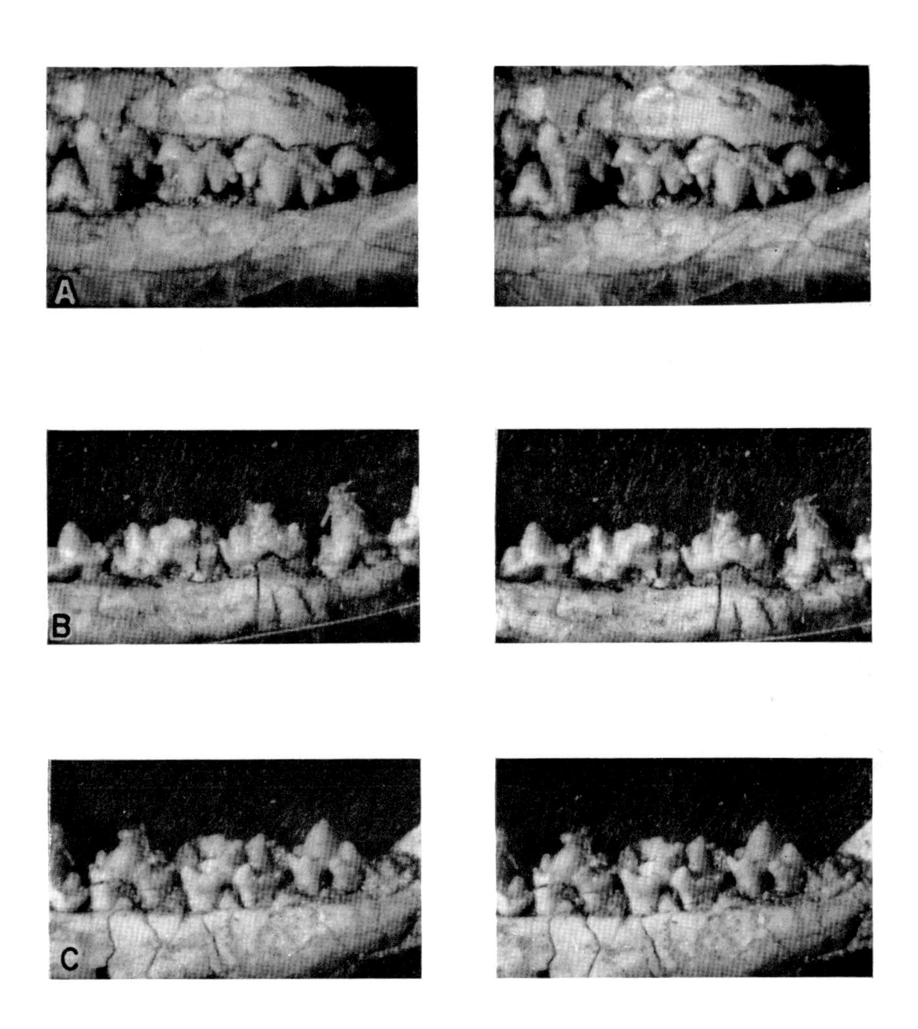

J. R. E. MILLS

PLATE 2

Second molars of *Morganucodon*.
A. Buccal aspect of M_2.
B. Lingual aspect of the same tooth.
C. Buccal aspect of M^2.
D. Lingual aspect of same tooth. × 35.

PLATE 3

A, B. Lower molars of *Morganucodon* with attached alveolus, showing root form. × 35.
C, D. Alveoli of dentary showing roots of lower premolars in process of resorption and replacement by bone. In **D** the arrow indicates a point where bone is invading the root fragment. × 18.

PLATE 4

A. Upper canine and two upper premolars of *Morganucodon*. × 35.
B. Specimen from Ewenny showing apparently last milk molar and M_{1-2} of *Morganucodon*. × 18.
C. Lower first molar of *Morganucodon*, showing early wear. × 35.
D. Upper molar of *Morganucodon* showing advanced wear. Note deep grooving associated with presence of cusp E. × 35.

PLATE 5

Molar teeth of *Morganucodon oehleri*.
A. Upper left last premolar and M^{1-3} buccal aspect.
B. Buccal aspect of lower right M_{1-3}, with last and part of penultimate premolars.
C. Lingual aspect of lower right M_{1-3}, with erupting M_4 and part of last premolar. × 6.

The origin of the tribosphenic molar

A. W. CROMPTON

Department of Biology and Museum of Comparative Zoology, Harvard University, Cambridge, Massachusetts, U.S.A.

Molar occlusion in a series of mammals lying on or close to the phylogenetic line leading to mammals with tribosphenic molars and in Jurassic symmetrodonts and pantotheres is described and discussed. Primitive tribosphenic molars consist essentially of six matching shearing surfaces; three of these were present in the molars of the primitive therian *Kuehneotherium.* In the independent phylogenetic lines arising from Triassic therians and terminating in spalacotherian symmetrodonts, dryolestid and paurodont pantotheres, and in mammals with tribosphenic molars, the shearing surfaces were modified along parallel lines to increase their efficiency but, in addition, in the line leading to eutherians and metatherians further shearing surfaces were progressively added to the simple triangular molars in order to fully exploit transverse mandibular movements. The metacone of tribosphenic molars and cusp 'c' of the upper molars of Triassic therians do not appear to be homologous cusps. The protocone developed initially as a lingual extension which sheared against the posterior surface of the trigonid medial to the crista obliqua; only later were the protocone and the talonid basin developed to form the effective crushing and shearing mechanism which characterized so many of the Cretaceous and Tertiary non-'carnivore' mammals. It is argued that the contact between the paraconid and the lingual cingulum of the matching upper molar was not the precursor of the protocone.

CONTENTS

INTRODUCTION

The purpose of this paper is to review the principal morphological stages involved in the origin of the tribosphenic molar and to attempt to explain the changes involved in functional terms.

In 1956 Patterson fully reviewed the current literature on the evolution of mammalian molars and gave a comprehensive account of the origin of the tribosphenic molar. He concluded that dryolestid pantotheres were possibly ancestral to mammals with tribosphenic molars but in 1964 Mills demonstrated that the main line leading to mammals with tribosphenic molars included or lay close to the following forms: Rhaetic symmetrodonts, *Amphitherium*, *Peramus*, and the Albian therians. When Patterson and Mills wrote these papers only the lower molars of *Amphitherium*, *Peramus*, and the Albian therians had been adequately described. In recent years, however, our knowledge of the molars of Mesozoic therian mammals has been substantially increased. K. A. Kermack, Lees & Mussett (1965) have described from the early Cretaceous a lower molar (*Aegialodon dawsoni*) (Wealden) which clearly shows that tribosphenic molars were present at that time. In this paper they also gave the first adequate description of the oldest known therian molars from the Rhaetic of Wales, first discovered and described by Kühne (1950). Kühne's single specimen (Duchy 33) was discussed by Patterson (1956) and Mills (1964). Subsequently, D. M. Kermack, K. A. Kermack & Mussett (1968) have given a more detailed description of the jaws and teeth of these early therians and named the mammal of which they were part *Kuehneotherium praecursoris*. The molars of this form have also been described and discussed by Parrington (1967), K. A. Kermack (1967), Crompton & Jenkins (1967, 1968), Hopson & Crompton (1969), and Crompton & Hiiemäe (1970). The upper and lower molars of *Peramus* have been described in detail by Clemens & Mills (1970). Patterson (1956) gave a detailed description of mammalian molars from the Albian and recently additional material of the same age has been described and named (*Pappotherium* and *Holoclemensia*) by Slaughter (1965, 1968*a*,*b*).

Recently more attention has been paid to the functional aspects of molar occlusion in primitive mammals. Mills (1955) and Butler (1961) have stressed the importance of homologous wear facets on occluding molars; Patterson (1956) and Mills (1964, 1966, 1967*a*,*b*) have demonstrated the importance of lateral mandibular movements during occlusion; Crompton & Hiiemäe (1969*a*,*b*, 1970) have reported on cine-flourographic studies of mastication in the American opossum; Every (1970 and a series of personal communications) has discussed the relationship between wear facets and mandibular movements; and Crompton & Sita-Lumsden (1970) have discussed the functional significance of the therian molar pattern. It is in the light of this new information that a review of the evolution of the tribosphenic molar is attempted. Occlusal details are better known in forms with tribosphenic molars than in earlier therians with 'pre-tribosphenic' molars. For this reason the structure of the tribosphenic molars of a species of *Didelphodus* will be discussed first, and more primitive forms will be discussed in the following order: *Pappotherium* (Early to Middle Cretaceous); *Aegialodon* (Early Cretaceous); *Peramus* (Late Jurassic); *Amphitherium* (Middle Jurassic); and *Kuehneotherium* (Rhaetic).

MOLAR OCCLUSION IN *DIDELPHODUS*

In order to illustrate the occlusion of a primitive tribosphenic molar, a specimen of *Didelphodus* sp. (Yale Peabody Museum, YPM 23727) was chosen. The molars of this specimen are only slightly worn and wear facets are well preserved. The upper and lower molars were preserved in contact and separated for study.

The molar teeth consist essentially of six shearing surfaces on the vertical or near-vertical slopes of the cusps and ridges connecting the cusps. In Fig. 1 two occluding

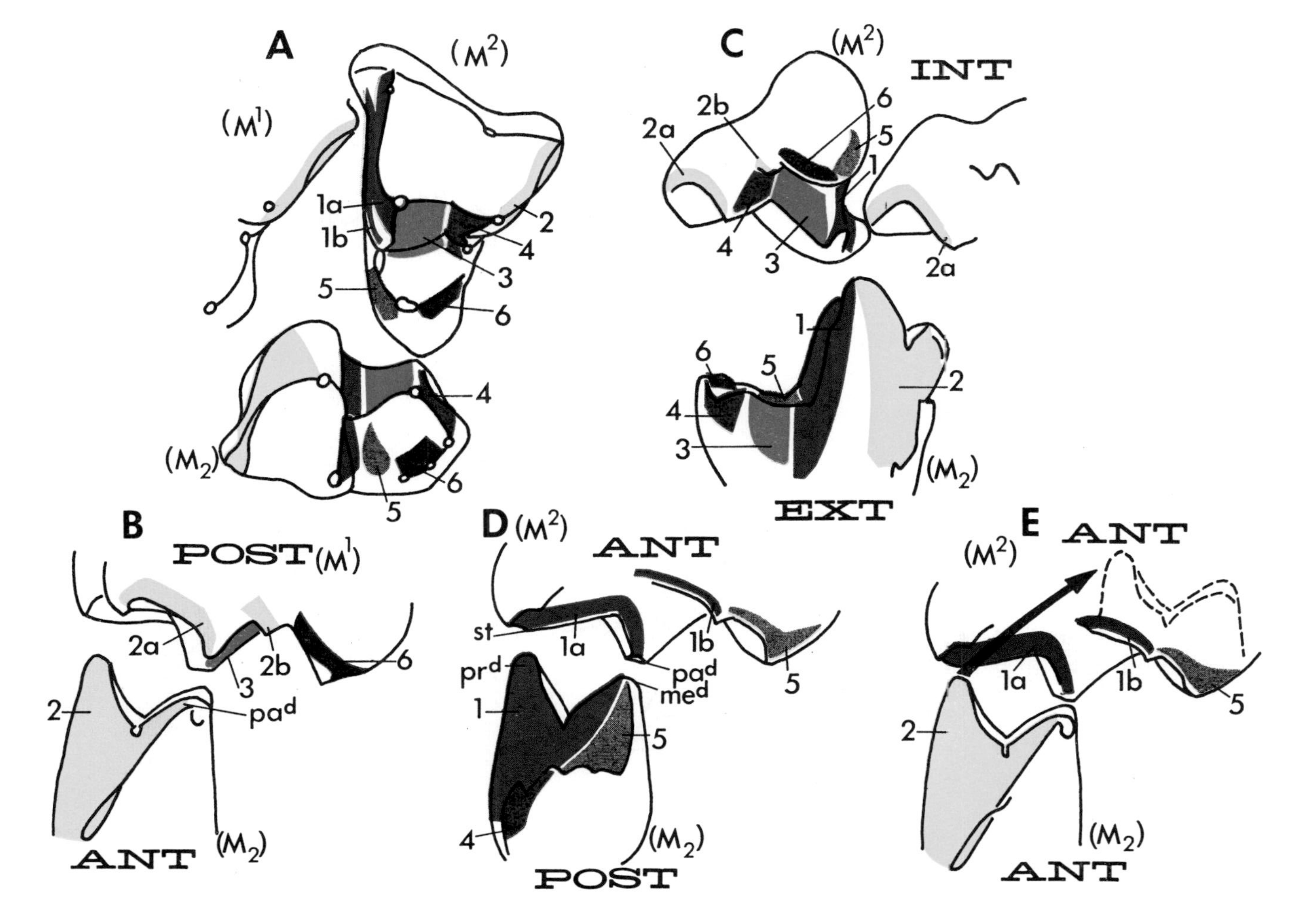

FIGURE 1. *Didelphodus* sp. Series of views of M^2, M^3, and M_2 to illustrate the position, extent, and interrelationship of six matching shearing surfaces on upper and lower molars. **E.** Direction of movement of the lower jaw during the final stages of occlusion.

molars are illustrated so as to show the extent and relationships of matching shearing surfaces on upper and lower teeth. The six shearing surfaces have been numbered *1* to *6* and the approximate extent of each shearing surface is indicated by a distinct colour. This colour code is also used for homologous shearing surfaces in more primitive molars illustrated in Figs 3 to 8. In Fig. 2**A, B** the crowns of opposing molars are shown in occlusal view at the beginning (**A**) of tooth contact and at the end (**B**) of the power stroke of the masticatory cycle. In Fig. 2**A** the cusps are labelled. The leading edges of the six shearing surfaces are labelled in Fig. 2**B**. These are the crests or ridges joining cusps or radiating from cusps. These are illustrated and labelled in terms of current nomenclature (Van Valen, 1969; Szalay, 1969) in Fig. 1**C**. In some cases a single shearing surface on a lower molar matches two shearing surfaces on the corresponding upper molar. As the teeth came into occlusion, the lower surface first

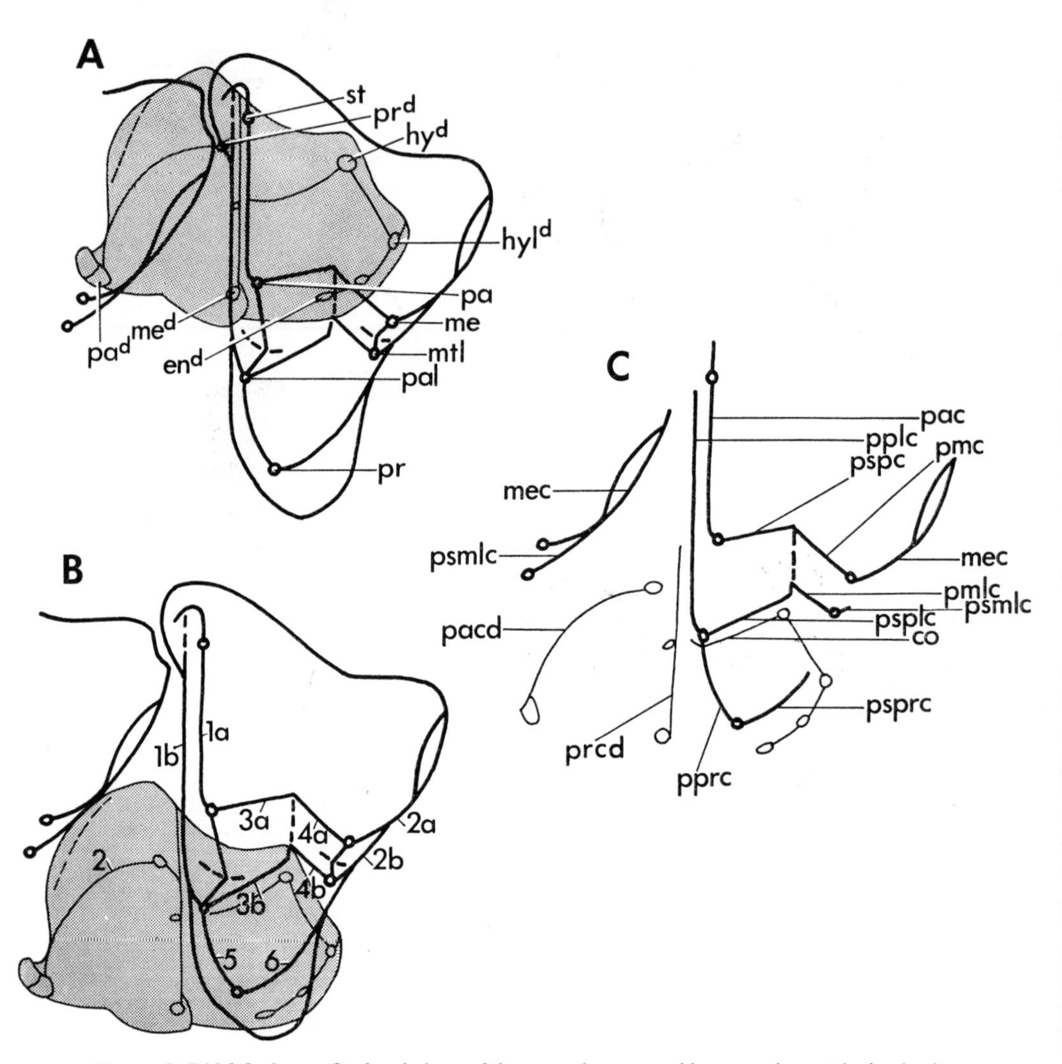

FIGURE 2. *Didelphodus* sp. Occlusal views of the second upper and lower molars at the beginning (**A**) of tooth to tooth contact and end (**B**) of the power stroke. **C**. Diagram of the cutting or leading edges of the six shearing surfaces.

sheared past one and then moved on to the next. In order to avoid confusion the double shearing surfaces have been designated *a* and *b*. For example, the matching surfaces for shearing surface *1* on the lower molar are shearing surfaces *1a* and *1b* on the corresponding upper molar.

The cutting edge of shearing surface *1a* is the crest joining the paracone and stylacone (paracrista, *pac*). In occlusal view (Figs 1**A** and 2**A**) it is linear and transversely orientated relative to the mid-line of the skull; the shearing surface itself is vertically orientated. In anterior view it is crescent-shaped (Fig. 1**D**). The matching shearing surface on the back of the trigonid has a leading edge which connects the protoconid to the metaconid (protocristid, *prc*^{d}) and when seen in posterior view is v-shaped (Fig. 1**D**). Because the lower jaw moved upwards and medially during occlusion (Fig. 1**E**), the inner border of shearing surface *1* on the lower molar (Fig. 1**D**) runs obliquely from the tip of the metaconid to the point where the crista obliqua joins the posterior surface of the trigonid. As the cutting or leading edges of shearing surfaces *1* approached one another, the food to be sheared was trapped in an ellipsoid space which decreased in size as the leading edges passed by one another (R. Every, pers. comm.; Crompton & Sita-Lumsden, 1970). Once the leading edges had passed one another, no further shearing was possible. However, the crescent-shaped ridge joining the paraconule and parastylar region, the preparaconule crista (*pplc*), is the cutting edge of shearing surface *1b* (Fig. 2**D**) and provided a second cutting edge for the protocristid. This shearing edge functioned towards the end of mandibular occlusion. Shearing surface *1* is the largest shearing surface on the molars of *Didelphodus;* it was the first shearing surface to come into contact at the beginning of occlusion and parts of the matching shearing surfaces remained in contact throughout active occlusion. For this reason, more than the other shearing surfaces, it appears to have acted as the principal guide for mandibular movement during the final stages of occlusion.

The cutting edges of shearing surface *2a* run from the metacone backwards to the metastylar region (metacrista, *mec*). The cutting edge of the matching surface on the lower molar joins the protoconid to the paraconid (paracristid, *pac*^{d}). In contrast to the leading edges of shearing surface *1*, those of shearing surface *2*, when seen in crown view (Figs 1**A** and 2**A**) are not transverse but diagonal relative to the mid-line of the palate. In addition they are not linear, as is shearing surface *1*, but uppers and lowers are curved in opposite directions.*

The shearing surface below the paracristid is not vertical, but oblique. This is the reason why in occlusal view (Fig. 1**A**) and buccal view (Fig. 1**C**) a large part of this shearing surface is visible on the lower molar. This is what would be expected if the matching surfaces were to remain in contact for a limited distance as the mandible moves upwards and inwards; this direction of movement would separate diagonal shearing surfaces which were vertically orientated (Crompton & Sita-Lumsden,

* We (Crompton & Sita-Lumsden, 1970) have claimed that this was an adaption to permit both the linear transverse and curved diagonal shearing surfaces of opposing teeth to be used simultaneously. However, Dr Butler has pointed out to us that if the curved surfaces are viewed directly in line with the direction of mandibular movement, the two opposing cutting edges appear as a straight line and not as curves. The shearing surfaces on the diagonal (postvallum and prevallid) are bevelled, or oblique, relative to the vertical axis of mandibular movement. It is this fact, and not the 'curvature in opposite directions', which enables both surfaces of the trigon and trigonid to be used simultaneously.

1970). If the cutting edges of shearing surfaces *2* are viewed from behind, in the case of the uppers, and in front, in the case of the lowers (Fig. 1**B**) they form an ellipsoid space which trapped the food to be sheared in a manner similar to that described for shearing surfaces *1*. In *Didelphodus* the metaconule is not large and a postmetaconule crista (*psmlc*) is reduced. However, in closely related forms from the Cretaceous, e.g. *Cimolestes*, this crista is well-developed and provided a second cutting edge for the paracristid. In contrast to shearing surface *1*, shearing surfaces *2* on the upper and lower molars did not remain in contact throughout occlusion and when the molars were in tight occlusion, they were separated.

Shearing surfaces *3* and *4* are related to the hypoconid and embrasure between the paracone and metacone. The leading edge of shearing surface *3a* is the postparacrista (*pspc*) which sheared past the crista obliqua (*co*) running forwards from the hypoconid (*hy*^{d}). The postparaconule crista (*psplc*, *3b*) duplicated the function of the postparacrista. The premetacrista (*pmc*, *2a*) and the premetaconule crista (*pmlc*, *2b*) sheared past the crest joining the hypoconid and hypoconulid. Shearing surfaces *3* and *4* are obliquely orientated relative to the vertical axis and lie parallel to the direction of jaw movement (upwards and inwards). The crest running externally and slightly forwards from the protocone, the preprotocrista (*pprc*) is the leading edge of a shearing surface (*5*). This surface sheared down the posterior surface of the trigonid medially to shearing surface *1* (Fig. 1**D**). Shearing surfaces *5* tend to be transversely orientated and only came into contact towards the end of active occlusion when the mandible was reaching its most medial position. Consequently, these shearing surfaces appear to have developed to exploit transverse mandibular movements and may be viewed as an extension of shearing surface *1b*. It is for this reason that the protocone, when seen in occlusal view in a primitive tribosphenic molar, lies on a transverse axis which is closer to the protocone than to the metacone.

The crest running backwards and externally from the protocone, the postprotocrista (*psprc*) is the leading edge of shearing surface *6*. The leading edge of the matching shearing surface on the talonid is the crest running forwards from the hypoconulid to the entoconid and on forwards to the base of the posterior surface of the trigonid (Fig. 1**A**,**B**). The protocone-talonid and hypoconid-trigon relationships provided an effective mechanism for crushing, puncturing, and shearing and are the regions which were so greatly modified in many of the more advanced non- 'carnivore' eutherian and metatherian mammals. In some Cretaceous mammals, for example, *Cimolestes* and *Kennalestes*, anterior and posterior cingula are present on the anterior and posterior surfaces of the protocone. These are absent in *Didelphodus* but they were possibly a primitive feature lost in this form. They provided additional cutting edges which initially duplicated the function of shearing surfaces *1b* and *2b*.

The position, orientation, and relative sizes of the shearing surfaces in *Didelphodus* were determined by the direction of mandibular movement and the type of food to be broken down. The teeth are constructed so that some of the shearing surfaces were used simultaneously while others were used successively as the lower jaw moved upwards and inwards as the molars moved into tight occlusion. The principal cusps of the molars provided the necessary elevation for the cutting edges of the shearing surfaces and consequently, the position of the principal cusps of the tribosphenic molar were primarily determined by the position and orientation of the shearing surfaces. Because of transverse mandibular movements, a single shearing surface on the lower molar could successively shear past several shearing surfaces of the upper molar. This is the reason why tribosphenic lower molars are narrower than the matching uppers and why in molar evolution during the Cretaceous the lower molar was the conservative element.

As a result of transverse mandibular movements, food that was sheared during occlusion was deposited into the oral cavity where it could be manipulated by the tongue and again presented to the teeth to be broken down by the multiple cutting blades of the molars (Crompton & Hiiemäe, 1969*a*, *b*, 1970). If the jaw movements were strictly orthal, food would be trapped between the opposing molars. Transverse mandibular movements appear to have been present to some extent in all early mammals and this feature may well prove to be unique to mammals and should perhaps be considered an additional diagnostic feature of Rhaetic mammals. Transverse mandibular movements do not appear to have been present in any of the advanced mammal-like reptiles.

MOLAR OCCLUSION IN *PAPPOTHERIUM*

The type of *Pappotherium pattersoni* is based upon an upper M^2 and M^3 but the lower molar (Shuler Museum of Paleontology, Southern Methodist University, Dallas, Texas, SMP-SMU 61726), also described by Slaughter (1965) from the Trinity, could possibly be a lower molar of *Pappotherium* or at least of a closely related form. Details of the molar occlusion of this genus are illustrated in Fig. 3. Basically the molars are similar to those of *Didelphodus*. Shearing surfaces *1a* and *1b* are transversely orientated and nearly vertical (Fig. 3 **A**). A wide gap separates the cutting edges of shearing surfaces *1a* and *1b* (Fig. 3 **E**). The parastyle and stylacone are large. The metacrista and paracristid (cutting edges of shearing surfaces *2*) are diagonally orientated and curved when seen in crown view. The metacrista supports an additional cusp ('c') half way along its length. The postmetaconule crista (*2b*) is poorly developed. Shearing surfaces *3* and *4* on either side of the hypoconid are present but *4* is relatively much smaller than in *Didelphodus*. Matching shearing surfaces are present on the slopes of the paracone and metacone. In *Pappotherium* the metacone is smaller relative to the paracone than in *Didelphodus* and in contrast to the latter genus, the second rank shearing surfaces *3b* and *4b* (the postparaconule crista and premetaconule crista) are absent. The protocone is small and pointed; consequently, shearing surfaces *5* and *6* are relatively smaller and less important than in *Didelphodus*. The border between shearing surfaces *1* and *5* (Fig. 3 **E**) on the posterior surface of the trigonid is closer to the medial border of the trigonid and the talonid basin is narrower than in *Didelphodus*. In crown view (Fig. 3 **A**) it can be seen that the protocone and the paracone lie on approximately the same transverse axis. Also in this view, the upper molar above the tip of the protocone is narrower (antero-posteriorly) than in *Didelphodus*. Consequently, it is unlikely that the posterior side of the protocone could contact the front surface of the paraconid of the succeeding lower molar. This is relevant when considering molar occlusion in *Aegialodon*.

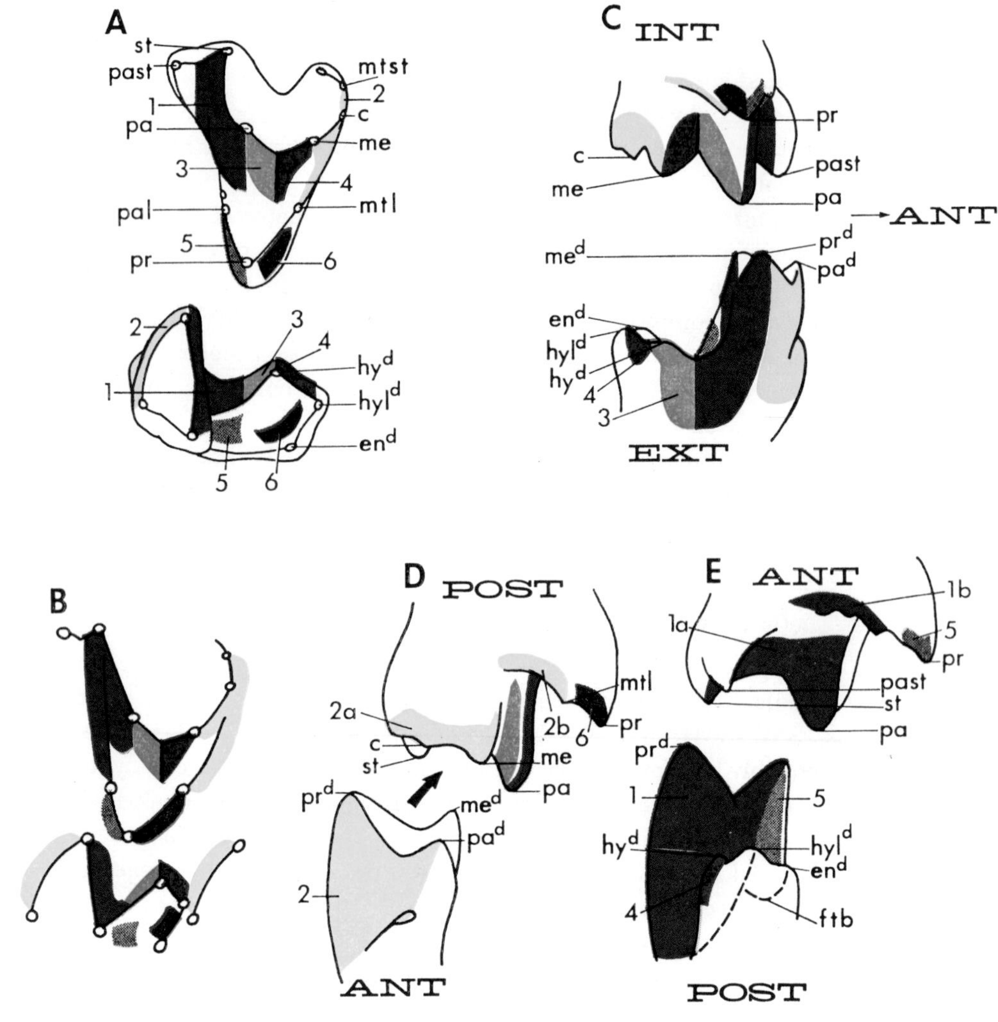

FIGURE 3. Series of views of M^2 of *Pappotherium pattersoni* and unidentified lower molar from the Trinity to illustrate matching shearing surfaces. It is suggested that the lower molar may be associated with the upper molar of *Pappotherium* or a closely related form. The leading edges of the six shearing surfaces are shown diagrammatically in **B**.

THE LOWER MOLARS OF *AEGIALODON DAWSONI*

This genus is based upon a worn, isolated lower molar (K. A. Kermack *et al.*, 1965) from the Lower Wealden (Early Cretaceous). In Fig. 4 an attempt has been made to reconstruct a matching upper molar and the lower molar has been reconstructed in this figure because the type is heavily worn. Shearing surface *1* is transverse but not as vertically orientated as in more advanced forms. It is broken up into two distinct facets (K. A. Kermack *et al.*, 1965, facets d_1 and d_2). This is indicated by a faint line in Fig. 4**E**. It is possible that the paracrista (*1a*) may have been responsible for the outer facet and that a small postmetaconule crista (*1b*) may have been responsible for the medial facet. Shearing surface *2* is well developed and this indicates that a distinct metacone and metacrista were present on the matching upper molar. K. A. Kermack

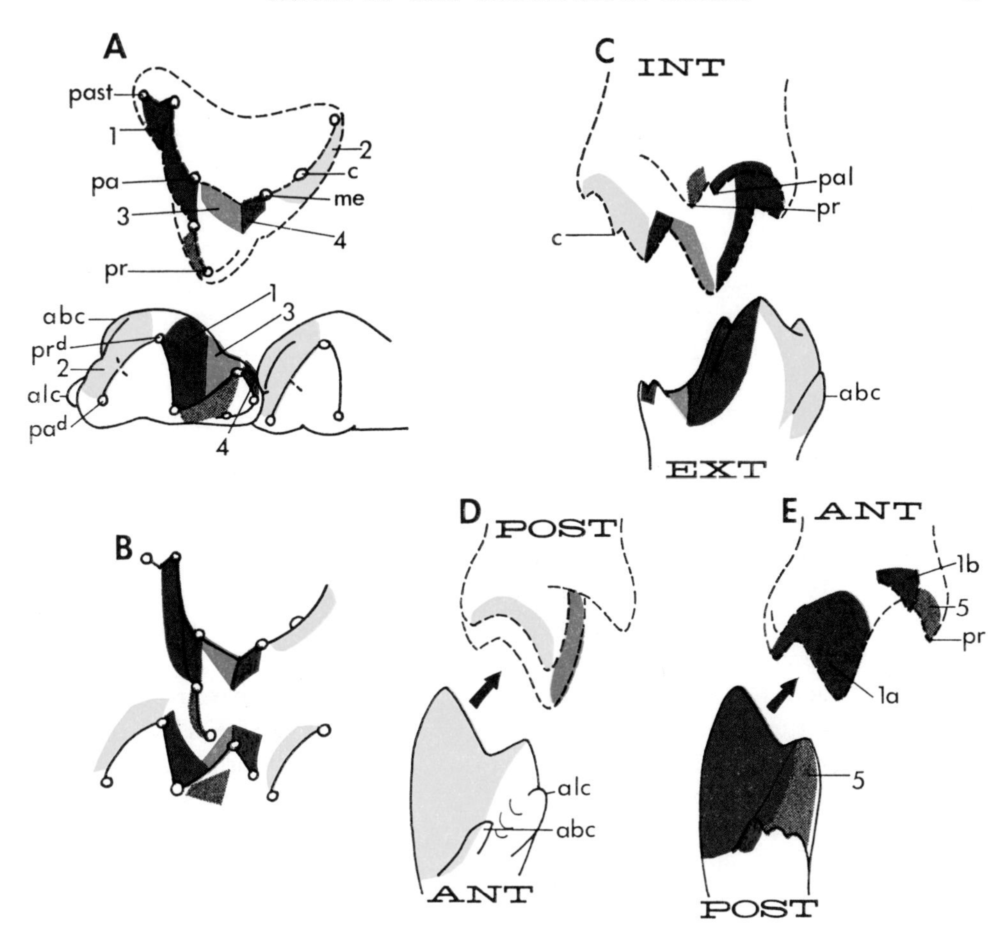

FIGURE 4. *Aegialodon dawsoni.* Series of views of a reconstruction of the type lower molar and a hypothetical matching upper molar to illustrate the position, extent, and interrelationships of five shearing surfaces.

et al. (1965) have suggested that *Aegialodon* had two hypoconids but Mr A. Sita-Lumsden and I have interpreted this to be an artifact resulting either from wear or damage. A small but pronounced shearing surface *3* is present. Its size indicates that the embrasure between the paracone and the metacone was shallow. Shearing surface *4* is very small and this is in agreement with a shallow metacone-paracone embrasure. The talonid basin is short (antero-posteriorly) fairly narrow, and the crista obliqua is directed towards the tip of the metaconid (Fig. 4**E**). However, a clear shearing surface (*5*) is present on the posterior surface of the trigonid medial to the crista obliqua. As has been suggested by K. A. Kermack *et al.* (1965) this was caused by a matching shearing surface on the front of the protocone. The entoconid is small and there is no clear evidence that shearing surface *6* was present. This suggests that the protocone was small and that the molar above the tip of the protocone was, relative to the size of the tooth, narrower than in *Pappotherium* or *Didelphodus*. This view has also been expressed by Clemens & Mills (1970). In forms more advanced than *Aegialodon*, shearing surface *5* enlarged and the medial border of shearing surface *1* on the posterior surface of the trigonid migrated laterally; the talonid basin was enlarged and moulded to fit the enlarging protocone and the ridges flanking the entoconid sheared against the postero-medial surface of the protocone (shearing surface *6*). K. A. Kermack *et al.* (1965) have described oblique wear facets on the antero-dorsal tip of the paraconid and protoconid above a continuous shearing surface on the antero-buccal surface of the trigonid. The latter appears to be homologous with shearing surface *2* of more advanced forms. They have suggested that the facet on the paraconid resulted from tooth to tooth contact between the paraconid and protocone. This relationship is seen in the molars of fairly specialized Mesozoic mammals such as *Gypsonictops*, which have a large protocone. K. A. Kermack *et al.*'s interpretation of the wear facets in *Aegialodon* therefore suggests that the protocone in this form must have been large. In view of *Aegialodon's* phylogenetic position it is unlikely that the protocone would be a large cusp; this cusp is small in *Pappotherium* and absent in *Peramus*. The apical facets on the paraconid and protoconid were probably the result of food abrasion (Crompton & Hiiemäe, 1969*a*, 1970) resulting from cusps being used to puncture food prior to occlusion and were not necessarily formed by tooth to tooth contact. Mills (1964, 1967*a*) has suggested that the precursor of the protocone was initially a contact between the tip of the paraconid and an internal cingulum on the uppers in the early therian, *Kuehneotherium*. K. A. Kermack *et al.* (1965), D. M. Kermack *et al.* (1968), and McKenna (1969) have accepted this view and have referred to the point of contact on the lingual cingulum of the upper molars of *Kuehneotherium* as an incipient protocone. Mills (1964, 1967*a*) has suggested that in forms more advanced than *Kuehneotherium* the contact on the cingulum developed into a distinct cusp (protocone). In a hypothetical form he suggests that the cingulum cusp increased in height and the paraconid reduced in height and that the contact was transferred from the paraconid to the talonid basin by a splitting of the paraconid into two sister cusps: one increasing in height to form the true paraconid and the other, the 'anterior cusp', progressively reducing in height until it was the same height as the talonid basin of the preceding tooth. The anterior cusp (referred to as the 'mesial cuspule' by K. A. Kermack *et al.* 1965 and the 'labial anterior basal cusp' by Mills & Clemens, 1970) is present in *Peramus* and *Aegialodon* (*alc*). Mills has suggested that when the protocone contact was transferrred from the 'anterior cusp' to the talonid basin the 'anterior cusp' was lost. In this way the protocone contact was transferred from the tooth behind to the talonid basin in front. However, the 'anterior cusp' was also present in *Kuehneotherium* and undoubted symmetrodont *Tinodon*, where a true protocone was certainly not present. This suggests that the 'anterior cusp' is a primitive feature and is not a remnant of an earlier protocone-paraconid contact. A contact between the paraconid and the back of the upper tooth above the protocone was present in Mesozoic mammals with large, well-developed protocones. This contact was apparently absent in *Pappotherium* because of the small size of the protocone. It is therefore unlikely that paraconid-protocone contact was present in early therians, lost in Albian therians, and redeveloped in some Late Cretaceous therians.

The lower molar of *Aegialodon* is characterized by an anterobuccal cingulum (*abc*, Fig. 4D). A shallow concavity is present between this cingulum and the antero-lingual cuspule (*alc* = Mills' 'anterior cusp' and K. A. Kermack *et al.*'s 'mesial cuspule').

The talonid of the preceding tooth fitted into this concavity (K. A. Kermack *et al.*, 1965). In later eutherians (e.g. *Pappotherium* and *Didelphodus*) the antero-lingual cusp is lost and the antero-buccal cingulum retained. In primitive metatherians, e.g. *Holoclemensia* and *Didelphis* (Crompton & Hiiemäe, 1970), on the other hand, the antero-buccal cingulum is enlarged and the antero-lingual cusp is retained as a distinct vertical ridge. This may prove to be an additional characteristic to help identify isolated lower molars of primitive metatherians.

MOLAR OCCLUSION IN *PERAMUS*

Clemens & Mills (1970) have described the upper molars of *Peramus* and discussed molar occlusion in this genus in detail. The details of the molars of *Peramus* illustrated in Fig. 5 are based primarily upon drawings made by Mr A. Sita-Lumsden and an excellent series of colour stereo photographs taken by Dr R. Every. Clemens & Mills

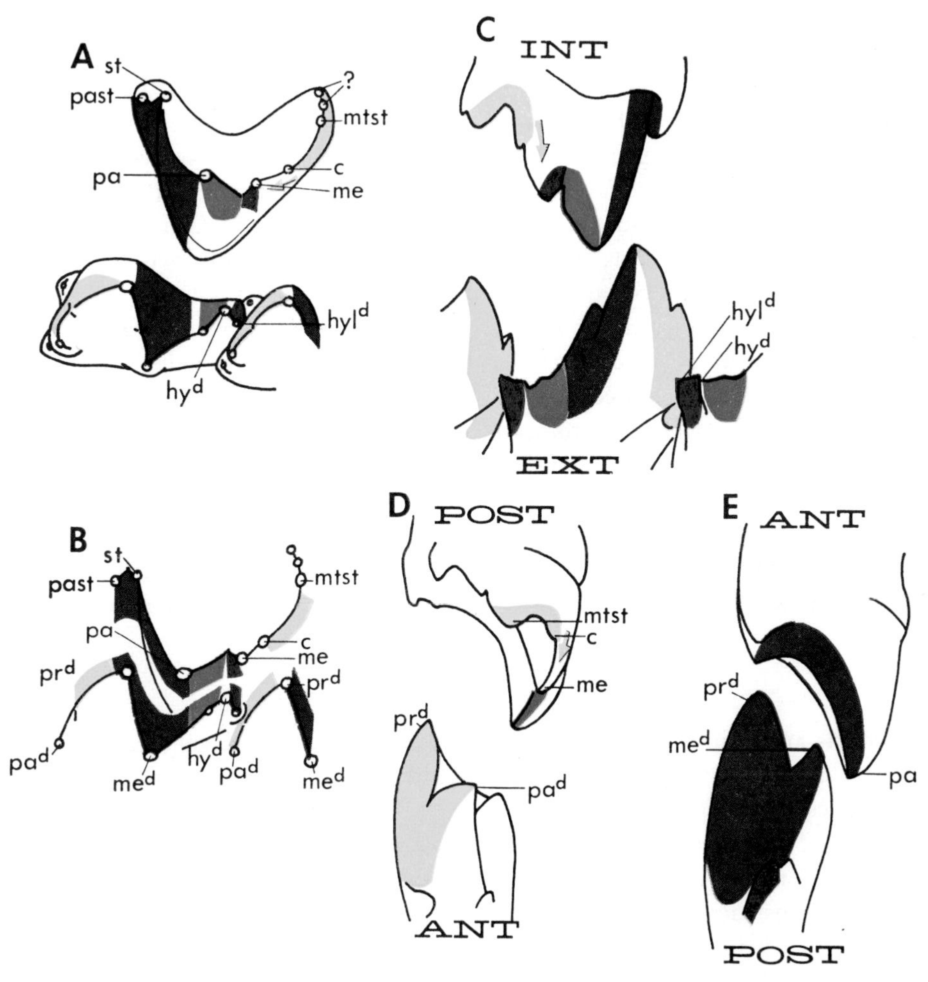

FIGURE 5. *Peramus tenuirostris*. Series of views of molars to illustrate the position, extent, and interrelationships of the four shearing surfaces.

(1970) have pointed out that in *Peramus* the stylacone, parastyle, and lingual cingulum are reduced and for this reason they do not consider it to be directly ancestral to *Aegialodon* and the Albian therians. However, *Peramus* is important because it illustrates an early stage in the development of the metacone and hypoconid.

Shearing surface *1a* on the upper molars (Fig. 5**E**) is prominent although the buccal portion of this surface is poorly developed because of the small size of the stylacone. The matching surface on the posterior surface of the trigonid is prominent (Fig. 5**E**). A paraconule and a preparaconule crista (*1b*) are absent. Shearing surface *1* is transversely orientated and nearly vertical. The metacrista which runs back from the metacone to the metastylar region supports several small cusps. The first, which is present on M^3, is extremely small and has been labelled 'c'. Clemens & Mills (1970) do not recognize this as a distinct cusp and it is apparently absent on M^2. This cusp appears to be homologous with the cusp in the center of the metacrista in *Pappotherium*. The metastylar region supports three small cusps. A distinct wear facet is present above the crest between cusp 'c' and the most buccal of the metastylar cusps. This appears to be homologous with at least part of shearing surface *2a* of more advanced forms. In M^2 this facet extends upwards and lingually to the tip of the metacone. Clemens & Mills (1970) have shown that the metacone of M^2 and M^3 lies close to the paracone and have suggested that a metacone is absent from M^1. The cusp distal to the paracone in M^1, they suggest, is homologous to cusp 'c' of *Kuehneotherium*. The shearing surface in front of the trigonid (*2*) below the paracristid is prominent. Consequently, the orientation of shearing surfaces *1* and *2* are practically identical to those of more advanced forms. The talonid basin is small and a distinct hypoconid and hypoconulid are present. Shearing surface *3* is well developed but shearing surface *4* is minute. The reason for this is that the metacone lies close to the paracone and the embrasure between these cusps is shallow. Also, the hypoconid is close to the hypoconulid. As was suggested by Dr R. Every (pers. comm.) before the upper molar of *Peramus* was discovered, a protocone is absent in this genus as are several of the features correlated with a protocone. For example, shearing surfaces *5* and *6* are absent and the medial border of shearing surface *1* (Fig. 5**E**) lies close to the medial border of the trigonid; the talonid is narrow, is not basined, and the entoconid is either absent or extremely small. The dorsal surface of the talonid, when viewed from behind (Fig. 5**E**), is widely separated from the tips of the protoconid and metaconid.

THE MOLARS OF *AMPHITHERIUM*

Clemens and Mills (1970) have also discussed the lower molars of *Amphitherium* and have concluded that this form could not have been directly ancestral to *Peramus*. Nevertheless, *Amphitherium* is of great interest because it represents a stage preceding the origin of a hypoconid and, possibly, a metacone as well.

The upper molars of *Amphitherium* are not known but an attempt has been made in Fig. 6 to reconstruct an upper molar. Unlike *Peramus* and more advanced therians, a single talonid cusp is present and the talonid overlaps the succeeding lower molar (Fig. 6**A, C**). This is a specialized feature of *Amphitherium*. Shearing surface *1* is the dominant shearing surface and, judging by its structure, the lower jaw moved orthally

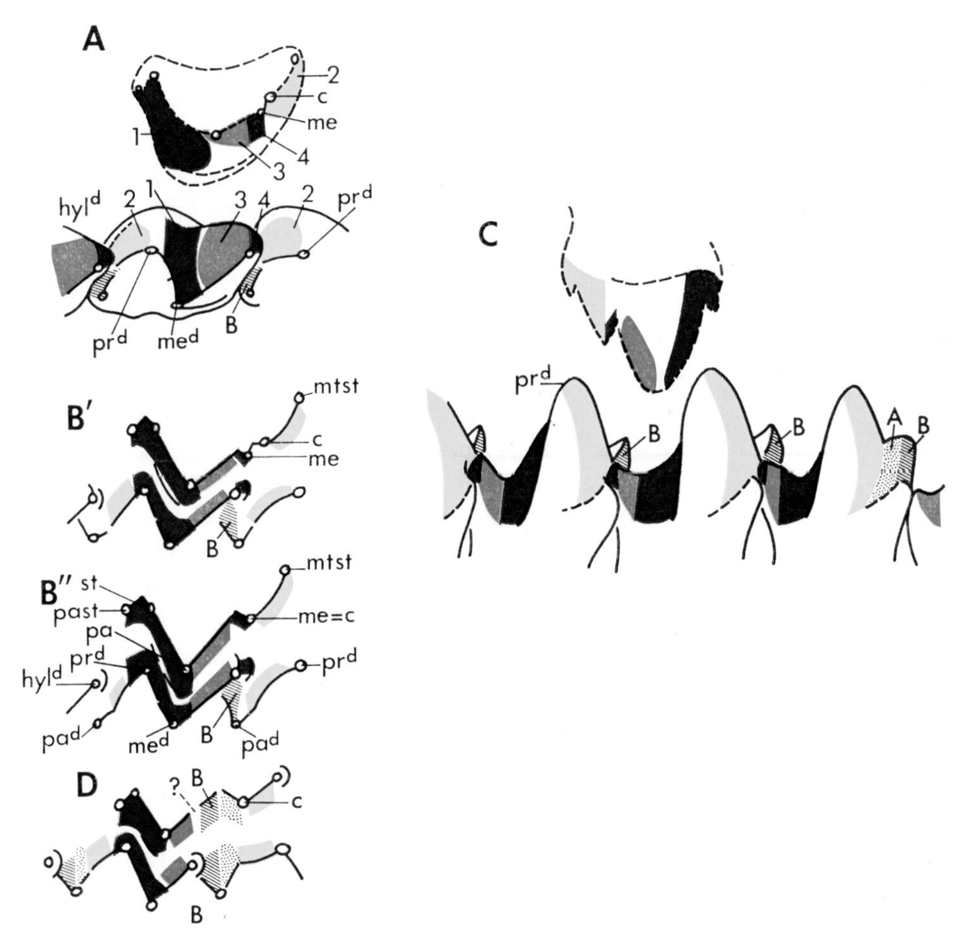

FIGURE 6. *Amphitherium prevostii.* Occlusal and external views of the lower molars and reconstruction of a hypothetical upper molar; **B′** and **B″**, alternate possibilities of cusps and cutting edges of shearing surfaces on the upper molars are illustrated diagrammatically. **D**. Matching shearing surfaces on *Kuehneotherium* molars.

and slightly medially during occlusion. In contrast to more advanced forms, a single continuous shearing surface is not present on the front surface of the trigonid. In the worn posterior molars two distinct facets can be recognized. One lies below the crest running from the protoconid to the notch between this cusp and the paraconid. This shearing surface appears to be homologous with at least the buccal part of shearing surface *2* of more advanced forms. The matching shearing surface on the upper probably lay above the crest joining cusp ‘c’ to the metastylar region. In the anterior molars of *Amphitherium* cusp ‘c’ has worn a distinct groove down the anterior surface of the trigonid (Mills, 1964) but this is absent in the posterior molars. Shearing surface *3* is present below the crest running forwards from the talonid cusp (hypoconulid ?) and a matching shearing surface was presumably present on the postero-lingual surface of the paracone. In worn molars of *Amphitherium* this facet is continuous with a small facet on the antero-buccal surface of the paraconid. This shearing surface is not present in advanced forms and in order to distinguish it from shearing surfaces

1 through *6*, it is referred to as shearing surface *B*. Mills (1964) has described a minute facet on the posterior surface of the talonid cusp.

There appear to be two possible ways to account for the wear facets on the lower molars of *Amphitherium*. It can be argued that both cusp 'c' and an incipient metacone were present (Fig. 6**A**,**B′**) or that cusp 'c' and the metacone were homologous cusps (Fig. 6**B″**,**C**). If the former is correct, a very short premetaconule crista was present in *Amphitherium*, the metacone was small, and cusp 'c' was probably larger than the metacone. It is possible that in forms more advanced than *Amphitherium* a distinct hypoconid arose antero-buccally to the hypoconulid and fitted into the shallow embrasure between the metacone and paracone. As the hypoconid migrated buccally, the length of the crest linking it to the hypoconulid (cutting edge of shearing surface *4*) was increased. Coupled with this, the embrasure between the metacone and paracone was deepened so that the length of the premetaconule crista was increased. The metacone increased progressively in size until it equalled the size of the paracone. As the metacone increased in height, it is probable that cusp 'c' decreased in size and, as a result, shearing surface *2* extended beyond cusp 'c' on to the postero-lingual surface of the metacone and on the lower molars, the matching surface could spread on to the antero-buccal surface of the paraconid. With the development of a hypoconid and metacone and enlargement of shearing surface *3*, the postero-lingual surface of the paracone was separated from the paraconid (Figs 5**A** and 3**B**) and, as a result, shearing surface *B* was lost. This hypothesis on the origin of the metacone is highly tentative and does not rule out the possibility that cusp 'c' of *Kuehneotherium* is homologous with the true metacone of *Peramus* and more advanced forms. An alternate possibility is that in forms transitional between *Kuehneotherium* and *Amphitherium* the talonid grew backwards until its posterior border contacted the antero-lingual surface of the metacone = cusp 'c' (Fig. 6**B″**). This contact could have been enlarged by deepening the embrasure between the paracone and the metacone and splitting the talonid cusp to form a hypoconid and hypoconulid. This point will be returned to below when discussing molar occlusion in *Kuehneotherium*.

In the anterior lower molars of *Amphitherium* (Fig. 6**C**) shearing surface *2* is separated from shearing surface *B* by a narrow facet which has been labelled *A*. A similar shearing surface is present in *Kuehneotherium* (Fig. 6**D**) and the matching surface on the upper molars lies on the antero-lingual surface of cusp 'c'. As shearing surface *A* is not present in the posterior lower molars, it suggests that cusp 'c' of the uppers was reduced in importance and that this was related to the development of an incipient metacone.

THE MOLARS AND OCCLUSION IN *KUEHNEOTHERIUM*

The reconstructions of the molars of *Kuehneotherium* given in Fig. 7 are based upon illustrations in the paper of K. A. Kermack *et al.* (1965), D. M. Kermack *et al.* (1968), and a collection of the molars of this genus at the Museum of Zoology, Cambridge University. Unfortunately, no specimen with the molars *in situ* in the dentary of maxilla has been described.

The principal and most conspicuous shearing surface is *1*. It is essentially transverse and nearly vertical and striations on its surface indicate that the lower jaw moved

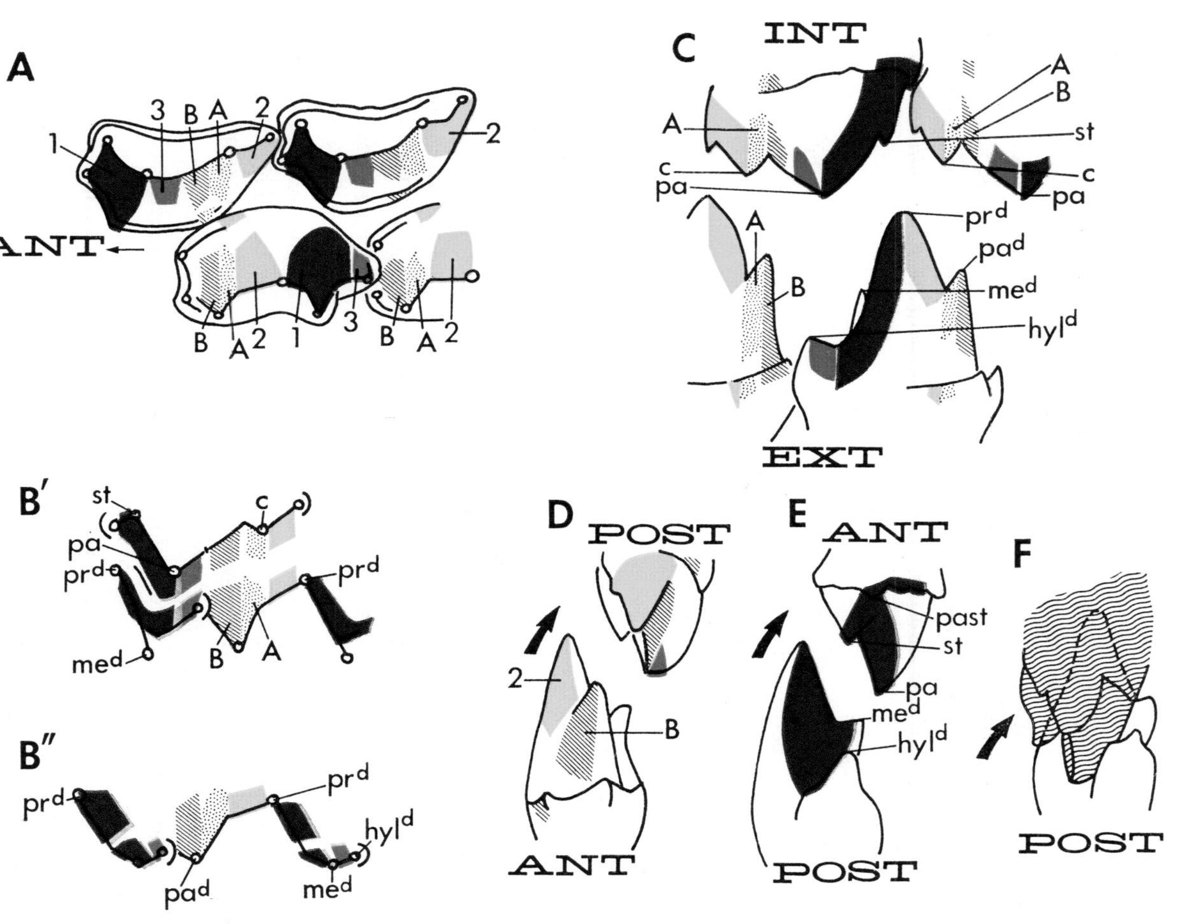

FIGURE 7. *Kuehneotherium praecursoris*. Series of views to illustrate the position and extent of matching shearing surfaces on upper and lower molars. In **B′** the cusps and cutting edges of shearing surfaces are shown diagrammatically and compared with *Tinodon* (**B″**), a Late Jurassic symmetrodont.

upwards and slightly medially during occlusion (Fig. 7**E,F**). As in the case of *Eozostrodon* (= *Morganucodon*), (Parrington, 1967; Crompton & Jenkins, 1968) shearing surfaces were formed by extensive modelling of the structure of the crown by wear, i.e. matching shearing surfaces were not present on freshly erupted upper and lower molars. Wear started on the anterior cingulum of the uppers and extended until a flat vertical and transversely orientated shearing surface was formed. In more advanced therians less of the crown had to be worn away to produce matching shearing surfaces. As in later dryolestids and paurodont pantotheres, *Peramus*, and mammals with tribosphenic molars the trigonid is skewed with the protocristid more transverse than the paracristid. In the upper molars the paracrista is more transverse than the crest joining the paracone, cusp 'c', and the metastylar region. The basic adaption of therian molars, i.e. triangular molars with one shearing surface transverse and vertical and the other diagonal and oblique, was therefore present in *Kuehneotherium* and this feature allies this genus with therian mammals and clearly separates it from nontherian mammals.

Cusp 'c'* is widely separated from the paracone and the paraconid sheared up a v-shaped embrasure between these two cusps while cusp 'c' sheared down a v-shaped embrasure between the protoconid and paraconid. Consequently, unlike *Peramus*, the postero-lingual face of the trigon and the antero-buccal face of the trigonid are divided into three distinct matching shearing surfaces. The shearing surface above the crest running backwards from cusp 'c' to the metastylar region appears to be homologous with the outer part of shearing surface *2* in more advanced forms and a matching surface is present on the antero-buccal surface of the protoconid. Matching shearing surfaces are present on the antero-lingual surface of cusp 'c' of the uppers and the postero-buccal surface of the paraconid. This shearing surface has been labelled *A* and is also present in the anterior molars of *Amphitherium*. Shearing surface *B* on the antero-buccal surface of the paraconid is prominent and a matching shearing surface is present on the postero-lingual surface of the paracone (Fig. 7**A,D**). Shearing surface *3* on the outer surface of the talonid is present, as is the matching surface on the postero-lingual surface of the paracone. In worn molars shearing surfaces *B* and *3* are continuous on the postero-lingual surface of the paracone and this presupposes that the matching surfaces, which on the lowers lay on successive molars (Fig. 7**A**), were also continuous. This may have been the reason why in forms transitional between *Kuehneotherium* and *Amphitherium* the talonid could migrate posteriorly alongside the external surface of the succeeding lower molar without introducing a functional problem. On occasional lower molars of *Kuehneotherium* a minute wear facet is present on the back of the talonid (e.g. specimen No. 54 in the Cambridge Museum of Zoology) behind the talonid cusp (hypoconulid) and in one molar (No. 59), a small facet is present between shearing surfaces *3* and *B* on the postero-lingual surface of the paracone. This may indicate the beginning of shearing surface *4* of more advanced forms and the first stage in the development of a metacone and hypoconid. If this assumption is correct, the metacone arose posteriorly to this small facet and because it

* I have referred to the cusp postero-internally of the paracone as cusp 'c' and have not followed K. A. Kermack *et al.* (1968) in calling it a metacone because there is a possibility that cusp 'c' of *Kuehneotherium* is not homologous with the metacone of *Peramus* and more advanced forms.

lies near the tip of the paracone, it may explain the close proximity of the paracone and the metacone in M^2 and M^3 in *Peramus* and the considerably larger distance separating cusp 'c' from the metacone on M^1. The progressive increase in the distance between the paracone and the metacone in the early therians appears to have been associated with an enlargement of shearing surface *4*. *Peramus* and later therian molars are characterized by a single shearing surface (*2*) on the antero-external surface of the trigonid and postero-internal surface of the trigon whereas *Kuehneotherium* is characterized by multiple facets on these surfaces.* In an earlier paper (Crompton & Hiiemäe, 1970) we figured an occlusal view of the molars of *Kuehneotherium*. At that time we did not realize that the shearing facets on the anterior and posterior surfaces of the trigonid and trigon, respectively, were multiple. *Amphitherium* appears to represent a transitional stage; shearing surface *A* was absent in the posterior molars but not in the anterior molars because cusp 'c' was probably reduced in size in the posterior molars.

Kuehneotherium upper molars are characterized by a prominent external cingulum. In the early stages of wear a small v-shaped groove was worn in the cingulum. This resulted from the paraconid shearing past the upper cingulum and, in worn specimens, the sides of this notch in the cingulum are continuous with facets *A* and *B* (Fig. 7**A, C**). Mills (1964), K. A. Kermack *et al.* (1965), and D. M. Kermack *et al.* (1968) have interpreted the region of the cingulum where the notch is worn by the paraconid as an incipient protocone or as indicating the site where the protocone would develop in advanced forms. D. M. Kermack *et al.* (1968) have claimed that the cingulum 'acted as a stop for the tip of the paraconid to prevent over closure and is the homologue of the protocone'. For reasons discussed above, it is considered unlikely that the protocone-talonid contact of tribosphenic molars was preceded by a protocone-paraconid contact which migrated on to the talonid. It is also unlikely that the paraconid-lingual cingulum contact of *Kuehneotherium* acted as an occlusal stop because the v-shaped notch in the upper cingulum indicates that the protoconid sheared past the cingulum rather than being stopped by it. The lingual cingulum of a *Kuehneotherium* upper molar above the paracrista was also totally obliterated by wear in old teeth. Substantial modelling of the tooth by wear, such as wearing away at the inner cingulum of the uppers to produce shearing surfaces, has also been observed in *Eozostrodon* (Crompton & Jenkins, 1968). The large lingual cingulum is present in the upper molars of both *Eozostrodon* and *Kuehneotherium* and is probably a primitive feature.† It is reduced in *Peramus*. I think it is more probable that the protocone arose in conjunction with the talonid in order to develop a shearing surface on the posterior surface of the trigonid medial to the crista obliqua and a mechanism for effective crushing and puncturing. The development of a hypoconid with shearing surfaces below the cutting edges forming by the crests directed postero-lingually and antero-lingually from this cusp resulted in a widening (bucco-lingually) of the talonid. Once this had taken place it was possible to develop a talonid basin and an effective crushing and shearing mechanism between this basin and the protocone.

* The single facet is associated with a metacone and metacrista and the multiple facets were possibly associated with a large cusp 'c'.

† It is lost in Jurassic dryolestid pantotheres, but redeveloped in Early Cretaceous pantotheres (Henkel & Krebs, 1969).

OCCLUSION IN SYMMETRODONTS AND PANTOTHERES

In Fig. 7**B**″ the principal shearing surfaces on the lower molars of *Tinodon*, a Late Jurassic symmetrodont, are shown diagrammatically. The shearing surfaces in this form are almost identical to those of *Kuehneotherium*. This is especially true for the front of the trigonid where facets *2*, *A* and *B* are all present. Facets *B* and *2* are particularly well shown in Dr R. Every's photograph of the anterolingual view of *Tinodon* molars (Crompton & Jenkins, 1968, plate 3, fig. A). The talonid of *Tinodon* molars is relatively slightly smaller than those of *Kuehneotherium* molars. *Spalacotherium*, a Late Jurassic symmetrodont, is characterized by molars in which the protocristid is transverse and the shearing surface below it vertical, whereas the paracristid is diagonal and the shearing surface below it, oblique. However, unlike *Tinodon* and *Kuehneotherium*, the anterior surface of the trigon forms a single shearing surface rather than the three distinct facets which characterize *Tinodon* and *Kuehneotherium*. The same is true for the posterior surface of the trigon of *Peralestes*. Simpson (1928) has suggested that the molars of *Spalacotherium* and *Peralestes* are from the same type of animal. If this is the case then the spread of shearing surface *2* antero-lingually on the front of the trigonid and posterior surface of the trigon was related to a decrease in the size of cusp 'c' in *Peralestes* and loss of shearing surface *A*.

The same general trend characterized the molars of dryolestid and paurodont pantotheres (Fig. 8**G**). The front surface of the trigon supports a single shearing surface and the leading edge for the matching surface on the upper runs postero-buccally from cusp 'c'. This shearing surface is apparently homologous with shearing surface *2* described for *Peramus* and more advanced forms. In these pantotheres these single shearing surfaces are obtained because cusp 'c' has moved relative to the paraconid so that both cusps lie on the same transverse axis. Therefore, cusp 'c' occupies the same position *vis-à-vis* the paraconid in dryolestid and paurodont pantotheres as the metacone in tribosphenic molars. Shearing surface *1* is large and well developed; *3* is small but this is related to the small talonid. Occlusion in dryolestid pantotheres has been discussed in an earlier paper (Crompton & Sita-Lumsden, 1970). Consequently, advanced symmetrodonts and pantotheres (dryolestids and paurodonts) parallel later therian mammals (*Peramus*, *Aegialodon Pappotherium*) in developing a single shearing surface on the front surface of the trigonid and posterior surface of the trigon behind cusp 'c'. This is possibly related to increased transverse mandibular movements and a more effective shearing mechanism. One large single shearing surface with a crescent-shaped leading, or cutting, edge was presumably more effective than several small shearing surfaces present in the same position.

SUMMARY AND CONCLUSIONS

The principal stages involved in the origin of the tribosphenic molar are discussed. These stages are shown diagrammatically in Fig. 8. In this figure only the main cusps and leading edges of the principal shearing surfaces (colour blocks) are illustrated. In the early therian mammal, *Kuehneotherium*, the lower jaw moved dorso-medially during active occlusion. The evolution of the tribosphenic molar from those of a form

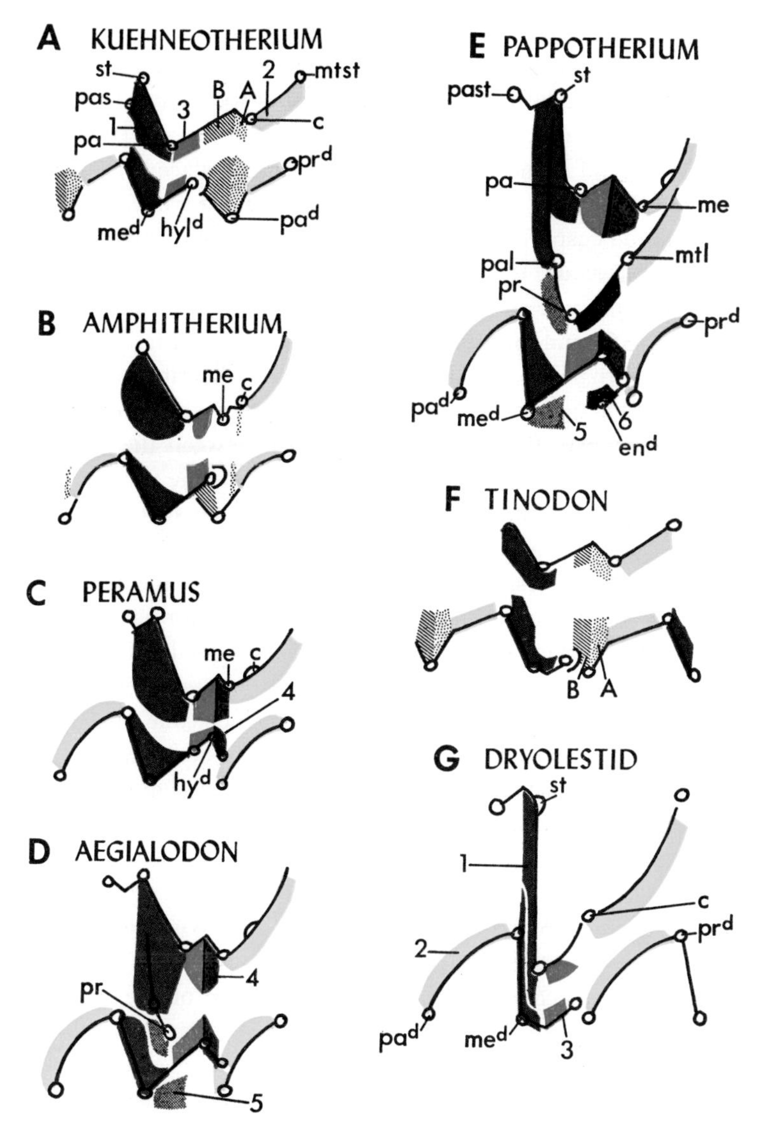

FIGURE 8. **A–E**. Morphological stages in the evolution of the therian molar with cusps, shearing surfaces, and cutting edges shown diagrammatically. **F**, **G**. Diagram of the shearing surfaces in a symmetrodont (*Tinodon*) and a dryolestid.

such as *Kuehneotherium* consisted essentially of increasing the transverse component of mandibular movement, increasing the size and efficiency of some of the shearing surfaces present in the earliest therians, and adding shearing surfaces that were used successively as the lower molars moved upwards and medially. This resulted in a widening of the upper molars and the development of a three-cusped talonid basin.

In *Kuehneotherium* the leading edge of the principal shearing surface (*1*) is transversely orientated when seen in crown view. This shearing surface retains this orientation in tribosphenic molars. The paraconid wears a groove down the posterior surface of the trigon and cusp ‘c’ of the uppers wears a groove down the front face of the trigonid. As a result, the front of the trigonid and posterior surface of the trigon are divided up into distinct facets: *2*, *A*, and *B* on the lowers and *2*, *A*, *B*, and *3* on the uppers. The leading edges of these shearing surfaces were short and the front of the trigonid and posterior surface of the trigon did not form a very effective shearing mechanism. The talonid is short and a small shearing surface (*3*) was present on the external surface of the talonid and the postero-lingual surface of the paracone. In some individuals the talonid cusp wore a small groove in the postero-lingual face of the trigon near the tip of the paracone. It is concluded that this indicates the position where the metacone developed in more advanced forms.

In *Peramus* (Fig. 8**C**) and, presumably, in *Amphitherium* (Fig. 8**B**) the upper molar is transversely wider than the matching lower molar. Shearing surface *1* in *Peramus* is well developed and longer on the uppers than the lowers. In *Peramus* and perhaps in *Amphitherium* a new cusp, the metacone, appeared between the paracone and cusp 'c'. In *Peramus* shearing surface *2* spread forwards on to the paraconid on the lower and on to the metacone on the upper obliterating shearing surface *A*. The metacone and hypoconid developed simultaneously and the crest running between the hypoconid and hypoconulid on the lowers, and antero-buccally from the metacone on the uppers, formed the cutting edges of a new shearing surface (*4*). In forms more advanced than *Peramus* this shearing surface increased in size and, coupled with this, there was an increase in the size of the metacone, deepening of the embrasure between the paracone and the metacone, a separation of the paracone and the metacone, and a widening of the talonid. This separated shearing surfaces *3* on the postero-lingual surface of the paracone from shearing surface *B* on the front face of the paraconid; as a result, shearing surface *B* was lost. The widening of the talonid preceded the development of a protocone.

In *Aegialodon* (Fig. 8**D**) shearing surface *4* increased in size. A new cusp, the protocone, arose on the uppers roughly on the same transverse axis as the paracone. This cusp further increased the width of the uppers relative to the lowers. The anterior surface of this cusp sheared down the posterior surface of the trigonid lingual to shearing surface *1*. This cusp therefore provided a new shearing surface which functioned at the end of active occlusion when the lower molars reached their most lingual position. In addition, the sharp tip of the protocone and the widened talonid provided an excellent mechanism for crushing and puncturing of food.

In *Pappotherium* (Fig. 8**E**) the trigonid is basined and a new cusp, the entoconid, added on the postero-lingual border of the talonid. The crests radiating from the entoconid are the leading edges of a new shearing surface (*6*) which sheared against the posterior surface of the widened (antero-posteriorly) protocone. Shearing surfaces *1* and *2* ceased to be effective for the breakdown of food once the cutting edges (paracrista and protocristid, and metacrista and paracristid) had passed one another. *Pappotherium* was characterized by the development of a second rank of cutting edges, the post-paraconule crista and premetaconule crista. These duplicated the functions of the protocrista and metacrista. The protocristid and paracristid sheared past this second rank of shearing surfaces towards the end of active occlusion. Because the lower jaw moved dorsomedially, the second rank shearing surfaces lies dorso-medially of the first rank. For effective shearing it is essential that the leading edges of the shearing surfaces (when viewed from a position perpendicular to the plane of the shearing surfaces) be concave. The leading edges of opposing shearing surfaces then form an ellipsoid space in which food to be sheared can be trapped. This appears to be the reason why cusps are present at both ends or at the anterior end of the cutting edges of the principal shearing surfaces; paracone and stylacone for *1a*, paraconule for *1b*, metacone and metastyle for *2a*, metaconule for *2b*, and protocone for *5* (see Fig. 1**B**,**D**).

In mammals with tribosphenic teeth, but more advanced than *Pappotherium*, shearing surfaces *3* and *4* are also duplicated and the postparaconule and premetaconule crista form the leading edges of a second rank of shearing surfaces (Fig. 2).

It is concluded that the contact point between the paraconid and lingual cingulum of the upper molars of *Kuehneotherium* did not represent the precursor of the protocone as suggested by Mills (1964) and D. M. Kermack *et al.* (1968). Therefore, it is unlikely that in the evolution of the tribosphenic molar the protocone migrated forwards relative to the paracone. The protocone, it is concluded, arose medially of the paracone and the widening of the talonid preceded the development of the protocone.

It is shown that the pattern of wear facets on the molars of *Kuehneotherium* is essentially the same as that of the Jurassic symmetrodont, *Tinodon*, and distinct from the pattern present in the Dryolestidae, Paurodontidae, Peramuridae, Spalacotheriidae, and therian mammals with tribosphenic molars. It is concluded that the single shearing surface on the front of the trigonid arose independently and by different routes in the lines leading to (1) *Peramus* and forms with tribosphenic molars; (2) Spalacotheriidae; and (3) Paurodontidae and Dryolestidae. In the first line the cutting edge of the matching shearing surface of the upper molar runs from the metacone to the metastylar region; in the second line, from the paracone to the metastylar region; and in the third line, from cusp 'c' to the metastylar region. The single shearing surfaces appear to be related to increased transverse mandibular movement and the development of a more effective shearing mechanism. In all these lines the transverse width of the upper molar relative to the lower was increased.

ACKNOWLEDGEMENTS

I wish to thank Dr F. R. Parrington for his help, advice, and permission to study, photograph, and draw several of the *Kuehneotherium* molars in the collection of the Museum of Zoology, Cambridge; Drs W. A. Clemens and J. R. E. Mills for permission to study and draw the upper molars of *Peramus* while they were still studying this specimen; Dr K. A. Kermack for permission to study the type molar of *Aegialodon;* Dr B. H. Slaughter for permission to study both his described and undescribed collection of therian molars from the Trinity; and Dr H. W. Ball, Keeper of Palaeontology at the British Museum (Natural History), for permission to study the collection of Mesozoic mammals housed in the British Museum. I am particularly grateful to Dr R. G. Every for copies of stereo photographs of the molars of *Peramus*, *Amphitherium*, and some dryolestids and for informative discussions.

Mr A. Sita-Lumsden studied and prepared drawings of several of the specimens discussed in this paper, Mr C. Schaff prepared the specimens of *Didelphodus* and dryolestid pantotheres, Miss M. Estey prepared the drawings, and Mrs G. Dundon typed the manuscript. To all these co-workers I am extremely grateful.

REFERENCES

Butler, P. M., 1961. Relationships between upper and lower molar patterns. International colloquium on the evolution of lower and non-specialized mammals. *Kon. Vlaamse Acad. Wetensch. Lett. Sch. Kunsten België*, Part 1, 117–126. Brussels.

Clemens, W. A. & Mills, J. R. E., 1971. Review of *Peramus tenuirostris* Owen (Eupantotheria Mammalia). *Bull. Br. Mus. Nat. Hist.* (Geol.), **20**, No. 3, 89–113.

Crompton, A. W. & Hiiemäe, K., 1969*a*. Functional occlusion in tribosphenic molars. *Nature, Lond.*, **222**: 678–679.

Crompton, A. W. & Hiiemäe, K., 1969*b*. How mammalian molar teeth work. *Discovery* (Magazine of the Peabody Mus. Nat. His., Yale Univ.), **5**: 23–34.

Crompton, A. W. & Hiiemäe, K., 1970. Molar occlusion and mandibular movements during occlusion in the American opposum, *Didelphis marsupialis* L. *Zool. J. Linn. Soc.*, **49**: 21–47.

Crompton, A. W. & Jenkins, F. A., Jr., 1967. American Jurassic symmetrodonts and Rhaetic 'Pantotheres'. *Science, N.Y.*, **155**: 1006–1009.

Crompton, A. W. & Jenkins, F. A., Jr., 1968. Molar occlusion in late Triassic mammals. *Biol. Rev.*, **43**: 427–458.

Crompton, A. W. & Sita-Lumsden, A., 1970. Functional significance of the therian molar pattern. *Nature, Lond.*, **227**: 197–199.

Every, R. G., 1970. Sharpness of teeth in man and other primates. *Postilla*, **143**: 1–30.

Henkel, S. & Krebs, B., 1969. Zwei Säugetier-Unterkiefer aus Uña (Prov. Cuenca, Spanien). *Neues Jb. Geol. Paläont. Mh.*, **8**: 449–463.

Hopson, J. A. & Crompton, A. W., 1969. Origin of mammals. In Th. Dobzhansky *et al.* (Eds), *Evolutionary biology*, **3**: 15–72. New York: Appleton-Century-Crofts.

Kermack, K. A., 1967. The interrelations of early mammals. *J. Linn. Soc.* (*Zool.*), **47**: 241–249.

Kermack, D. M., Kermack, K. A. & Mussett, F., 1968. The Welsh pantothere *Kuehneotherium praecursoris*. *J. Linn. Soc.* (*Zool.*), **47**: 407–423.

Kermack, K. A., Lees, P. M. & Mussett, F., 1965. *Aegialodon dawsoni*, a new trituberculosectorial tooth from the lower Wealden. *Proc. R. Soc.* (B), **162**: 535–554.

Kühne, W. G., 1950. A symmetrodont tooth from the Rhaeto-Lias. *Nature, Lond.*, **166**: 696–697.

McKenna, M. C., 1969. The origin and early differentiation of therian mammals. *Ann. N.Y. Acad. Sci.*, **167**: 217–240.

Mills, J. R. E., 1955. Ideal dental occlusion in primates. *Dent. Practnr, Bristol*, **6**: 47–61.

Mills, J. R. E., 1964. The dentitions of *Peramus* and *Amphitherium*. *Proc. Linn. Soc. Lond.*, **175**: 117–133.

Mills, J. R. E., 1966. The functional occlusion of the teeth of Insectivora. *J. Linn. Soc.* (*Zool.*), **47**: 1–25.

Mills, J. R. E., 1967*a*. Development of the protocone during the Mesozoic. *J. dent. Res.*, **46**: 883–893.

Mills, J. R. E., 1967*b*. A comparison of the lateral jaw movements in some mammals from wear facets on the teeth. *Archs oral Biol.*, **12**: 645–661.

Parrington, F. R., 1967. The origin of mammals. *Advmt Sci., Lond.*, **24**: 1–9.

Patterson, B., 1956. Early Cretaceous mammals and the evolution of mammalian molar teeth. *Fieldiana, Zool.*, **13**: 1–105.

Simpson, G. G., 1928. *A catalogue of the Mesozoic Mammalia in the Geological Department of the British Museum* (*Natural History*). London: Br. Mus. (Nat. Hist.).

Slaughter, B. H., 1965. A therian from the lower Cretaceous (Albian) of Texas. *Postilla*, **93**: 1–18.

Slaughter, B. H., 1968*a*. Earliest known marsupials. *Science, N.Y.*, **162**: 254–255.

Slaughter, B. H., 1968*b*. *Holoclemensia* instead of *Clemensia*. *Science, N.Y.*, **162**: 1306.

Szalay, F. S., 1969. Mixodectidae, Microsyopiday, and the insectivore-primate transition. *Bull. Am. Mus. nat. Hist.*, **140** (4): 193–330.

Van Valen, L., 1969. Deltatheridia, a new order of mammals. *Bull. Am. Mus. nat. Hist.*, **132**: 1–126.

ABBREVIATIONS USED IN FIGURES

abc antero-buccal cuspule (cingulum)
alc antero-labial cuspule
c cusp 'c'
co crista obliqua
en^d entoconid
ftb floor of talonid basin
hy^d hypoconid
hyl^d hypoconulid
me metacone
mec metacrista
me^d metaconid
mtl metaconule
mtst metastyle
pa paracone
pac paracrista
pac^d paracristid
pa^d paraconid
pal paraconule
past parastyle
pmc premetacrista
pmlc premetaconule crista
pplc preparaconule crista
pprc preprotocrista
pr protocone
pr^d protoconid
prc^d protocristid
psmlc postmetaconule crista
pspc postparacrista
psplc postparaconule crista
psprc postprotocrista
st stylacone
1a, 1b–6 matching shearing surfaces

Evolution of the mandible and lower dentition in dryolestids (Pantotheria, Mammalia)

BERNARD KREBS

Lehrstuhl für Paläontologie, Freie Universität, West Berlin, Germany

The lower jaws of the oldest known dryolestids (from the Kimmeridgian in Portugal) are here compared with those of the youngest known dryolestids (from the Lower Cretaceous in Spain). The Kimmeridgian forms still possessed a coronoid and a splenial, as well as a Meckelian cartilage which persisted in the adult. These reptilian elements are missing in the dryolestids of the Lower Cretaceous.

The presence of eight molars, which remains constant into the Cretaceous, is interpreted as a primitive feature inherited from the therapsids of the Triassic. The pattern of the lower molars is so specialized that the dryolestids are disqualified as ancestors of the modern Theria. They have, however, perfected the mechanism for sharpening the teeth (*thegosis*, R. G. Every) parallel to the modern Theria. The interlocking of the molars in the dryolestids of the Lower Cretaceous is related to this process.

Der Unterkiefer der bisher ältesten Dryolestiden (aus dem Kimmeridgien von Portugal) wird mit jenem der bisher jüngsten Dryolestiden (aus der Unteren Kreide von Spanien) verglichen. Die Formen des Kimmeridgien besassen noch ein Coronoid und ein Spleniale sowie einen zeitlebens persistierenden Meckel'schen Knorpel. Diese reptilischen Elemente sind bei den Dryolestiden aus der Unteren Kreide verschwunden.

Das bis in die Kreide konstante Vorhandensein von acht Molaren wird als primitives, von Therapsiden der Trias übernommenes Merkmal gedeutet. Die Gestalt der unteren Molaren ist dagegen so spezialisiert, dass die Dryolestiden als Ahnen der modernen Theria ausscheiden. Sie haben aber parallel zu den modernen Theria den Schärfungsmechanismus der Zähne (*Thegosis* nach R. G. Every) vervollkommnet. Die gegenseitige Verzahnung der Molaren bei den Dryolestiden der Unteren Kreide ist eine Folge dieses Prozesses.

CONTENTS

INTRODUCTION

The excavations for Mesozoic mammals by Professor W. G. Kühne and his staff (Lehrstuhl für Paläontologie der Freien Universität Berlin) have yielded the oldest as well as the youngest dryolestids known up to the present time. The oldest finds come from the Guimarota lignite mine, near Leiria in Portugal. The age of this locality was fixed with the help of characeans and ostracods as Kimmeridgian. In addition to multituberculates and docodonts, the mammals are represented in the Guimarota mine by pantotheres—namely, by paurodontids, peramurids, and dryolestids. The upper jaws, lower jaws, and isolated teeth of dryolestids are present.

The youngest finds come from the Uña lignite mine in the province of Cuenca, Spain. Unfortunately, the age of this locality can at present only be fixed as Lower Cretaceous. The coal-layers have been placed by Spanish geologists in the Albian, although the ostracodes, for which no direct material for comparison is known, seem rather to indicate a somewhat older dating. The digging at Uña is presently being carried on further, and a geological-stratigraphical investigation is under way. To date, the mammal-finds at Uña include two complete dryolestid lower jaws, along with isolated teeth of multituberculates and pantotheres.

Since only the lower jaws are present from these latest dryolestids, and the examination of the upper jaws from the Guimarota mine has not yet been completed, only the lower jaws will be considered in this report.

THE LOWER JAW

The finds from the Kimmeridgian

The dryolestids from the Guimarota mine may be related to several taxa; and the correlation of the various types of upper and lower jaws, which were always found separated, has not yet been completely successful. These dryolestids are different from the already-known, somewhat younger forms of the uppermost Jurassic from Swanage (South England) and Como Bluff (Wyoming) (Simpson, 1928*a*, 1929). The taxonomical problems will be treated in a monographical description of pantotheres from the Guimarota mine to appear in the *Memorias dos Serviços Geologicos de Portugal.* They are unimportant here, since all Guimarota dryolestids agree in the characteristics to be discussed in this report.

The lower jaw of dryolestids from the Guimarota mine (Fig. 1) bears, on a relatively long alveolar part, four incisors, one canine, four premolars, and eight molars, as is characteristic of this group. The alveolar border lies, in the area of the molars, noticeably deeper on the outside than on the inside. The dentition will be discussed later in more detail. Directly behind the last molar rises the high coronoid process, which exhibits an apophysis at the back. The condylar process slants upward and rearward. A well-developed angular process projects backward. The large pterygoid fossa is visible on the inside of the jaw under the coronoid process. At the extreme anterior end of the pterygoid fossa begins the inner groove, which gradually approaches the ventral border and extends forward to the symphysis. The symphysial surface shows that both mandibular rami moved against each other.

The area around the pterygoid fossa deserves special attention. The fragment pictured in Plate 1**A** belongs to a shattered lower jaw; the remaining pieces and the teeth prove that it belonged to a dryolestid. Above the pterygoid fossa, at the rear end of the alveolar border, at the foot of the coronoid process, an approximately triangular field may be seen, which, in comparison with the other surfaces of the jaw, is a little deeper and rough. This is precisely the place where, in reptiles and particularly in

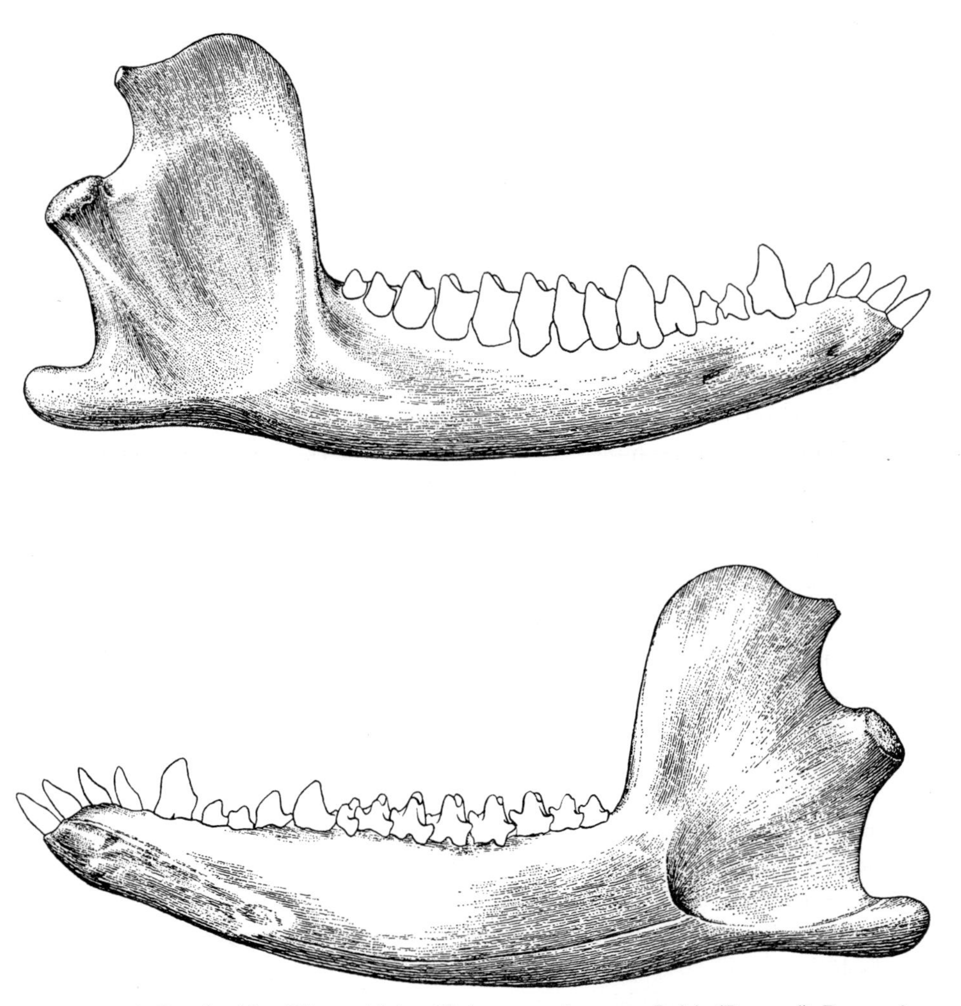

FIGURE 1. Dryolestidae. Kimmeridgian. Guimarota mine, near Leiria (Portugal). Restoration of a right lower jaw, based mainly on specimen 'Guimarota 6'. ×4½. Above, external view. Below, internal view.

therapsids, the coronoid is found. There is no doubt that this dryolestid also possessed a coronoid bone. The coronoid was only a small, thin plate of bone inserted in the dentary, and was lost after the post-mortem decomposition of the connective tissue. Thus, only the sutural surface for the coronoid may be observed on the fossil lower jaw.

The area directly before the pterygoid fossa may be seen clearly on the fragment reproduced in Plate 1**B**; further fragments and teeth prove here again that this piece belongs to a dryolestid. Above and below the posterior part of the inner groove runs, on each side, a finer groove. These two smaller grooves approach each other gradually toward the front, become indistinct, and disappear approximately under the last molar. Toward the back they curve away from each other to join with the borders of the pterygoid fossa. The area between these grooves, below the coronoid and before the insertion-point of the adductor muscles—here the pterygoid fossa—corresponds to the place where the splenial lies in reptiles, and suggests that the dryolestids of the Guimarota mine also possessed a splenial bone. However, one must not imagine the splenial sunken in the dentary, like the coronoid, but rather lying like a scale on the inside of the dentary. Thus no surface suture exists; the splenial was probably fastened to the dentary only at its upper and lower edges with connective tissue. The observed small grooves would correspond to the place of insertion of the connective tissue in the dentary. The rear border of the splenial coincided with the front border of the pterygoid fossa. The front of the splenial, which ends in a point, must not have extended far beyond the end of the grooves, since the gently descending sides of the inner groove indicate no trace of an overlying bone.

The sutural surface for the coronoid and the attachment-grooves of the splenial may be observed on all dryolestid lower jaws from the Guimarota mine on which this section is visible (another example is presented in Plate 2**A**). The individual differences in the development of the coronoid and splenial are a typical feature of rudimentary, functionless elements.

The presence of coronoid and splenial bones raises the question whether still more rudimentary reptilian dermal bones were present in the lower jaw. One would look for them on the rear part of the medial side. However, the surface of the pterygoid fossa, as well as that of the condylar process, is smooth and offers no indication of such accessory elements.

At the front end of the pterygoid fossa opens the dental foramen, through which the inferior alveolar artery and nerve enter the jaw. The inner groove begins at exactly the same level. This close relationship between the location of the inner groove and the dental foramen is noticeable in all dryolestids. Doubtless the inner groove held the mylohyoid artery and nerve, which branch away from the inferior alveolar artery and nerve just before these latter enter the dental foramen. Accordingly, the inner groove would correspond to the sulcus mylohyoideus of modern mammals, as Owen assumed as early as 1871.

Since, however, the mylohyoid artery and nerve supply the mouth floor, it remains puzzling, why the inner groove runs all the way to the symphysis. This fact suggests that the inner groove held, along with the blood-vessel and nerve, a Meckelian cartilage, a possibility considered by Simpson (1928*b*). The strongest argument in favour of this is that the inner groove penetrates into the symphysis region and leads into the cavities there, corresponding to the extension of the embryonic Meckelian cartilage of recent mammals. It is unlikely that the inner groove may be explained as the remainder of a Meckelian cartilage which is reabsorbed in the adult stage, since, in that case, the groove would have had to disappear (like, in adults, the alveolus of a tooth which has fallen out). The dryolestids of the Guimarota mine had a Meckelian cartilage which persisted throughout their lifetime. The persistence of a Meckelian cartilage accords well with the possession of a splenial bone, since this was the element which originally covered the Meckelian cartilage on the inside of the jaw.

Contrary to the opinion of Bensley (1902), who interpreted the inner groove of Mesozoic mammals as exclusively a sulcus primordialis, the present finds show that the inner groove held a Meckelian cartilage as well as the mylohyoideal artery and nerve. This blood-vessel and nerve stand also in close embryological relation to the

Meckelian cartilage. A further proof of the double function of the inner groove is its occasional division into two grooves running alongside each other (Plate 1**A**).

The persistence of a Meckelian cartilage does not necessarily mean that these dryolestids possessed a reptilian jaw articulation as well. It could be conceivable, that the proximal part of the cartilage becomes the malleus of the middle ear, while the distal part, which separated early, continues to exist in the lower jaw.

Figure 2 shows an attempted restoration of the area around the pterygoid fossa of a dryolestid from the Guimarota mine. The lower border of the pterygoid fossa forms an edge which ventrally limits the insertion-surface of the internal pterygoid muscle. The edge of the pterygoid fossa, where it projects sharply at the level of the dental foramen, corresponds (like the lingula mandibulae of modern mammals) to the insertion-point of the sphenomandibular ligament. The notch to be found in the border of the pterygoid fossa above this point served to allow the passage of the lingual nerve.

The finds from the Lower Cretaceous

The two lower jaws from Uña are characterized as dryolestids by the general form, the dental formula (four incisors, one canine, four premolars, and eight molars), and the tooth pattern. They have been described and named *Crusafontia cuencana* by Henkel & Krebs (1969) (Fig. 3). Aside from the vertical position of the articular process, several things have changed with respect to the dryolestids of the Kimmeridgian, especially in the area of the pterygoid fossa.

Between the pterygoid fossa and the angle formed by the alveolar border and the anterior border of the coronoid process, a small bump projects from the inner surface of the jaw. This is the place where the sutural surface for the coronoid is found in the forms of the Guimarota mine. The bump shows no roughness on its surface. It is thus not a sutural surface, but probably the coronoid itself. Since the bump is not bounded by a suture, the coronoid must be fused early in ontogeny with the dentary. Here, in the adult animal, the coronoid as an independent element has disappeared.

Yet another difference from the dryolestids of the Kimmeridgian lies in the fact that the small grooves above and below the inner groove are missing. The splenial has by this time completely disappeared. In addition, the inner groove itself is only weakly pronounced and does not extend to the symphysis but only to about under the penultimate molar. Consequently, a Meckelian cartilage was no longer present. Only these Cretaceous dryolestids would be, by definition, complete mammals.

THE DENTITION

The dental formula

The large number of molars raises the question whether the dental formula of the dryolestids is primitive or specialized, as against other pantotheres such as *Peramus* or the paurodontids. The fact that *Peramus*-precursors and paurodontids are present in the Guimarota mine which possess more molars than the younger forms from Swanage and Como Bluff proves that a larger number of molars is the starting-point of the evolution. An earlier pantothere which possesses eight molars, like *Amphitherium* of the Bathonian, may be considered the ancestor not only of the dryolestids, but

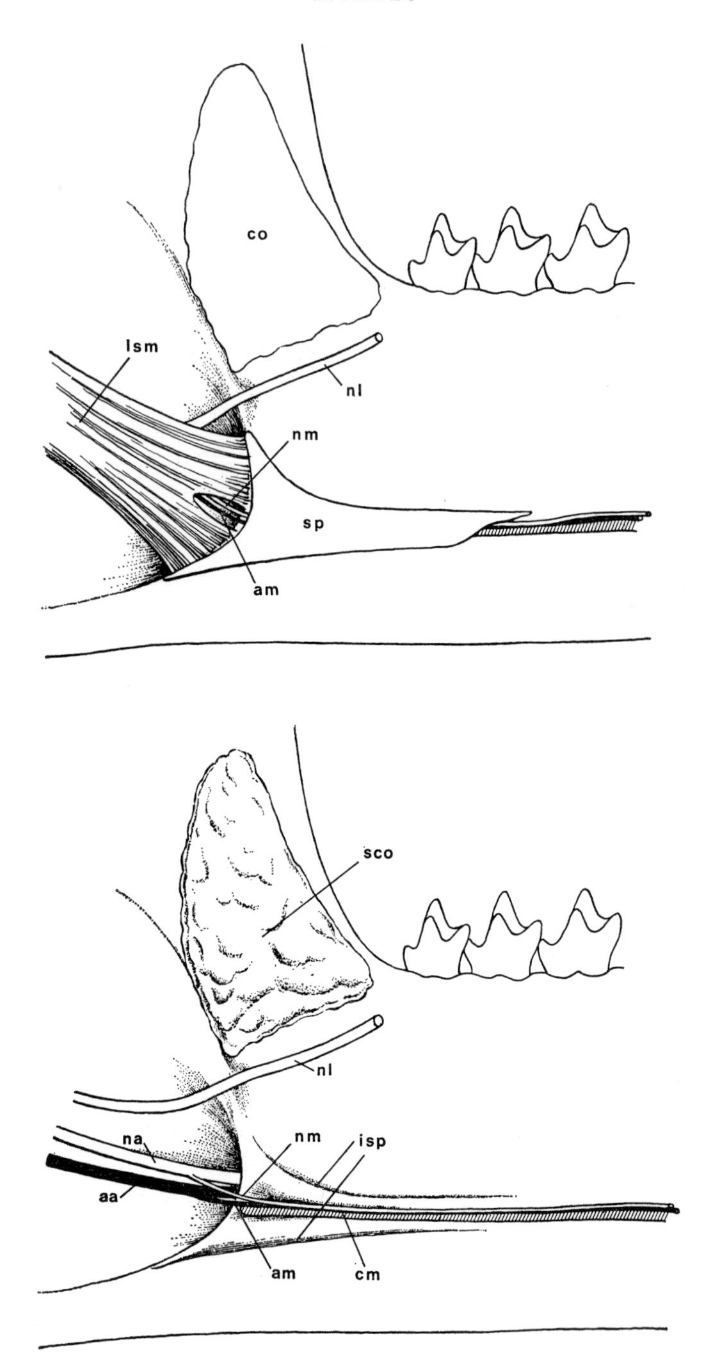

FIGURE 2. Dryolestidae. Kimmeridgian. Guimarota mine, near Leiria (Portugal). Attempted restoration of the posterior part of a left lower jaw in internal view. Above, with coronoid, splenial and sphenomandibular ligament *in situ*. Below, with coronoid, splenial and ligament removed.

aa, Arteria alveolaris inferior; am, arteria mylohyoidea; cm, cartilago Meckeli; co, coronoid; isp, insertion-grooves of the splenial; lsm, ligamentum sphenomandibulare; na, nervus alveolaris inferior; nl, nervus lingualis; nm, nervus mylohyoideus; sco, sutural surface of the coronoid; sp. splenial.

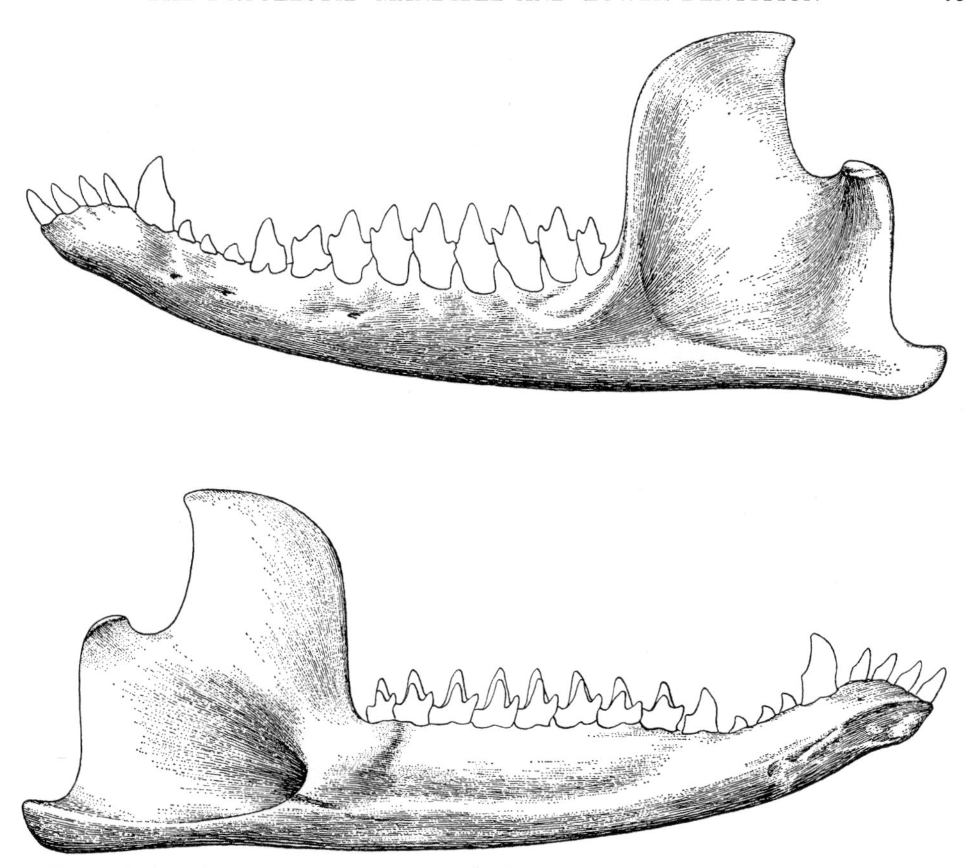

FIGURE 3. *Crusafontia cuencana* Henkel & Krebs (Dryolestidae). Lower Cretaceous. Uña (prov. Cuenca, Spain). Restoration of a left lower jaw, combination of both finds from Uña. ×6½. Above, external view. Below, internal view.

of the paurodontids and peramurids as well. As regards their dental formula, the dryolestids are not specialized, but conservative. This conservatism is emphasized, in that eight molars were always present in the finds from Uña as well.

Whether *Amphitherium* has more teeth than its Rhaetic ancestors seems to remain an open question. The few mammals from the Rhaetic whose dental formula is known have, to be sure, fewer molars. With the absence of any transitional form from the Lower Jurassic, it is in no way certain that *Amphitherium* necessarily traces back to one of these forms. Many therapsids exhibit a dozen postcanine teeth. It is quite imaginable that in the Rhaetic, along with the already-mentioned forms, other mammals lived who kept the large number of molars and passed them on through *Amphitherium* to the dryolestids. (Perhaps these missing links would also show fewer contradictions than the known forms in other features.)

The molars

While, then, the dryolestids are conservative in their dental formula, the pattern of their molars is specialized. In *Amphitherium*, the lower molars consist of a trigonid with three cusps ordered in the form of an equilateral triangle, a relatively large, single-cusped talonid, and two equally large roots. This pattern continues in the paurodontids, and also in the ancestors of the modern Theria, where new characteristics appear only gradually and quite late. In the dryolestids from the Guimarota mine, the molars have already changed. They are mesio-distally shortened, the labial angle of the trigonid has become very sharp, and the talonid is small. The front root is considerably larger than the rear one, and extends labially, so that only the front root is visible from the outside. This specialization in the pattern of the molars is the factor which above all disqualifies the dryolestids as ancestors of the modern Theria. However, the dryolestids do display certain features which also appear in the line of the modern Theria—be it as the further development of a common heritage, or as parallel evolution.

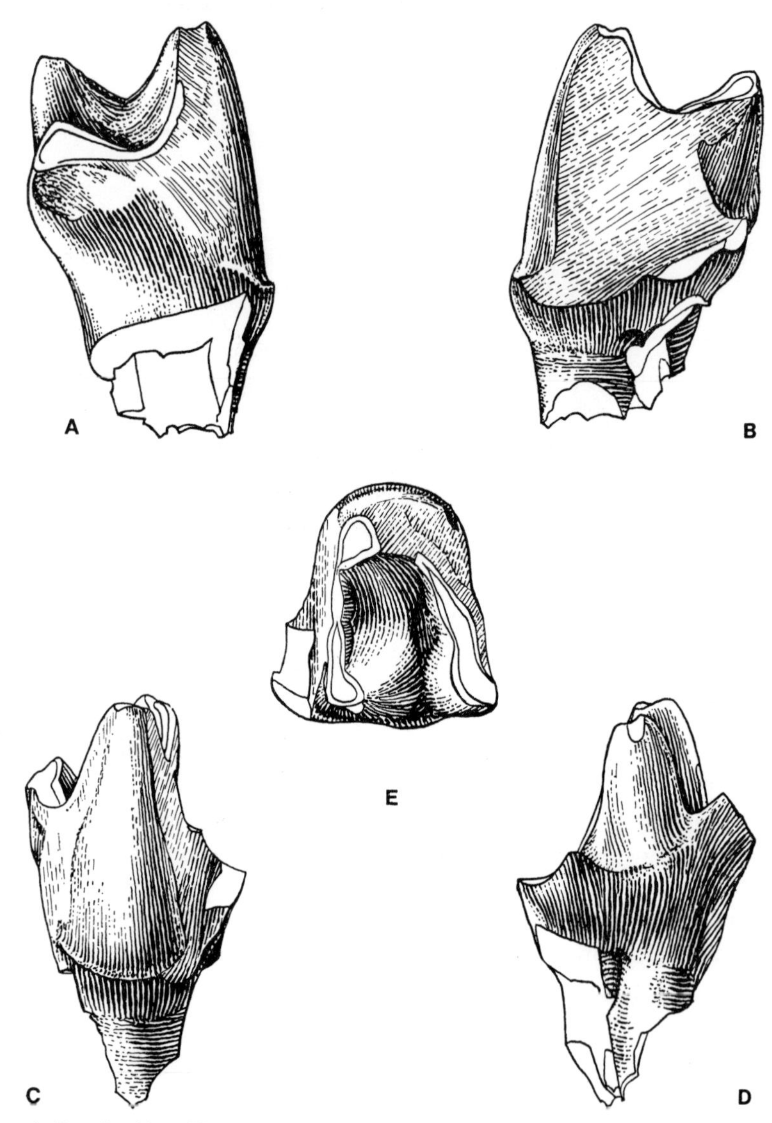

FIGURE 4. Dryolestidae. Kimmeridgian. Guimarota mine, near Leiria (Portugal). Left lower molar of specimen 'Guimarota 112', separated during preparation. ×30.
A, Anterior view; **B**, posterior view; **C**, external view; **D**, internal view; **E**, occlusal view.

One such feature is the mechanism which serves for sharpening the teeth. This process was first observed by Dr R. G. Every, who gave it the name of *thegosis*. It is explained in Every & Kühne (1971) in this volume, and is here only briefly summarized. Through the mastication of food, the cusps and crests of the teeth become worn and dull. Thegosis is a process which counteracts this wear. In the absence of food, the edges of the upper and lower teeth are whetted against each other, so that they both become sharp again. The wear caused by chewing results in arched, smooth, indistinctly bounded surfaces. Through thegosis, even, sharply bordered facets come into existence along the cutting edges. These facets exhibit fine parallel striations, which indicate the direction of the grinding movement. Thegosis occurs in modern mammals and in the pantotheres; in the dryolestids the process reached a special degree of perfection.

On the lower molars of dryolestids, one thegosis-facet is found on the front and on the rear flank of each trigonid (Figs 4 and 5). They border on the two cutting edges

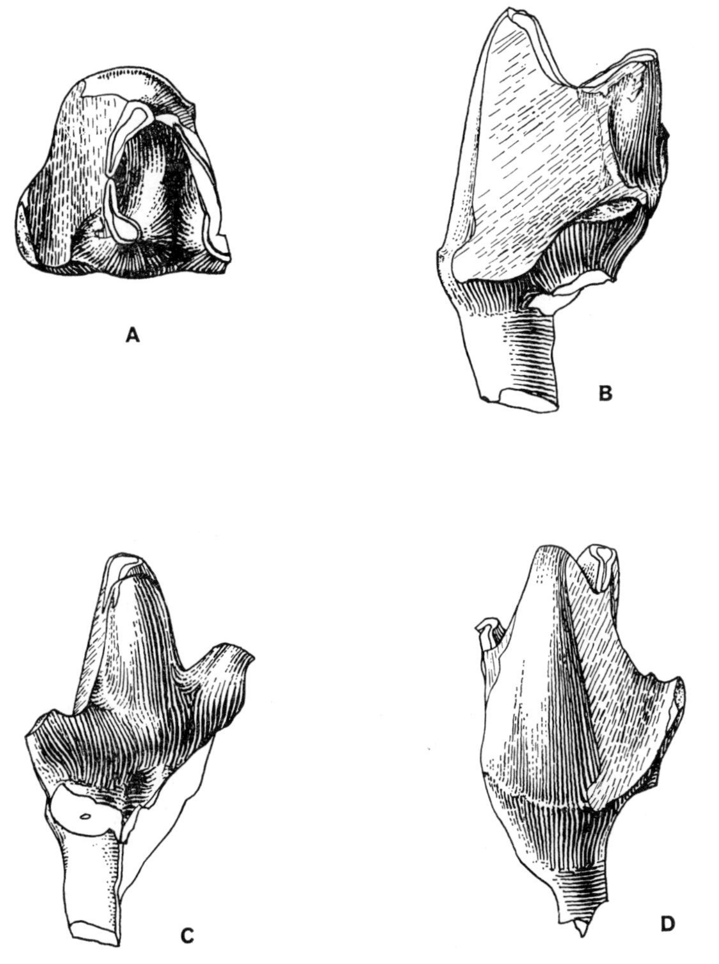

FIGURE 5. Dryolestidae. Kimmeridgian. Guimarota mine, near Leiria (Portugal). An other left lower molar of specimen 'Guimarota 112', separated during preparation. ×30.
A, Occlusal view; **B**, posterior view; **C**, internal view; **D**, external view.

which join the protoconid and paraconid on the one side, and the protoconid and metaconid on the other. Their completely flat surfaces stand out against the more or less rounded, unused tooth surface. On the downwardly and outwardly directed slope of the talonid, a facet may be observed in the form of a groove with a rounded bottom. In heavily-worn molars, this groove-like facet extends laterally into the facet on the rear flank of the trigonid. The striations of the thegosis-facets run at an angle of about 45° outwards and downwards. From this, one can see that in the dryolestids, to sharpen the worn edges, the trigonids of the lower molars were pulled diagonally between the trigons of the upper molars. In this process, the front surface of a lower molar rubs over the rear surface of an upper molar, and at the same time, the rear surface of the same lower molar rubs against the front surface of the next following upper molar. The main cusp of the upper molar glides, as if on a rail, over the groove of the talonid, which slopes labially downward, with the result that this groove is more and more deeply engraved. (Accordingly, the protoconid cusp of the lower molar glides in a groove in the talon of the same upper molar, which is directed inwards and upwards.)

Inside and above, the facet of the talonid extends always up to the inside surface of the tooth; there is never an unused space between the lingual end of the facet and the inside surface of the tooth. This indicates that the main cusp of the upper molar moved beyond the lingual end of the talonid. The whetting action ended thus at the lingual end of the talonid groove. Had it begun there, the danger would have existed that the main cusp of the upper molar could have come a little too deeply in contact with the talonid and been broken off. The grinding movement of the lower jaw must therefore have been exclusively from above and inside in a downwards and outwards direction, probably alternating to the left for the molars on the left side, and to the right for the molars on the right side. Whetting of the teeth was effected by opening the mouth—thus, in the direction of its movement, it is a process opposite to chewing.

In this connection, one may remember the difference in the height of the alveolar border on the inside and on the outside of the lower jaw. At grinding, a lingually directed pressure is brought against the lower molars. The elevated alveolar border works against this pressure. (In the upper jaw, accordingly, the alveolar border, which projects ventrally on the outside, protects the molars against labially directed pressure.)

Such a sharpening mechanism demands extraordinary precision in the succession of the molars. In particular, after the upper and lower molars have ground into one another, their relative positioning must remain unchanged. In the modern Theria, the proper spacing of the molars is assured, in that they stand under mesio-distal pressure (cf. Every & Kühne, 1971). Such a pressure also affected the spacing of the molars of the dryolestids from the Guimarota mine. At the point where the talonid of one molar strikes against the front surface of the next, a worn surface due to this pressure is noticeable. Worn surfaces on the molars of dryolestids from the Guimarota mine caused by mastication, thegosis and interdental pressure are depicted schematically in Fig. 6.

The process of the safe relative positioning of the molars was brought to perfection in the dryolestids of the Lower Cretaceous. The lower molars of *Crusafontia* (Figs 7 and 8) possess a cingulum which starts from the talonid, runs around the base of the protoconid, and rises again at the forward flank of the trigonid. It ends in a small point at the front under the paraconid. Thus the front end of the cingulum appears as a counterpart to the talonid. Since the cingulum does not reach to the antero-lingual edge of the trigonid, a notch exists there, into which the talonid of the next forward molar is inserted. The rear surface of the talonid displays, correspondingly, an indentation into which the cingulum-point of the following molar fits (Fig. 9).

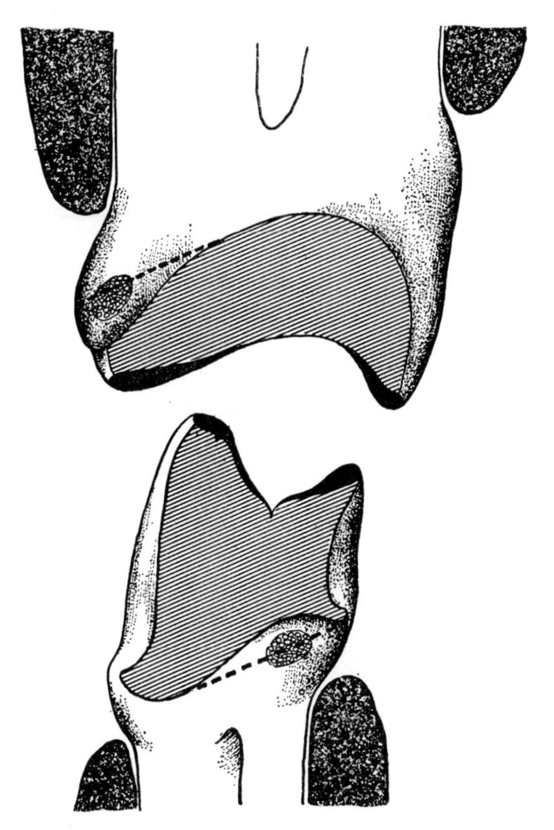

FIGURE 6. Diagram of the wear-surfaces on the molars of dryolestids, based on specimens from the Guimarota mine. On the lower molar the posterior side is visible; on the upper—transparently, as it were—the corresponding anterior side.

Black, wear due to mastication. Oblique hatching, thegosis-facets (the hatching corresponds to the direction of the striation). Circles, wear due to interdental pressure. Broken line, floor of the talonid or talon-groove. See p. 98.

Such a mutual interlocking assures not only the correct spacing of the molars, but also their relative transverse positioning. Beyond this, it insures each single molar against lingual breakage during grinding.

The interlocking of the molars is well known in some modern mammals—like *Didelphis*—and in their Mesozoic precursors. It is missing, however, in all previously known dryolestids from the Upper Jurassic. Apparently the Cretaceous dryolestids have acquired this characteristic feature through parallel evolution.

These observations show once more that the evolution of mammals progressed only gradually and like a mosaic. Beside the persistence of reptilian lower jaw elements,

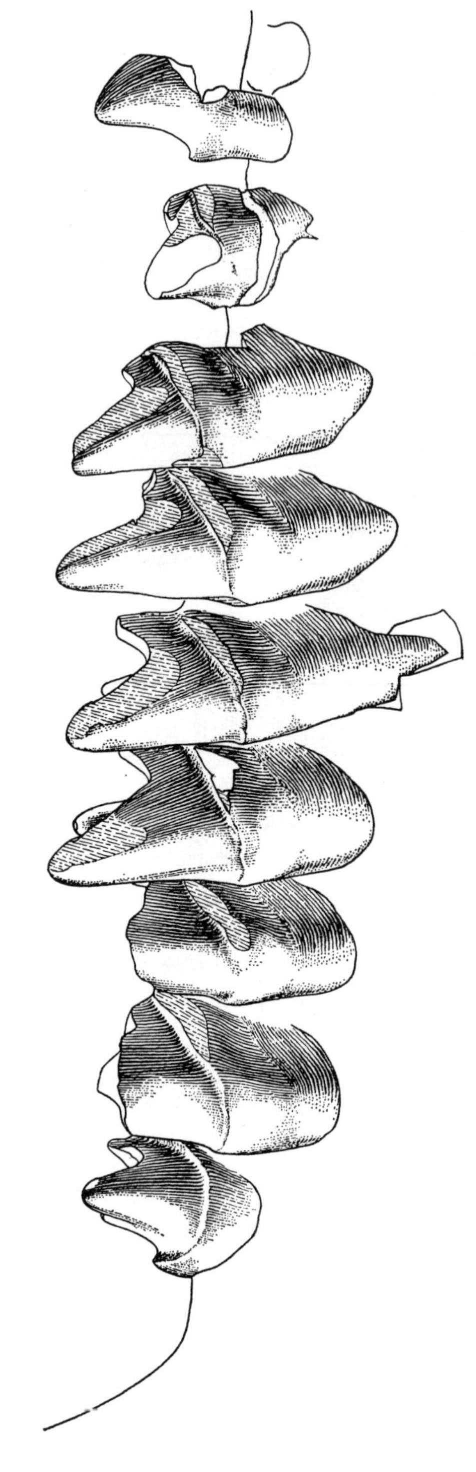

FIGURE 7. *Crusafontia cuencana* Henkel & Krebs (Dryolestidae). Lower Cretaceous. Uña (prov. Cuenca, Spain). Dentition of the right lower jaw in external view, showing thegosis-facets. Specimen 'Uña 2'. ×24.

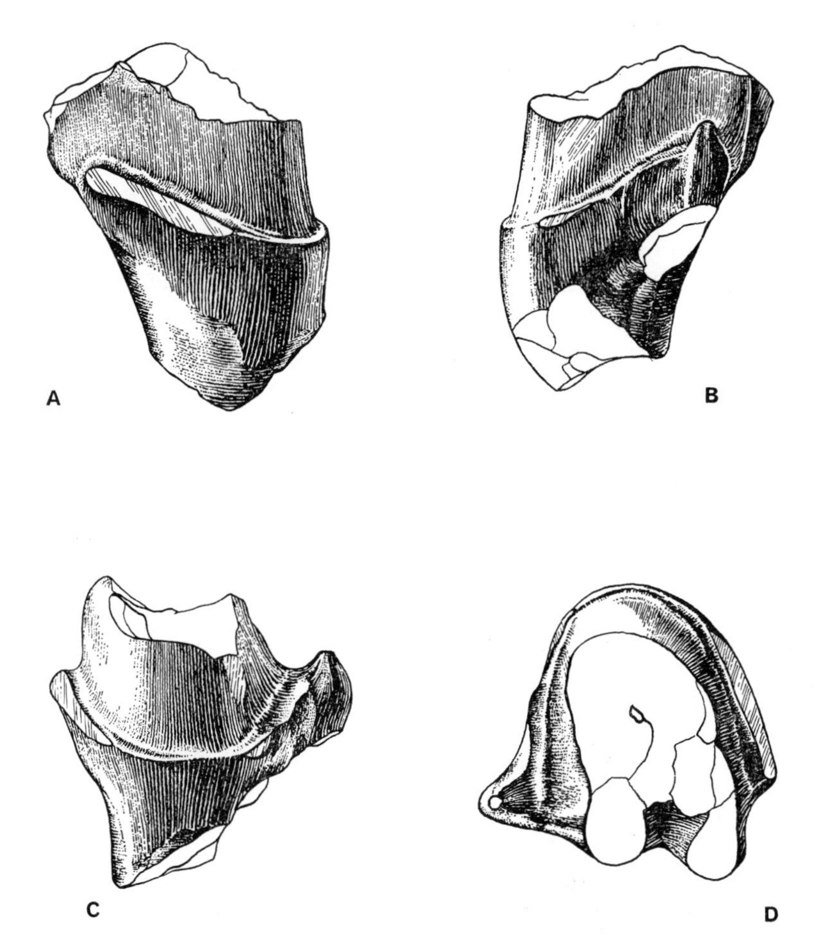

FIGURE 8. *Crusafontia cuencana* Henkel & Krebs (Dryolestidae). Lower Cretaceous. Uña (prov. Cuenca, Spain). Isolated left lower molar. The tooth, whose cusps are missing, shows the cingulum. ×30.

A, Anterior view; **B**, posterior view; **C**, external view; **D**, occlusal view.

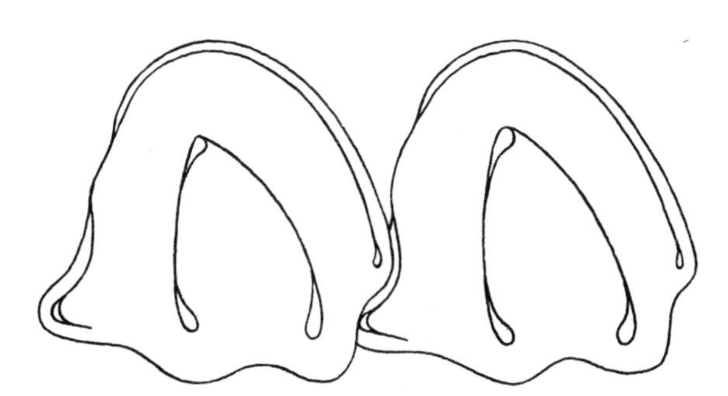

FIGURE 9. Diagrammatic occlusal view of two lower molars of *Crusafontia*, showing their interlocking.

we find the most advanced features in the tooth-functioning. The history of the dryolestids contributes to illustrate how evolution toward the modern mammals was achieved in parallel lines—of which many, however, proved to be blind alleys.

ACKNOWLEDGEMENTS

I wish to express my sincere gratitude to Dr R. G. Every, Dr K. A. Kermack, and Professor Dr W. G. Kühne for their valuable suggestions and useful discussions. I am especially grateful to Mr P. Berndt for his understanding help with the preparation of the drawings, and to Mr Gene Moore for the English translation. Thanks are also due to the Deutsche Forschungsgemeinschaft, which very generously supported our field-work. In Spain the work is organized in collaboration with the Instituto Provincial de Paleontología, Sabadell (Barcelona); we are much indebted to its Director, Professor Dr M. Crusafont Pairó.

REFERENCES

Bensley, B. A., 1902. On the identification of meckelian and mylohyoid grooves in the jaws of Mesozoic and Recent Mammalia. *Univ. Toronto Stud. biol. Ser.*, **3**: 1–9.
Every, R. G. & Kühne, W. G., 1971. Bimodal wear of mammalian teeth. In Kermack, D. M. & Kermack, K. A. (Eds), *Early Mammals. Zool. J. Linn. Soc.* **50**, Suppl. 1: 23–27.
Henkel, S. & Krebs, B., 1969. Zwei Säugetier-Unterkiefer aus der Unteren Kreide von Uña (Prov. Cuenca, Spanien). *Neues. Jb. Geol. Paläont. Mh.*, 1969: 449–463.
Owen, R., 1871. Monograph of the fossil Mammalia of the Mesozoic Formations. *Palaeontogr. Soc.* [*Monogr.*], **24**: vi + 115 pp.
Simpson, G. G., 1928*a*. *A catalogue of the Mesozoic Mammalia in the Geological Department of the British Museum (Natural History)*. London: Br. Mus. (Nat. Hist.).
Simpson, G. G., 1928*b*. Mesozoic Mammalia. XII. The internal mandibular groove of Jurassic Mammals. *Am. J. Sci.*, **15** (5): 461–470.
Simpson, G. G., 1929. American Mesozoic Mammalia. *Mem. Peabody Mus. Yale*, **3**: xv + 235 pp.

EXPLANATION OF PLATES

Plate 1

Dryolestidae. Kimmeridgian. Guimarota mine, near Leiria (Portugal).
A. Fragment of left lower jaw, internal view. Specimen 'Guimarota 112'. ×20 approx.
The sutural surface for the coronoid is located at the posterior end of the alveolar border, at the foot of the coronoid process. The inner groove begins at the anterior end of the pterygoid fossa, and branches in this specimen (shortly before the fracture-point). Above and below the inner groove run the insertion-grooves for the splenial (here only weakly pronounced).
B. Fragment of left lower jaw, internal view. Specimen 'Guimarota 18'. ×20 approx.
The sutural surface for the coronoid is limited above by the fracture. Clearly visible here is the beginning of the inner groove at the level of the dental foramen, as well as the insertion-grooves for the splenial.

Plate 2

A. Dryolestidae. Kimmeridgian. Guimarota mine, near Leiria (Portugal). Fragment of right lower jaw, internal view. Specimen 'Guimarota 33'. ×20 approx.
This specimen shows the satural surface for the coronoid and the insertion-grooves for the splenial above and below the inner groove. The anterior border of the pterygoid fossa is broken out.
B. *Crusafontia cuencana* Henkel & Krebs (Dryolestidae). Lower Cretaceous. Uña (prov. Cuenca, Spain). Detail of left lower jaw, internal view. Specimen 'Uña 1'. ×20 approx.
The coronoid, which has fused with the dentary is still recognizable as a bump at the posterior end of the alveolar border. The inner groove and the insertion-grooves for the splenial are missing.

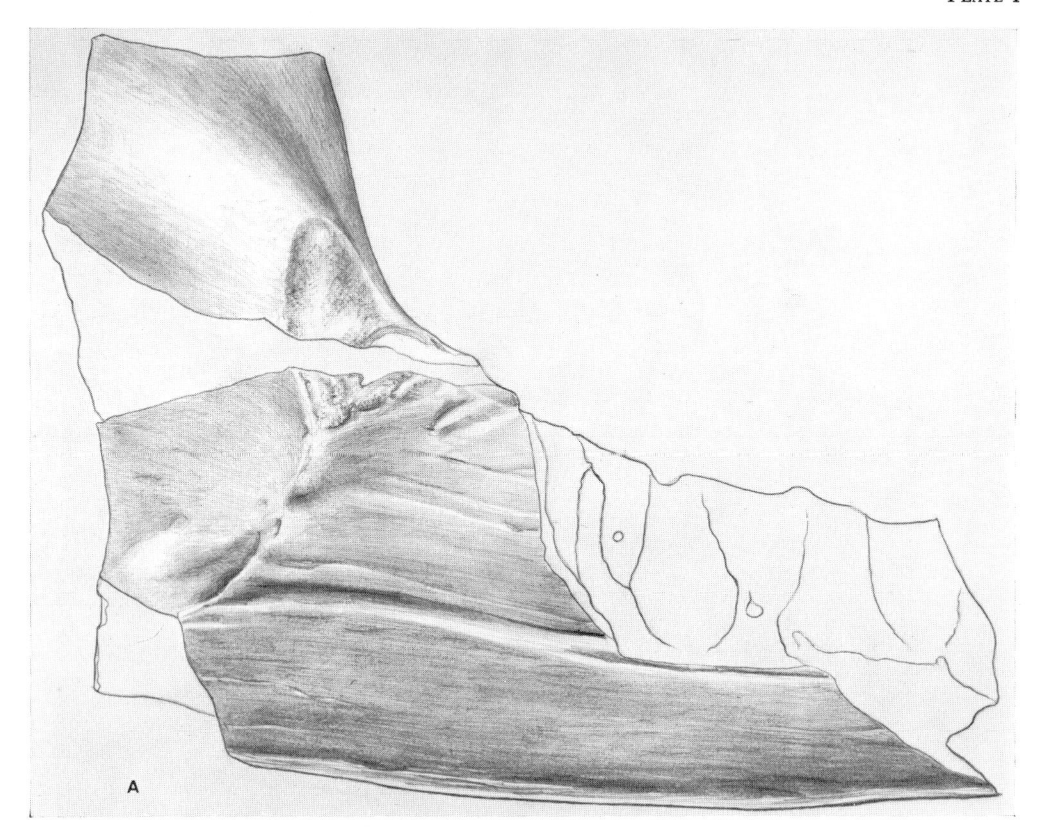

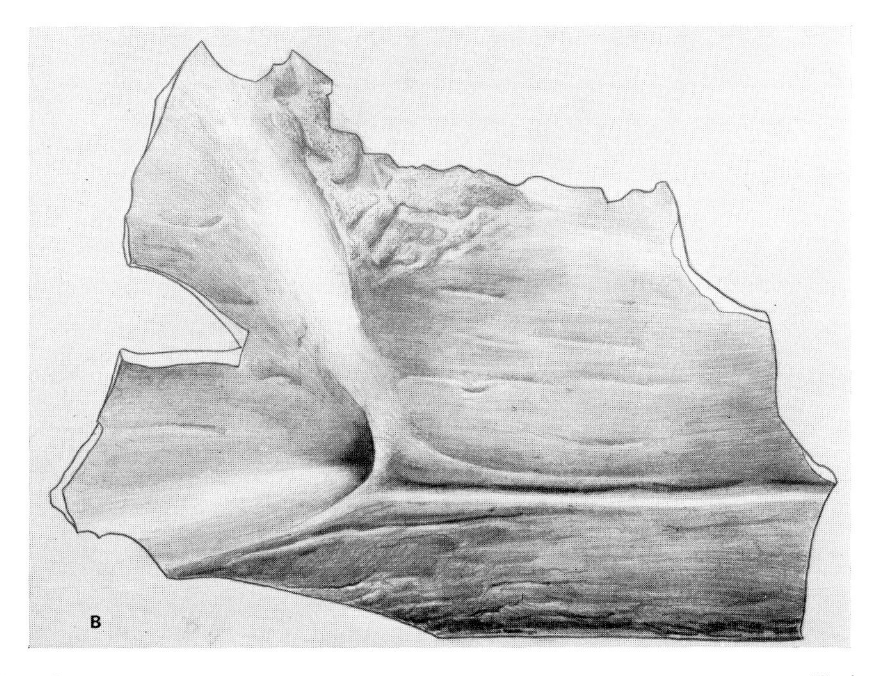

B. KREBS

(*Facing p.* 102)

B. KREBS

Therian and non-therian mammals

K. A. KERMACK

Department of Zoology, University College, London

AND

Z. KIELAN-JAWOROWSKA

Zakład Paleozoologii, Polska Akademia Nauk, Warsaw, Poland

The Mammalia form two main groups; the Theria with the lateral braincase wall formed by alisphenoid and squamosal, and the non-therian mammals with the lateral wall formed by the anterior lamina of the petrosal. The latter comprise the Docodonta, the Triconodonta, the Multituberculata and the Monotremata.

Recent work on multituberculates confirms this and enables us to discuss the inter-relations of the non-Theria.

Multituberculata and Monotremata have in common: (1) greatly reduced alisphenoid, (2) jugal absent, (3) lacrimal usually reduced or absent, (4) multituberculate ectopterygoid resembling the monotreme 'Echidna pterygoid'. Multituberculata and Monotremata are clearly related.

In contrast *Morganucodon* (the only adequately known docodont) has: (1) a well developed—rather bauriomorph-like—alisphenoid, (2) conspicuous jugal and lacrimal. There is a possible ectopterygoid.

Braincase and dentition are similar in Triconodonta and Docodonta thus the former must have evolved from the latter. Together they form a natural group within the non-therian mammals.

The time gap between the first Docodonta and the first Multituberculata makes it possible to derive both from a common mammalian ancestor.

CONTENTS

INTRODUCTION

In spite of the great adaptive variety of the placental mammals, the structure of the braincase is essentially the same in all. The same pattern is found in marsupials, although they separated from the placentals not later than the early Upper Cretaceous (Clemens,

1968). Thus the braincase is a conservative part of the mammalian skeleton, and so a good character to indicate the relationships of the major groups of mammals.

The braincase of the fossil non-therian mammals was almost unknown until recently, and our ideas of their interrelations were based exclusively on the structure of their dentition. The study of the dentition of Mesozoic mammals led Kermack & Mussett (1959) and Simpson (1959) to the conclusion that the Mammalia is a group of polyphyletic origin. According to Simpson (1959) several lines, possibly as many as nine and not less than four, independently crossed the reptilian-mammalian boundary in the Late Triassic.

This opinion was regarded as a well established fact by all the students of the group. It is amusing that the discussion initiated by Simpson's paper was not whether the Mammalia are a group of a monophyletic or polyphyletic origin, but how to avoid the situation—inconvenient from the viewpoint of systematics—that the most important class of vertebrates is not a natural one (Olson, 1959; Van Valen, 1960; Reed, 1960; Simpson, 1960, 1961; Brink, 1963; Crusafont-Pairó, 1962; MacIntyre, 1967). Two main trends in this discussion may be recognized. According to the first proposal, either the Therapsida (Van Valen, 1960) or both the Therapsida and the Pelycosauria immediately ancestral to them (Reed, 1960) should be included into the Mammalia. According to the second (MacIntyre, 1967) only the Theria should be called mammals.

Of the Mesozoic mammals the Symmetrodonta and Pantotheria belong to the therian line (Patterson, 1956). The non-therian mammals include the Triconodonta (possibly including the Amphilestidae),* the Docodonta (including *Morganucodon*), the Haramiyidae, the Multituberculata and the Monotremata (Kermack, 1967). One of us (K.A.K., 1967) accepted *Diarthrognathus* formally as a mammal, since it has a rudimentary squamosal-dentary articulation. He, however, made it clear that this was its only claim to mammalian status and that its other characters were more cynodont than mammalian. In particular it has a typically reptilian braincase (Crompton, 1958; Hopson & Crompton, 1969). If the criterion of a dentary-squamosal joint is to be used without admitting any exceptions, not only *Diarthrognathus* but also the cynodont *Probainognathus*, from the Middle Triassic of South America (Romer, 1969) would have to be classified as a mammal, although Romer himself rejects this. It would be most convenient to have a single character to discriminate between reptiles or mammals, but the example of *Diarthrognathus* and *Probainognathus* shows that the pattern of therapsid evolution is too complicated to allow us to do this.

In the present paper we shall discuss the braincase structure of the early mammals and the light this throws on the relationships of the various groups. Nothing is known on the braincase in the amphilestid triconodonts, nor in the haramiyids. The latter are only known by isolated teeth. Parrington (1967) and Hahn (1969) are of the opinion that these teeth show a close resemblance to those of the Multituberculata. We shall not discuss the amphilestids and the haramiyids further here.

The groups of non-therian mammals in which some part, at any rate, of the braincase is known are:

(1) *Monotremata*—here the braincase is well known and has been reasonably well understood since Watson (1916).

* Kermack (1967) points out that the amphilestids may be perhaps Theria.

(2) *Triconodonta*—four petrosals are known, one almost complete, and part of the basicranium, including the basisphenoid. The material comes from the Upper Jurassic of Purbeck and was described by Simpson (1928) and Kermack (1963).

(3) *Docodonta*—it has been possible to make reconstruction of the braincase of one member of the group—*Morganucodon*—on the basis of an almost complete skull from the Chinese Lufeng beds and isolated bones from Welsh fissures of the same age—Upper Trias. This reconstruction has been made by one of us (K.A.K.) along with Mrs F. Mussett and Father H. Rigney. By the courtesy of these workers we are able to discuss some of their findings here.

Coeval with *Morganucodon* are the closely related *Erythrotherium* and *Megazostrodon* from the Red Beds of South Africa. Skulls of these animals are known (Crompton, 1964; Crompton & Jenkins, 1968), but nothing has yet been published of the braincase. One specimen of *Sinoconodon rigneyi*, from the Late Triassic, (Lufeng Series) of China is a fragmentary braincase (Patterson & Olson, 1961). This closely resembles that of *Morganucodon*. Kermack (1967) considered it probable that this braincase was of *Morganucodon*, which is also found in the Lufeng beds; but we now consider it more likely that Patterson & Olson (1961) were correct in referring this specimen to *Sinoconodon*. In other ways *Sinoconodon* is close to *Morganucodon*, being a morganucodont showing triconodont features in its dentition. These resemblances may indicate affinity between *Sinoconodon* and the Triconodonta, as Patterson & Olson (1961) originally suggested. But we shall here classify *Sinoconodon* with *Morganucodon* in the Docodonta. *Eozostrodon problematicus*, known by a single, fragmentary tooth from a fissure at Holwell, England (Parrington, 1941) may be allied to *Sinoconodon*. Parrington (1967) and Hopson & Crompton (1969) regard *Morganucodon* as a synonym of *Eozostrodon*. In our opinion these two forms are not congeneric.

(4) *Multituberculata*—until recently the only reasonably complete multituberculate skulls* were those of the genera *Ptilodus* and *Taeniolabis*, described by Simpson (1937). Unfortunately in *Taeniolabis* it was impossible to make out any detail in the braincase. In *Ptilodus* Simpson was able to make a reconstruction of the braincase. This was essentially correct as far as the identification of the foramina in the basicranial region is concerned, and the overall shape of the braincase. But due to the obliteration of all sutures he reconstructed the braincase on essentially therian lines—postulating that the squamosal makes a large contribution to the lateral wall of the braincase, with an alisphenoid of therian pattern. The occipital plate was not preserved either in *Ptilodus* nor in *Taeniolabis*.

In the collection of Cretaceous multituberculates from Mongolia—studied by one of us (Z.K.-J.) there are several complete multituberculate skulls, two of which, belonging to the genera *Kamptobaatar* and *Sloanbaatar* (Figs 1 and 2; Plates 1 and 2), have well preserved braincases (Kielan-Jaworowska, 1970*a*, *b*, 1971).

In 1967 one of us (K.A.K.) suggested that on their braincase structure the orders Triconodonta, Docodonta, and Multituberculata formed a natural group with the Monotremata. This group he called the non-therian mammals to contrast them with the Theria, which clearly formed a natural group. His opinion was accepted by Hopson

* There has been a complete skull of a multituberculate for many years at Princeton. This was collected by Professor Jepsen from the Polecat Bench Formation, but unfortunately has never been described.

& Crompton (1969) and Hopson (1970). K.A.K. had reached his conclusions mostly on the study of the braincases of the triconodonts and docodonts, since the only multituberculate material available to him was that described by Simpson (1937). But Z.K.-J. studying the Cretaceous multituberculates from Mongolia came independently to the same conclusions.

The aim of the present paper is: (i) to combine the conclusions which each of us reached independently when studying the braincase structure of different groups of

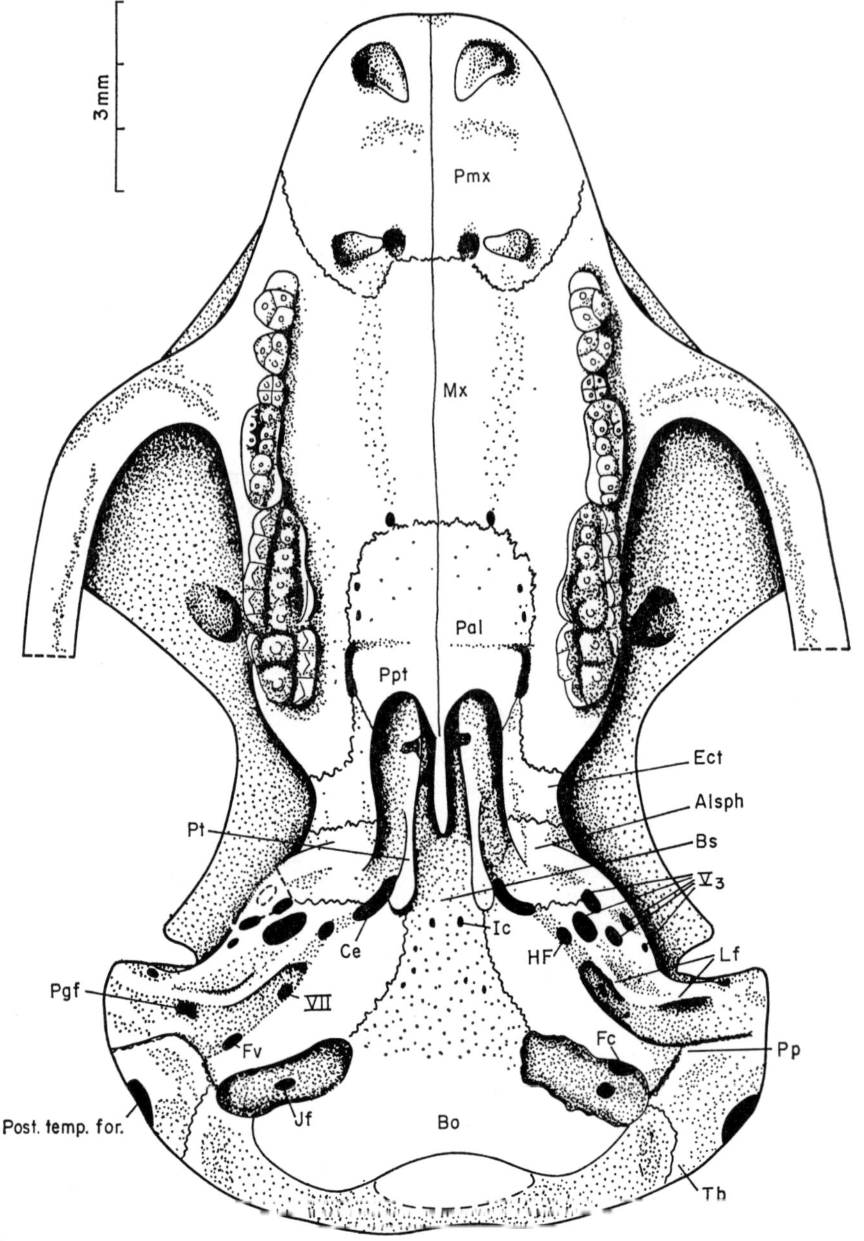

FIGURE 1. *Kamptobaatar kuczynskii:* reconstruction of the skull in palatal view (after Kielan-Jaworowska, 1970*b*). Abbreviations see p. 115.

non-therian mammals, (ii) to compare the general pattern of the braincase structure of the non-therian mammals and (iii) to discuss their relationships.

THE THERAPSID BRAINCASE

The most primitive pattern of the therapsid braincase is found in the gorgonopsids. Here the lateral wall of the braincase proper is formed by the prootic and opisthotic

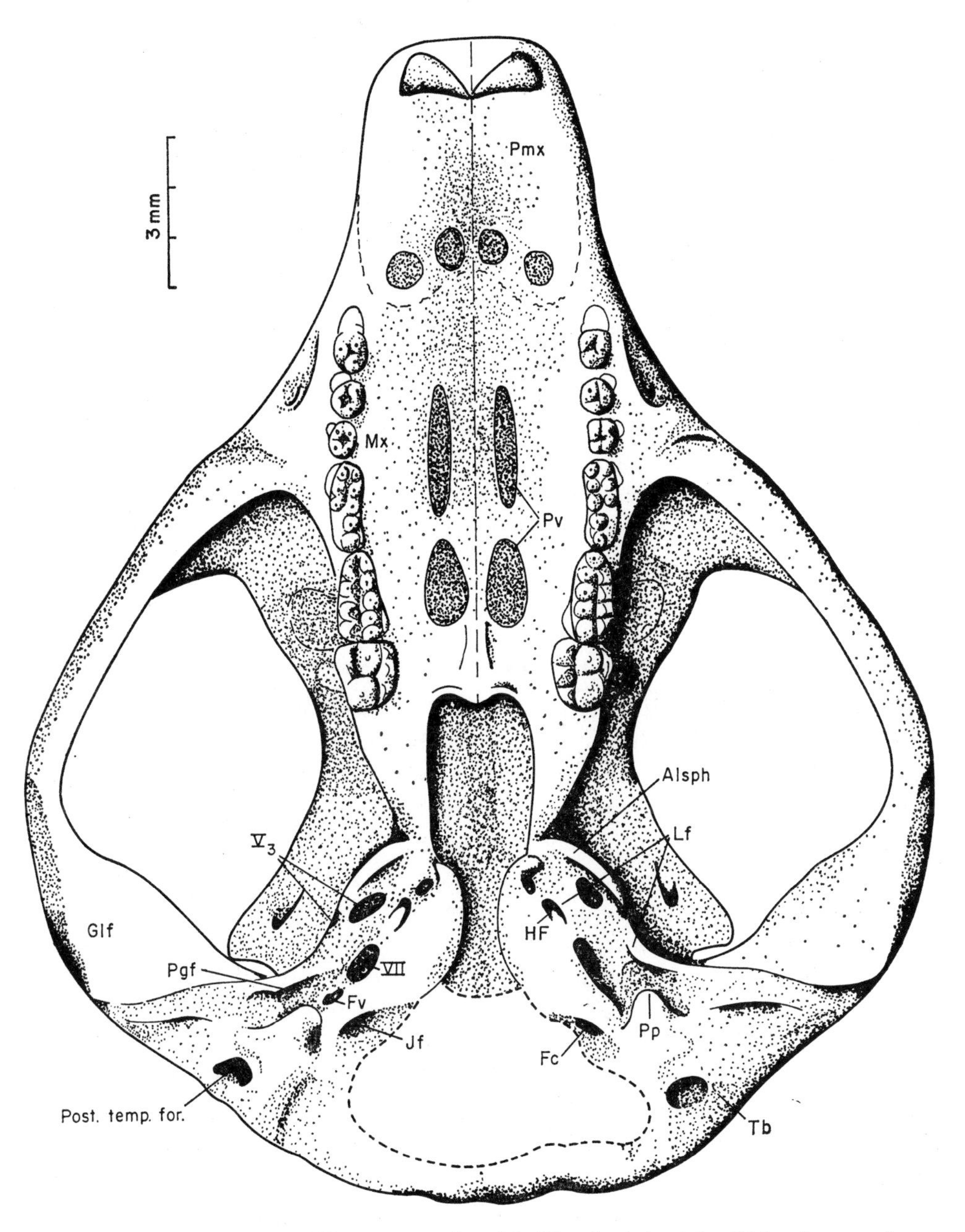

FIGURE 2. *Sloanbaatar mirabilis:* reconstruction of skull in palatal view (after Kielan-Jaworowska, 1971). Abbreviations see p. 115.

bones. The trigeminal (V) nerve leaves the braincase by a notch in the front of the prootic (the incisura prootica). The alisphenoid (epipterygoid) is a simple structure lying outside the area of the braincase altogether and acting as a strut connecting the palate to the skull roof. The space between the braincase and the alisphenoid is the cavum epiptericum in which lies the ganglion of the Vth nerve—the semilunar ganglion.

In what may be called the main line of therapsid evolution—that passing through the Therocephalia and their successors the Cynodontia—the trend is to broaden the alisphenoid. This trend culminates in the later cynodonts, where the ascending process of the alisphenoid has greatly broadened and is suturally connected with a lateral lappet on the prootic. Thus the alisphenoid covers in lateral view the anterior part of the prootic, being separated from it by the cavum epiptericum. The trigeminal nerve still leaves the braincase by the prootic incisure and the semilunar ganglion still occupies the cavum epiptericum. The ganglion is, however, now covered laterally by the broad alisphenoid. After leaving the semilunar ganglion the mandibular and possibly the maxillary branch of the trigeminal nerve (Simpson, 1933) pass through a foramen between the prootic and the ascending process of the alisphenoid.

The structure of the lateral wall of the braincase in the Tritylodontia and in *Diarthrognathus* (Crompton, 1958) is identical with that of the later cynodonts.* This is a good reason for considering *Diarthrognathus* a reptile.

In the other line of therapsid evolution, passing from the Therocephalia to the Bauriamorpha, the great increase in breadth of the ascending process of the alisphenoid does not occur. The bone remains in much the same condition as in the Gorgonopsia, and its ascending process does not come into contact with the anterior edge of the prootic.

THE DOCODONT BRAINCASE

The only member of this order in which the braincase is known is *Morganucodon*. Here the lateral wall of the braincase is formed by the anterior lamina of the petrosal. Through this lamina passes the mandibular nerve (V) by a single foramen pseudovale. Anterior to this the maxillary branch of the trigeminal (Vth) passes through the anterior lamina by the foramen pseudorotundum.

The alisphenoid lies in front of the petrosal. The ascending process of the alisphenoid does not contact the anterior lamina, but there is a clear space between this part of the two bones (Kermack, Mussett & Rigney, 1970). Both in this character, and in the width of the ascending process, *Morganucodon* resembles the condition in bauriomorphs—or for that matter in gorgonopsids—much more than it does the condition in typical cynodonts.

The petrosal has an anterior flange which articulates with the quadrate process of the alisphenoid as in cynodonts or tritylodonts. Between the quadrate ramus of the alisphenoid and the petrosal there is a small cavum epiptericum. The cochlea is almost straight.

* The only difference is that the tritylodonts have a single bone—the petrosal surrounding the inner ear, instead of the opisthotic and prootic of typical reptiles (Kühne, 1956). The condition in *Diarthrognathus* is similar (Crompton, 1958).

The jugal and lacrimal are conspicuous, and there may be a small ectopterygoid lying mostly anterior to the lesser palatine foramen.

THE MONOTREME BRAINCASE

The identification of the bones in the monotreme skull is difficult as the sutures between the bones are obliterated early in life. The first reconstruction of the braincase structure of the monotremes by Van Bemmelen (1901) was made in therian terms, so that he called the extensive bone in front of the squamosal the alisphenoid. A more thorough embryological study of the skull structure of both *Echidna* and *Ornithorhynchus* led Watson (1916) to a different interpretation.

According to him almost all the lateral wall of the braincase is formed by the extensive anterior lamina of the petrosal. The squamosal makes only a small contribution while the alisphenoid is reduced to a narrow, ventral element. The alisphenoid abuts ventrally on a small bone, called by Gaupp (1908) the 'Echidna pterygoid'. The homology of this is not clear. It was considered by Broom (1914) to be the ectopterygoid. This seems plausible. The post-temporal fossa is present.

The cranial nerves II, III, IV, V_1, V_2 and VI are transmitted through a comparatively large sphenorbital fissure between the orbitosphenoid and the anterior lamina of the petrosal. In *Ornithorhynchus* there may be a separate foramen pseudorotundum for V_2. The foramen for V_3 (foramen pseudoovale) is large. It is situated between the alisphenoid, the anterior lamina and the ventral body of the petrosal.

THE MULTITUBERCULATE BRAINCASE

The Late Cretaceous (Coniacian or Santonian) genera *Kamptobaatar* and *Sloanbaatar*, from Mongolia (Kielan-Jaworowska, 1970*b*, 1971) give new information on the multituberculate braincase. Its lateral wall is almost entirely formed by the expanded anterior lamina of the petrosal. The squamosal contributes little. Due to the great breadth of the basicranium in multituberculates, first noted by Simpson (1937), part of the floor of the braincase is formed by the anterior lamina and the foramen pseudoovale lies in this ventral part of the anterior lamina. Apparently typical of the multituberculates is the division of the foramen pseudoovale into two or more parts (Simpson, 1937; Kielan-Jaworowska, 1970*b*).

It seems characteristic of the multituberculates that the bones of the braincase fuse early and the sutures between them become obliterated. Fortunately *Kamptobaatar* (Z. Pal. No. MgMI 33) is an immature specimen and some of the sutures of the braincase can be made out. In particular this specimen suggests that the prominent ridge, lying just lateral and anterior to the foramen pseudoovale (Fig. 1) was formed by a bone separate from the petrosal, although there are places where two bones are already fused. This bone can only be the alisphenoid. This is confirmed by the examination of an undescribed specimen (Z. Pal. No. MgMI 49). The alisphenoid is a small bone, which can be seen forming a limited part of the floor of the orbit, but not really part of the lateral wall of the braincase.

In ventral view, an ectopterygoid bone is present in front of the alisphenoid. The ectopterygoid tapers at the palatonasal foramen (Tatarinov, 1969), its anterior end

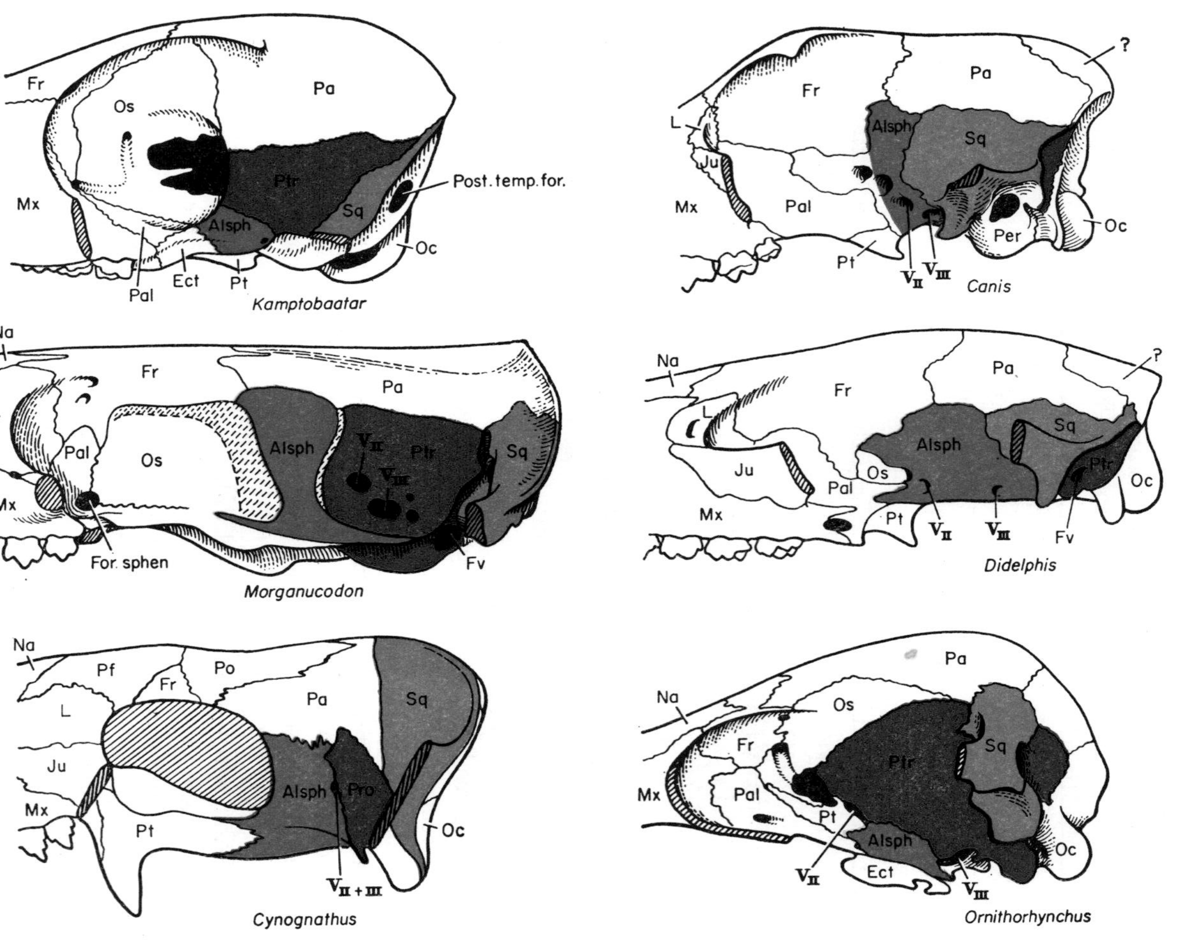

FIGURE 3. Diagram to show the structure of the lateral wall of the braincase in: (1) an advanced mammal-like reptile—*Cynognathus*, a cynodont; (2) *Morganucodon;* (3) *Kamptobaatar;* (4) *Ornithorhynchus;* (5) *Didelphis;* and (6) *Canis.* (5) and (6) are Theria; (2), (3) and (4) are non-therian mammals. Abbreviations see p. 115.

being inserted between the palatine and the maxilla. We do not know if the ectopterygoid is exposed on the interorbital wall.

In the skull of *Kamptobaatar*, a small cavum epiptericum is retained between the pterygoid, alisphenoid and petrosal. In *Kamptobaatar*, the region of the internal nares is preserved—it is unknown in the other genera—and shows a structure of three parallel ridges. The pterygoids are not situated along the lateral walls of the choanae, but are placed in the middle of each choanal channel, dividing it into an inner channel —between the vomer and the pterygoid, which served as the air passage, and an outer trough which probably served as an area for muscle attachment. The lateral wall of the outer trough is formed by the ectopterygoid.

The middle ear is completely open, and there is a crista parotica, formed by the anterior part of the paroccipital process, just as in monotremes. There is also a well developed post-temporal fossa. The jugal is absent and a well-developed lacrimal is only known in *Paulchoffatia* (Hahn, 1969).

THE THERIAN BRAINCASE

The braincase of the Theria is of a completely different pattern from that found in any of the groups of non-therian mammals. There is no anterior lamina of the petrosal its place is taken by the alisphenoid and the squamosal. The mandibular and maxillary branches of the trigeminal nerve leave the skull through two foramina in the alisphenoid —the mandibular branch through the foramen ovale and the maxillary branch through the foramen rotundum. There is no post-temporal foramen and no ectopterygoid. Finally there is a spiral cochlea—this is unknown in any of the non-therian groups where the cochlea is, at most, slightly curved. It must be noted that in talking about the therian braincase, we are talking about the eutherian braincase—found in the marsupials and placentals. Nothing is known of the pantotherian braincase beyond what can be inferred from that of their descendants—the Eutheria.

DISCUSSION

The facts presented above show that the non-therian braincase is constructed on quite a different pattern from that of the Theria.

In non-therian mammals the lateral wall of the braincase is mostly formed by an extensive anterior lamina of the petrosal, in front of which there is a comparatively small alisphenoid. Unfortunately the representatives of various groups of non-therian mammals upon which we have to base the comparisons derive from very different geological periods. This may make the differences in structure appear greater than they really were. In *Morganucodon*, which is of Late Triassic age, the alisphenoid is largest in all known non-therian mammals. In this respect *Morganucodon* is closer to the therapsid reptiles than any other of the non-therian group. In the multituberculates and monotremes the anterior lamina of the petrosal is larger than in *Morganucodon*, and correspondingly the alisphenoid is reduced. In this respect *Kamptobaatar* from the Upper Cretaceous is intermediate between the *Morganucodon* and the monotremes. In Cretaceous multituberculates the anterior lamina of the petrosal is more extensive than in *Morganucodon*, but less so than in *Ornithorhynchus*, while the alisphenoid is less extensive than in *Morganucodon*, but greater than in *Ornithorhynchus*.

In all the non-therian mammals the squamosal contributes practically nothing to the lateral wall of the braincase. The cranial part of the squamosal is narrow and situated far posteriorly.

The structure of this region in Theria is quite different. The anterior lamina of the petrosal is absent. The lateral wall of the braincase is formed both by the cranial part of the squamosal, which may be extensive, and the alisphenoid.

The position of the foramen for mandibular branch of the trigeminal nerve is quite different in non-therian and therian lines. In therians it is situated within the alisphenoid and it is called the foramen ovale. In non-therian mammals it either pierces the petrosal to form the foramen pseudovale or lies along the junction of the petrosal and alisphenoid.

There are further differences in the structure of the basicranial region. In all the Mesozoic non-therian mammals the cavum epiptericum seems to be present. It was lost in the monotremes and it does not exist in therian mammals. In the various groups of non-therian mammals different 'reptilian' bones are present, which do not exist in therian mammals. A septomaxilla is retained in monotremes and in docodonts (Patterson & Olson, 1961) while the ectopterygoid is present in the multituberculates. On Broom's (1914) interpretation the ectopterygoid persists in the living monotremes as the 'Echidna pterygoid' and we have been informed (Kermack *et al.*, 1970) that there is possibly an ectopterygoid in *Morganucodon*. These are simply examples of the retention of primitive reptilian characters in the non-therian groups of mammals, and cannot be used to indicate relationship in the same way as the structure of the lateral wall of the braincase. Similar reptilian survivals are the retention of a post-temporal fossa in those groups in which the occiput is known (not known in triconodonts) and the presence of large tabulars in the multituberculates and *Morganucodon*. In all groups the paroccipital process seems to be formed from the petrosal—as in reptiles.

The non-therian mammals can be divided up into two groups—the members of each group being clearly more closely related to each other than to the members of the other group.

The first group comprises the Multituberculata and the Monotremata. In both groups the braincase has become widened, consequent on the increase in size of the brain. This means that part of the anterior lamina of the petrosal faces ventrally. This part of the anterior lamina contains the foramen pseudovale, which consequently also faces ventrally in these two groups.

In both the alisphenoid is greatly reduced and occupies a ventral position in the skull. The alisphenoid lies close to the foramen pseudovale in multituberculates and forms actually part of its border in monotremes. This is due to the great size of the foramen pseudovale in the latter group, and probably has no other significance.

The ectopterygoid in multituberculates occupies a posterior position resembling that of the 'Echidna pterygoid' in monotremes. Both multituberculates and monotremes lack a jugal. The lacrimal seems to be absent in all except the earliest members of the Multituberculata (Hahn, 1969; Kielan-Jaworowska, 1970*a*) and is not present in the Monotremata. From what is known of the multituberculate post-cranial skeleton (Granger & Simpson, 1929) it does not much resemble that of the monotremes, but nor does it show any real resemblance to that of the Theria. The description of some of the new material from Mongolia may resolve this problem.

The one reasonably complete petrosal of a triconodont (*Trioracodon ferox*) closely resembles that of *Morganucodon* (Kermack, 1963) and the only known basicranium of a triconodont (*Triconodon mordax* B.M.47763) is very like that of *Morganucodon*. Both have the same system of three longitudinal ridges, differing from the ridges of the Multituberculata in being further back—not involving the internal nares, and being formed rather differently. There are also similarities between the teeth of triconodonts and morganucodonts (Mills, 1971) and there is not much doubt that the former must have evolved from the latter. Together they form a natural group within the non-therian mammals.

The time gap between the first Docodonta and the first Multituberculata makes it possible to derive both the major groups of the non-therian mammals from a common ancestor which was itself a mammal. Even if this were not so, and they crossed the reptile-mammal boundary independently, their common reptilian ancestor would not have lived long before the transition.

Kermack (1967) has given the reasons for considering the Theria and the 'non-Theria' to be quite distinct. Theria and non-Theria would have competed for the various ecological niches open to them. By the end of the Jurassic the Theria had established themselves as the dominant insectivorous group (Eupantotheria and Symmetrodonta) with the non-therian mammals the herbivores (Multituberculata) and carnivores (Triconodonta). *Docodon* and its Upper Jurassic relatives were an unsuccessful attempt by the non-Theria to produce the equivalent to a tribosphenic tooth.

The Docodon experiment had failed by early in the Cretaceous. The Triconodonta were eliminated at the end of the Mesozoic by competition from carnivorous insectivores and perhaps early creodonts. The Multituberculata survived until the end of the Eocene, when they were finally exterminated by therian competition—mainly from rodents (Jepson, 1940). The only non-Theria to survive to the present day are the Monotremata—highly specialized forms living in the Australian regions. This phenomena of the failure of one group to stand up to the competition of another is paralleled by the general failure of the marsupials to maintain themselves against placental competition.

ACKNOWLEDGEMENTS

We thank the University of London for a grant from the Central Research Fund to enable one of us (K.A.K.) to visit Warsaw and examine the multituberculates there. The work on which his part of this paper is based has been made possible by grants from the National Environmental Research Council.

REFERENCES

BRINK, A. S., 1963. The taxonomic position of the Synapsida. *S. Afr. J. Sci.*, **59**: 153.

BROOM, R., 1914. On the origin of mammals. *Phil. Trans. R. Soc.* (Ser. B), **206**: 1.

CLEMENS, W. A., 1968. Origin and early evolution of marsupials. *Evolution, Lancaster, Pa.*, **22**: 1.

CROMPTON, A. W., 1958. The cranial morphology of a new genus and species of ictidosaurian. *Proc. zool. Soc. Lond.*, **130**: 183.

CROMPTON, A. W., 1964. A preliminary description of a new mammal from the Upper Triassic of South Africa. *Proc. zool. Soc. Lond.*, **142**: 441.

CROMPTON, A. W. & JENKINS, F. A., 1968. Molar occlusion in Late Triassic mammals. *Biol. Rev.*, **43**: 427.

CRUSAFONT-PAIRÓ, M., 1962. Constitución de una nueva clase (Ambulatilia) para los llamodos 'Reptiles mamiferoides'. *Notas Comun. Inst. geol. min. Esp.*, **66**: 259.

GAUPP, E., 1908. Zur Entwicklungsgeschite und vergleichenden Morphologie des Schadels von *Echidna aculeata* var. *typica*. In R. Semon (Ed.), *Zool. Forsch-reis. in Australien*, **3**, **II** (2). *Denkschr. med.-naturw. Ges. Jena*, **VI** (2): 539.

GRANGER, W. & SIMPSON, G. G., 1929. A revision of the Tertiary Multituberculata. *Bull. Am. Mus. nat. Hist.*, **56**: 601.

HAHN, G., 1969. Beiträge zur Fauna der Grube Guimarota nr. 3-die Multituberculata. *Palaeontographia* (*Abt. A*), **133**: 1.

HOPSON, J. A., 1970. The classification of non-therian mammals. *J. Mammal.*, **51**: 1.

HOPSON, J. A. & CROMPTON, A. W., 1969. The origin of mammals. *Evolutionary Biology*, **3**: 15.

JEPSEN, G. L., 1940. Palaeocene faunas of the Polecat Bench Formation, Park County, Wyoming. *Proc. Am. phil. Soc.*, **83**: 217.

KERMACK, K. A., 1963. The cranial structure of the triconodonts. *Phil. Trans. R. Soc.* (Ser. B), **246**: 83.

KERMACK, K. A., 1967. The interrelations of early mammals. *J. Linn. Soc.* (*Zool.*), **47**: 311.

KERMACK, K. A. & MUSSETT, F., 1959. The jaw articulation in Mesozoic mammals. *Proc. XV Int. Congr. Zool.*, Sect. V, p. 442.

KIELAN-JAWOROWSKA, Z., 1970*a*. New Upper Cretaceous Multituberculate genera from Bayn Dzak, Gobi Desert. In *Results of the Pol.-Mong. Palaeont. Exped.*, Part II. *Palaeont. pol.*, **21**: 35.

KIELAN-JAWOROWSKA, Z., 1970*b*. Unknown structures in multituberculate skull. *Nature, Lond.*, **226**: 97.

KIELAN-JAWOROWSKA, Z., 1971. Skull structure and affinities of the Multituberculata. In *Results of the Pol.-Mong. Palaeont. Exped.*, Part III. *Palaeont. pol.*, **25**.

KÜHNE, W. G., 1956. *The Liassic therapsid* Oligokyphus. London: Br. Mus. (Nat. Hist.).

MACINTYRE, G. T., 1967. Foramen pseudovale and quasi-mammals. *Evolution, Lancaster, Pa.*, **21**: 834.

MILLS, J. R. E., 1971. The dentition of *Morganucodon*. In D. M. Kermack & K. A. Kermack (Eds), *Early Mammals. Zool. J. Linn. Soc.*, **50**, Suppl. 1: 29–63.

OLSON, E. C., 1959. The evolution of mammalian characters. *Evolution, Lancaster, Pa.*, **13**: 344.

PARRINGTON, F. R., 1941. Two mammalian teeth from the Lower Rhaetic of Somerset. *Ann. Mag. nat. Hist.* (Ser. 11), **8**: 140.

PARRINGTON, F. R., 1967. The origins of mammals. *Advmt Sci., Lond.*, **24**: 165.

PATTERSON, B., 1956. Early Cretaceous mammals and the evolution of mammalian molar teeth. *Fieldiana, Geol.*, **13**, no. 1.

PATTERSON, B. & OLSON, E. C., 1961. A triconodontid mammal from the Triassic of Yunnan. International colloquium on the evolution of lower and non-specialized mammals. *Kon. Vlaamse Acad. Wetensch. Lett. Sch. Kunsten België*, Part I: p. 129. Brussels.

REED, C. A., 1960. Polyphyletic or monophyletic ancestry of mammals, or: what is a class? *Evolution, Lancaster, Pa.*, **14**: 314.

ROMER, A. S., 1969. Cynodont reptile with incipient mammalian jaw articulation. *Science, N.Y.*, **166**: 881.

SIMPSON, G. G., 1928. *A catalogue of the Mesozoic Mammalia in the Geological Department of the British Museum* (*Natural History*). London: Br. Mus. (Nat. Hist.).

SIMPSON, G. G., 1933. The ear region and the foramina of the cynodont skull. *Am. J. Sci.*, **26**: 285.

SIMPSON, G. G., 1937. Skull structure of the Multituberculata. *Bull. Am. Mus. nat. Hist.*, **73**: 727.

SIMPSON, G. G., 1959. Mesozoic mammals and the polyphyletic origin of mammals. *Evolution, Lancaster, Pa.*, **13**: 405.

SIMPSON, G. G., 1960. Diagnosis of the classes Reptilia and Mammalia. *Evolution, Lancaster, Pa.*, **14**: 388.

SIMPSON, G. G., 1961. Evolution of Mesozoic mammals. International colloquium on the evolution of lower and non-specialized mammals. *Kon. Vlaamse. Acad. Wetensch. Lett. Sch. Kunsten Belgie*, Part 1, p. 57. Brussels.

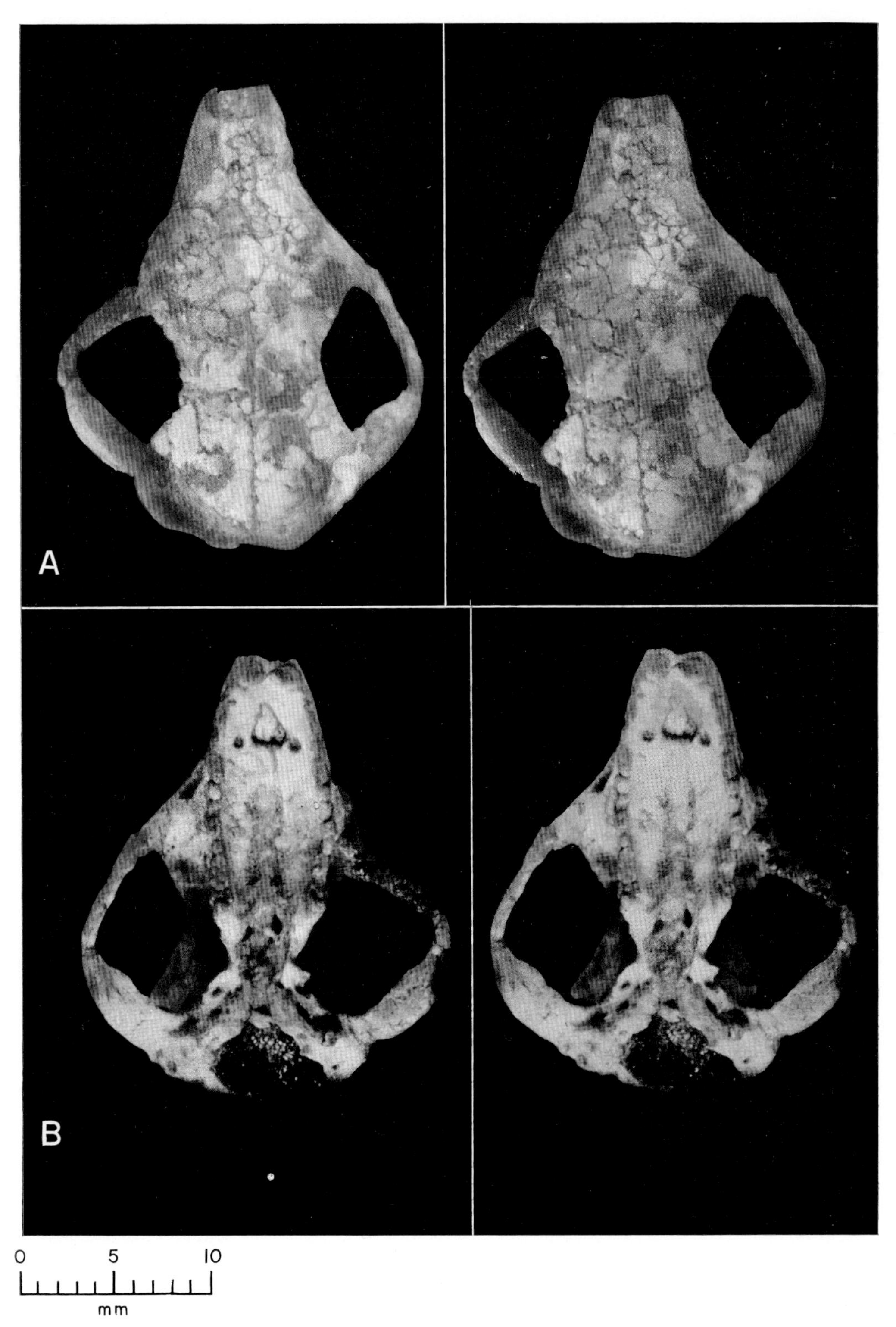

K. A. KERMACK AND Z. KIELAN-JAWOROWSKA

(*Facing p.* 114)

PLATE 2

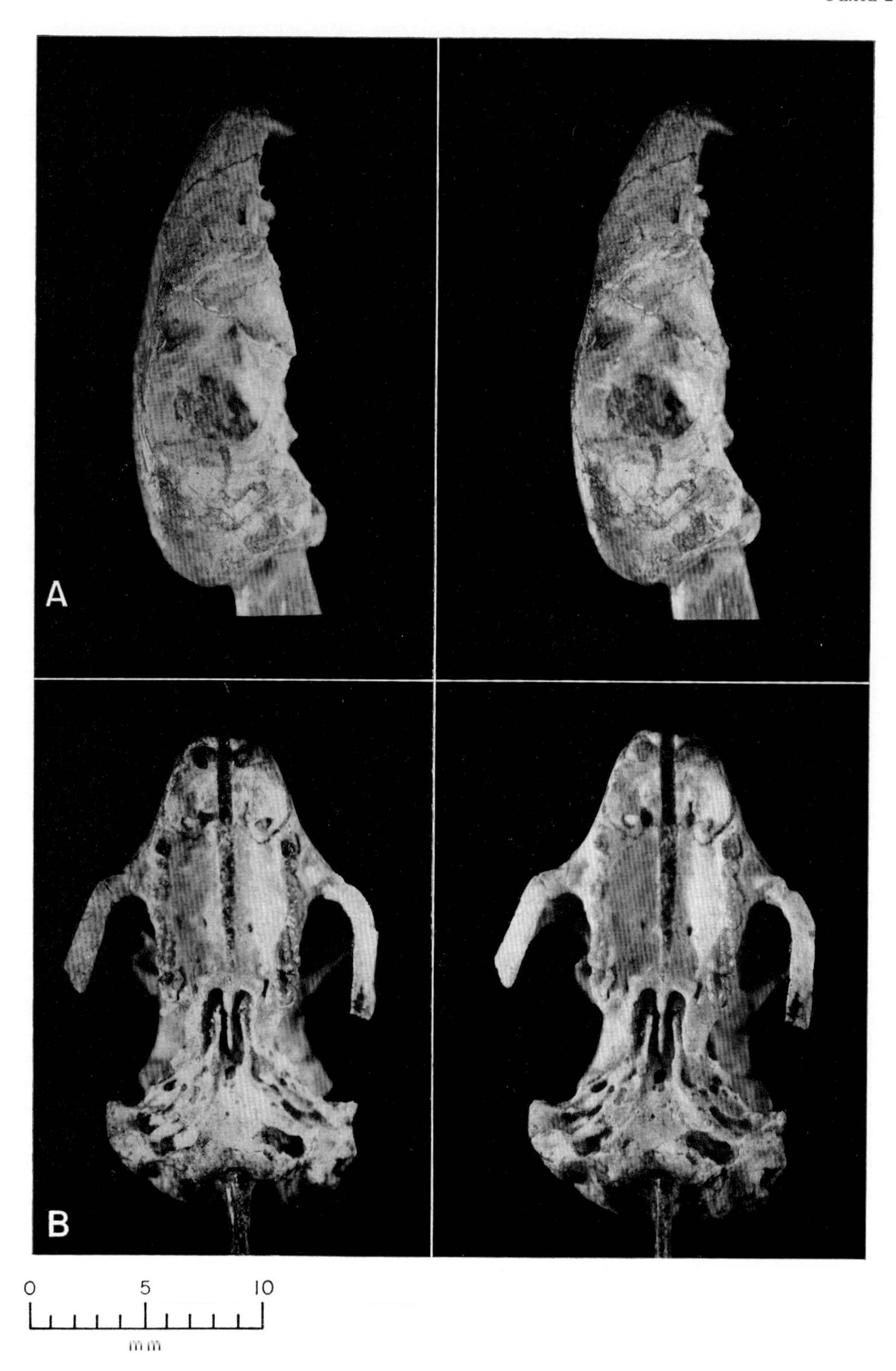

K. A. KERMACK AND Z. KIELAN-JAWOROWSKA

Tartarinov, L. P., (1969). Проблемы зволуци териодонтов. Автореферат диссертации.— Палеонт. Инст. А.Н. СССР. Моснва. [The problems of the evolution of the theriodonts.—Summary of thesis for doctorate. *Institute of Palaeontology, Academy of Sciences of U.S.S.R. Moscow.*]

Van Bemmelen, J. F., 1901. Der Schadelbau der Monotremen. In R. Semon (Ed.), *Zool. Forsch. in Australien* **3, II** (**1**). *Deukschr. med-naturw. Ges. Jena* **VI** (**1**): 729.

Van Valen, L., 1960. Therapsids as mammals. *Evolution, Lancaster, Pa.*, **14**: 304.

Watson, D. M. S., 1916. The monotreme skull: a contribution to mammalian morphogenesis. *Phil. Trans. R. Soc.* (Ser. B), **207**: 311.

ABBREVIATIONS USED IN FIGURES

Alsph	alisphenoid
Bo	basioccipital
Bs	basisphenoid
Ce	cavum epipterium
Ect	ectopterygoid
Fc	fenestra cochleae
For. sphen.	sphenopalatine foramen
Fr	frontal
Fv	fenestra vestibuli
Glf	glenoid
HF	hiatus canalis Fallopii
Ic	internal carotid foramen
Jf	jugular foramen
Ju	jugal
L	lacrimal
Lf	lateral flange
Mx	maxilla
Na	nasal
Oc	exoccipital
Os	orbitosphenoid
Pa	parietal
Pal	palatine
Per	tympanic
Pf	prefrontal
Pgf	lateral opening, of canalis prooticus
Pmx	premaxilla
Po	postorbital
Post. temp. for.	posterior temporal foramen
Ppt	pterygoid
Pro	prootic
Pt	pterygoid
Ptr	petrosal
Pv	palatal foramina
Sq	squamosal
Tb	tabular
V_3	foramina for madibular branch of trigeminal nerve
V_{II}	maxillary branch of trigeminal nerve
V_{III}	mandibular branch of mandibular nerve
VII	foramin for facial-nerve

EXPLANATION OF PLATES

Plate 1

Dorsal (**A**) and Ventral (**B**) views of skull of *Sloanbaatar mirabilis* (after Kielan-Jaworowska, 1970*a*).

Plate 2

Lateral (**A**) and Ventral (**B**) views of skull of *Kamptobaatar kuczynskii* (after Kielan-Jaworowska, 1971).

A review of English Early Cretaceous mammals

WILLIAM A. CLEMENS, JR.

Department of Paleontology, University of California, Berkeley, California, U.S.A.

AND

PATRICIA M. LEES

Department of Zoology, University College, London, England

Mammalian fossils have been discovered in Early Cretaceous strata cropping out in the Weald, south-eastern England. In the last ten years application of collecting techniques involving disaggregation of the sediments in dilute acid and concentration of the fossils by flotation in heavy liquids, has resulted in enlargement of the sample. This collecting technique has been improved to increase the efficiency of the separation and the safety of the operator. Wealden mammals, which are of Neocomian age, have been found in strata of the lower part of the Wadhurst Clay at two localities, Cliff End and Tighe Farm, and in the overlying Grinstead Clay at Paddockhurst Park. The mammalian fauna is known to include multituberculates, symmetrodonts, eupantotheres, and a therian closely related to early marsupials and placentals. A new spalacotheriid symmetrodont, ***Spalacotherium taylori***, and a new dryolestid eupantothere, ***Melanodon hodsoni***, are described.

CONTENTS

INTRODUCTION

The Early Cretaceous was a time of major change in the evolution of the Mammalia. Comparison of Late Jurassic faunas from the Purbeckian Lulworth Beds of England and the Morrison Formation of North America with the late Early Cretaceous (Albian) fauna found in Texas illustrates the magnitude of the change. However, few collections of intermediate age documenting the rates and directions of these evolutionary changes are known. Two mammals, *Manchurodon* and *Endotherium*, were recovered from deposits in what was Manchuria and are thought to be of Early Cretaceous age (Kermack, Lees & Mussett, 1965). Discovery of Early Cretaceous mammals at several localities in Spain was recently announced by Kühne & Crusafont-Pairó (1968). So far two mammals from these sites have been described: *Crusafontia* Henkel & Krebs 1969 is a dryolestid not represented in the collections from the Wealden of England. *Parendotherium* (Crusafont & Adrover, 1966) is probably a multituberculate (note McKenna, 1970).

The third area in which Early Cretaceous mammals have been discovered is the Weald of England. A historical resumé of collecting in this area of south-eastern England was published by Clemens (1963 *a*). Since the summer of 1961, when the last collections considered in that paper were made, groups from the Department of Zoology, University College, London directed by Dr K. A. Kermack have continued to prospect and collect in Early Cretaceous strata of the Weald. One of us (P.M.L.) has been responsible for the processing of the fossiliferous matrix obtained and discovery of the mammalian fossils described for the first time in this paper. We express our gratitude to Dr Kermack who has made it possible for us to present this second interim report on the continuing search for Wealden mammals.

MATERIAL AND METHODS

The search for Wealden mammals has involved particularly difficult challenges to the palaeontologist. To date mammalian fossils have been found only in strata that contain a high concentration of rock fragments of approximately the same size, thus preventing rapid concentration of the fossils by sorting on the basis of this character, or are well cemented and do not readily disaggregate. In the case of the Cliff End bone bed both conditions prevent rapid separation and concentration of the vertebrate fossils. Collecting techniques specifically designed to deal with these kinds of deposits have been developed independently at several institutions (for example, note Henkel, 1966).

The techniques used at University College, London, to process rock from the Early Cretaceous deposits of the Weald were described by Lees (1964). Since publication of this paper a few modifications of the technique increasing the safety of the operator have been instituted. These precautions may appear excessive but tetrabromoethane and, particularly, dibromoethane are highly toxic. In a large-scale project the operator will be exposed to the fumes from a considerable quantity of liquid over a prolonged period and damage to the respiratory system and liver can ensue. The modifications in technique instituted are as follows:

Because the fume cupboard in which the flotation is carried out is not fully sealed the operator wears a gas mask during this stage of the processing. A Siebe Gorman 'Puretha' respirator in conjunction with an absorption canister type O is used.

After flotation the concentrate is washed twice in 70% alcohol and then left in a fume cupboard until the alcohol has fully evaporated. At this stage each grain is still coated with a thin, strongly adherent layer of bromoethanes. Use of a vacuum oven for further cleaning, mentioned in the previous account (Lees, 1964: 305), has now been incorporated as a routine step in the processing. The concentrate is held in the oven at 120°–140°C for 12 hours and then allowed to cool to room temperature. After this treatment the concentrate will still have an odour of the brominated hydrocarbons. This can be removed by spreading out the concentrate in a fume cupboard and allowing it to stand for several days or, preferably, recycling it through the washing in alcohol and drying in the vacuum oven. Thorough cleaning of the concentrate greatly increases the safety of the operator who must sort it under a binocular microscope.

The toxic fumes produced during heating in the oven can be dealt with without danger to the operator. At University College the oven is exhausted by use of a single stage oil-filled pump. It is necessary to both limit the amount of brominated hydrocarbons released from the system and protect the oil in the pump from contamination. This is accomplished by introducing two liquid nitrogen filled traps between the oven and the pump.

The fossils described and mentioned here are preserved in the collections of the Department of Zoology, University College, London (catalogue number with prefix W); Department of Palaeontology, British Museum (Natural History) (catalogue number with prefix M); and the Yale Peabody Museum, Yale University (catalogue number with prefix Y.P.M.). All measurements are in millimetres.

LOCALITIES

Cliff End

Mammalian fossils were discovered at Cliff End, East Sussex, by Revs P. Teilhard de Chardin and F. Pelletier, who were working with C. Dawson, and described by A. Smith Woodward (1911). No additional mammalian fossils were discovered until 1960, when blocks of the bone bed were prepared at the Department of Zoology, University College, London. The fossils obtained during the academic year 1960–61 were described by Clemens (1963*a*). Subsequently Kermack *et al.* (1965) described a mammalian lower molar, type of *Aegialodon dawsoni*, and announced the discovery of eight mammalian teeth. The fossil sharks from the Cliff End, Tighe and Paddockhurst bone beds were studied by C. Patterson (1966).

All the mammalian teeth collected since 1960 have been obtained from isolated blocks of the bone bed found in the shingle or on the wave-cut platforms below the cliffs at Cliff End. The collecting area is indicated in Clemens, 1963*a*, text-fig. 1. It was thought that only the Ashdown Sand is exposed in the cliffs between Cliff End and Haddock's Cottages. Although exposed in the cliffs south-west of Haddock's Cottages, the lower unit of the Ashdown Beds, the Fairlight Clays, does not crop out above low tide level in the immediate vicinity of the collecting area. Allen (1960: 11) stated, '. . . rapid erosion of the cliffs has re-exposed the celebrated bone bed . . .' and described it as being '. . . beneath a massive sandstone yielding most of the larger joint blocks seen on the present shore'. On field trips in the autumn of 1960 and spring of 1961 we could not find an outcrop of the bone bed. It was assumed to be hidden beneath the shingle and part of the Ashdown Sand.

In the spring of 1962 Mr R. P. Murphy found a large block of the bone bed lying on a ledge more than ten feet above the shingle. The size and position of the block suggested it had fallen from the cliff face, however a search of the accessible areas of the cliff did not reveal an exposure of the bone bed. Continued collecting at Cliff End showed that blocks of the bone bed were available in quantity only after winter and spring storms. Not finding a source for these blocks in the cliff or on the platform below it Kermack *et al.* (1965) suggested they were derived from a bone bed in the Fairlight Clays cropping out on the sea floor off Cliff End. Patterson (1966) accepted this lower stratigraphic position.

The geology of the Cliff End area has been studied since 1965 by Dr E. R. Shephard-Thorn of the Institute of Geological Sciences, who will present an analysis of the stratigraphy of the bone bed elsewhere. For the purposes of placing the mammalian fossils described subsequently in the proper stratigraphic context he has prepared the following summary of his findings that significantly modify earlier interpretations.

> 'Careful study of the cliff sections, and examination of material brought down in cliff falls and exposed in temporary sections between 1965 and 1970 has prompted a revision of their stratigraphy. Briefly this starts from the recognition of a 1-m band of dark shales and thin interlaminated siltstones with a 10-cm clay ironstone band at the top as the equivalent of the basal Wadhurst Clay (traces of the Top Ashdown Pebble Bed (Allen, 1949*b*) occur beneath it). A massive unit of well-jointed, fine white sandstone, 10 m thick, forms the upper part of the cliff at Cliff End; it is capped by another thin pebble bed. Dark grey shales with thin siltstones, calcareous Tilgate Stone and nodular clay ironstone occur above, in a densely wooded and landslipped "undercliff"; the bone bed lies within these shales about 2·5 m above the top of the 10-m sandstone. The bone bed is lenticular in horizontal extent and appears to be well-developed between the faults at Haddocks Cottages and Cliff End.
>
> 'Cliff falls are fairly frequent at this locality and usually originate from the undermining of joint-blocks of the massive 10-m sandstone by marine erosion. The overlying shales containing the bone bed are often involved in these falls or subsequent slips and thus provide the intermittent supply of slabs of the bone bed to the beach, from which they have been previously recorded erroneously as *in situ*.
>
> 'Support for the revised stratigraphy outlined above has come from examination of the ostracod faunas of shale samples, collected at Cliff End, by Dr F. W. Anderson of the Institute of Geological Sciences. He confirms the equivalence of the 1-m band of shales with siltstones and ironstones to the basal Wadhurst Clay and assigns the bone-bed to the approximate horizon of the Lydd "S" phase of the *Cypridea tuberculata* Beds (*C. paulsgrovensis* Zone) of the zonal scheme he has recently put forward for the Wadhurst Clay (Anderson, Bazley & Shephard-Thorn, 1967).' (E. R. Shephard-Thorn, pers. comm.)

Allen (1967) has noted that the Cliff End bone bed and the Telham pebble bed can be distinguished lithologically from all other Wealden bone and pebble beds. Both are '... dominated by unstained sub-angular pebbles of "vein" quartz and grey or white quartzite, including types not known elsewhere (Allen, 1967: 262)'. The lithological similarity suggests that the Cliff End bone bed was formed as part of the Telham pebble bed. However, the Telham pebble bed is a unit within the lower part of the Wadhurst Clay while the Cliff End bone bed was assumed to be part of the underlying Ashdown Sand or Fairlight Clay. Now Shephard-Thorn has accumulated considerable evidence indicating that the upper part of the cliff section at Cliff End is made up of a coarse facies of the Wadhurst Clay. The Cliff End bone bed is one of the uppermost strata in this coarse facies.

On the basis of this new evidence we now assume that all the fossiliferous blocks collected at Cliff End were derived from the bone bed intermittently exposed near the top of the cliff. Also we accept Allen's (1967) and Shephard-Thorn's interpretation that the Telham pebble bed and the Cliff End bone bed probably were formed during the same depositional event. In other areas of the Weald, deposition of the Wadhurst Clay is thought to have been a Valanginian event (Allen, 1965). Evidence collected by Shephard-Thorn in the Cliff End area supports assignment of this age to the Cliff End bone bed.

Paddockhurst Park

In the autumn of 1966 Dr K. A. Kermack and P.M.L. undertook extensive excavations at the clay pit in Paddockhurst Park where some mammalian fossils had been recovered (Clemens, 1963*a*). A few blocks of fossiliferous rock were found, one yielding a fragment of a mammalian caniniform tooth (W-10), but a continuation of the bone bed could not be located. It must have been a lenticular unit within the Grinstead Clay that has now been completely excavated. Taylor (1963) has described the lithology of the fossiliferous rock.

The Grinstead Clay overlies the Wadhurst Clay and the Paddockhurst Park fossil assemblage is clearly more recent in age than that found at Cliff End. Allen (1965) suggests deposition of the Grinstead Clay occurred in the Valanginian, but Anderson & Hughes (1964) have assigned a Hauterivian age.

Tighe Farm

The remains of a variety of terrestrial and aquatic vertebrates have been found in the Telham pebble bed and immediately subjacent sandstones (Allen, 1949*a*). As part of their search for Wealden mammalian fossils Dr K. A. Kermack and P.M.L. sampled the Telham pebble bed and other coarse units in the lower part of the Wadhurst Clay. One of these samples was taken from a grit thought to be the Telham pebble bed (Allen, 1949*a*) or a unit slightly higher stratigraphically (Shephard-Thorn, Smart, Bisson & Edmonds, 1966). This grit crops out in the eastern wall of an abandoned quarry which is approximately 200 yards south-west of the Tighe farmhouse and abuts Stone and $2\frac{1}{4}$ miles east of Wittersham, Isle of Oxney, Kent (grid co-ordinates 93632672). The following section was exposed in the eastern face of the quarry in 1969.

4+ feet.	Grey and white, iron-stained, friable, thinly and irregularly stratified siltstones and clays.
6–8 inches.	Pebble bed, quartz grit with clay partings, broken and abraded vertebrate fossils present, friable to heavily indurated with ferruginous cement.
1 foot.	Calcareous siltstone.
2 inches.	Grey, iron-stained clay.
1 foot.	Calcareous siltstone.
4+ feet.	Yellowish brown, poorly cemented, thinly stratified sandstone.

Large samples of the pebble bed are now being processed with the same technique used for blocks of the Cliff End bone bed and a few mammalian teeth have been recovered. Probably these fossils are approximately contemporaneous with those from Cliff End and can be provisionally assigned a Valanginian age.

ORDER MULTITUBERCULATA

Family Plagiaulacidae

Loxaulax valdensis (Woodward, 1911)

The genus was rediagnosed by Hahn (1969: 9) as follows: 'Eine Gattung der Plagiaulacidae mit folgenden Besonderheiten: Nur isolierte Zähne bekannt (untere m, vordere obere p, obere i). Untere m mit 2 Höcker-Reihen. Höcker der buccalen Reihe (4) im Gegensatz zum Bau bei den übrigen Plagiaulacidae subselenodont; Höcker der lingualen Reihe (3) konisch. Vordere obere p tricuspid; obere i (wahrscheinlich i^2) mit einer Haupt-Spitze und mehreren in einer Reihe stehenden hinteren Neben-Spitzen.'

In his extensive review of the Kimmeridgian multituberculates found at Guimarota, Portugal, and in a forthcoming paper Hahn (1969 and pers. comm.) questions the identification of some of the teeth allocated (Clemens, 1963*a*) to *Loxaulax valdensis*. With the Wealden multituberculates we are caught between the desire to refrain from recognizing new taxa until they can be adequately diagnosed and the realization that the Wealden mammalian fauna must have included more than one species of multituberculate. At present it seems best to simply describe the material under the rubric of one taxon and explore the alternative phylogenetic relationships.

The fossils identified with some reservations as teeth of Wealden multituberculates by Simpson (1928) were reviewed by Clemens (1963*a*). Of these two (M10480, the type of *Loxaulax valdensis*, and M10481) are incontrovertibly molars of Wealden multituberculates. The fragment of an incisor first described by Lydekker (1893) (M5691) is certainly mammalian but doubtfully part of the dentition of a Wealden multituberculate. The purported mammalian affinities of two other fossils found in Wealden strata (M20241, not included in Simpson, 1928, and M13134) are doubtful. Without qualification Van Valen (1967) accepted both M13134 and M5691 as Wealden mammalian fossils. These fossils and the circumstances of their discovery have been re-evaluated and no reason found to alter previous interpretations (Clemens, 1963*a*).

The type specimen of *Loxaulax valdensis*, M10480, is a left M_1 that has been thoroughly described by Simpson (1928: 49–50). Although slightly larger (length = 2·1 mm; width = 1·5 mm) and a little more abraded, W-6 (Plates 1**A**,**B** and 4**A**) from the Cliff End bone bed is almost a duplicate of M10480. The only difference worthy of note is the absence of a distinct apex for the fourth (distal) labial cusp that is represented by a distinct transverse ridge. Its absence might reflect post mortem damage. The only other element of the lower dentition of *L. valdensis** possibly represented in our collection is M21102 from Paddockhurst Park (Clemens, 1963*a*, text-fig. 8), which might be a lower incisor of this multituberculate.

If M10480 and W-6 are correctly identified as first lower molars the only other molar of *L. valdensis* represented in the collections is a second upper molar. One specimen, M21098, a right M^2 has already been described (Clemens, 1963*a*, text-fig. 2) but is refigured here (Plates 1**C**,**D** and 4**B**). Another tooth from the Cliff End bone bed, W-1 (Plates 1**E**,**F** and 4**C**) probably is a left M^2. Like M21098 the highest cusps of W-1 form the longest row. Another similarity is that the cusp of this row at the narrowest end of the crown, when seen in occlusal view, is separated from the adjacent cusp by a cleft that is wider and deeper than those separating the other cusps. Taking this row as the lingual and the more isolated cusp as the most distal, the deep indentation on the mesiolabial edge of the crown would mark the area where M^2 abutted against M^1. The two cusps of the labial row of W-1 are of comparable size and position to those of M21098. Of these two molars W-1 is more abraded, particularly along its labial margin where the enamel on the labial slopes of the cusps is breached. The deep pits on the mesial slope of the labiomesial cusp might also have been enlarged by post-mortem abrasion. Making allowance for post-mortem damage, W-1 differs from M21098 only in its smaller size that, in itself, is not great enough to warrant allocation of W-1 to a species other than *L. valdensis*.

	M21098	W-1
Length	2·4 mm	1·9 mm
Width	1·9 mm	1·2+ mm

No anterior upper premolars have been discovered since 1961. Hahn (pers. comm.) has noted that the four-cusped premolar, M21106 (Clemens, 1963*a*, text-fig. 4) shows greater similarities to premolars of paulchoffatiids than those of plagiaulacids. He also concludes that the upper incisor, M21100, described earlier (Clemens, 1963*a*, text-fig. 5) might also be part of the dentition of a paulchoffatiid. Although more heavily abraded W-4 (Plate 2**A**,**B**) also from the Cliff End bone bed and M21100 appear to be examples of the same kind of incisor except that they are derived from opposite sides of the dentition.

* Since the manuscript was prepared another multituberculate lower molar (W-18) was discovered in the concentrates from the Cliff End bone bed. It is the shortest (mesiodistally) multituberculate molar yet found (length = 1·9 mm, width = 1·6 mm). Its labial row of cusps differs from those of M10480 and W-6 in that only three cusps are present. On all three the mesiolabial cusp is large, preceded by a cingulum, and separated by a deep cleft from the distal cusps. On W-18 the two cusps distal to the first are not as well separated as those of W-6 and M10480 and there is no evidence of a fourth cusp. The lingual cusp row of W-18 resembles those of W-6 and M10480. The distinguishing characters of W-18 are of the kind that might differentiate an M_2 from M_1, but the possibility that W-18 is a small M_1, or less likely an M^1, cannot be ruled out.

ORDER SYMMETRODONTA

Family Spalacotheriidae

Spalacotherium tricuspidens Owen

The type of *Peralestes longirostris* Owen is on a maxilla found in the Purbeck Beds. *Spalacotherium tricuspidens* Owen is based on a mandible from the same locality. Simpson (1928) favoured the interpretation that these taxa are founded on different parts of the dentition of one species of symmetrodont but refrained from proposing synonymy until their association could be proven beyond doubt. This nomenclature was employed in an earlier report on the Wealden fauna (Clemens, 1963*a*). Subsequently additional information about dental formulae (Clemens, 1963*b*) and relative sizes of symmetrodont dentitions (Crompton & Jenkins, 1968) has been brought forth to suggest *Peralestes* is a synonym of *Spalacotherium*. Although Simpson's criterion of proof beyond any question has not been satisfied, the probabilities now heavily favour the synonymy and it is adopted here.

One symmetrodont upper molar, W-3 (Plates **2C**,**D**,**E** and **4D**) was found at Cliff End. It has been abraded and its roots are now missing. Greater length (1·5 mm) than width (0·8 mm) suggest it is an anterior molar. The parastyle forms a pronounced salient on the crown and probably was a higher, more distinct cusp before post-mortem abrasion. The low ridge along the labial margin of the crown is irregularly crenulated but lacks evidence of large, distinct cusps. Unlike the labial and distolingual sides of the crown, its mesiolingual margin is convex. A ridge links the apex of the paracone to the base of the parastyle and swings around a deep pit in the crown. From the paracone, the highest cusp of the crown, a crest extends distolabially to the single metastylar cusp and bears a low metacone (Patterson, 1956). There is no evidence of a lingual cingulum.

Of the upper molars of *S. tricuspidens* from Purbeck W-3 most closely resembles M^1 in size and proportions. It differs from the anterior molars (M^{1-3}) in the presence of a deep pit mesiolabial to the paracone and absence of a cusp on the crest linking paracone and parastylar region. The mesial crest of M^2 of the Purbeckian specimens is damaged but a cusp is present on this crest of M^3. Post-mortem abrasion has modified the crown of W-3 but when compared with the molars of the Purbeckian symmetrodont the paracone and cusp in the position of a metacone of W-3 appear to be lower and less massive.

If the positions in the dental arcade of W-3 and the type and only specimen of *Eurylambda* from the Morrison Formation (Crompton & Jenkins, 1967) are correctly identified, the Wealden tooth is clearly not referable to this American symmetrodont. The angle formed at the apex of the paracone by the mesial and distal crests of W-3, *c.* 125°, is more acute than that of *Eurylambda*, *c.* 160° (Crompton & Jenkins, 1967). Also both crests from the paracone of *Eurylambda* carry more cusps than those of W-3 and a lingual cingulum is present. The only other symmetrodont, whose upper dentition has been discovered is *Spalacotheroides bridwelli*, a member of an Albian North American fauna (Patterson, 1955, 1956). The three isolated upper molars of *Spalacotheroides* so far recovered are thought to be posterior molars. None exhibits any particular similarity to W-3 and the only complete molar (PM 1133, Patterson, 1956, fig. 1) is easily distinguished by the acute angle of its trigon and multiplicity of stylar cusps.

Of the symmetrodonts whose upper dentitions are known, *Spalacotherium*, *Spalacotheroides* and *Eurylambda*, W-3 is from an animal most closely related to *Spalacotherium*. The differences between W-3 and anterior upper molars of *S. tricuspidens* from Purbeck could be indicative of intraspecific variations, and we choose to allocate W-3 to *Spalacotherium tricuspidens*.

Spalacotherium taylori sp. nov.

Type. British Museum (Nat. Hist.) M21103, a left lower molar (Fig. 1).

Horizon. Grinstead Clay, a member of the Hastings Beds, Neocomian age.

Locality. An abandoned clay pit in Paddockhurst Park, south-west of East Grinstead, East Sussex, grid co-ordinates 53281334 (for additional description see Clemens, 1963*a*).

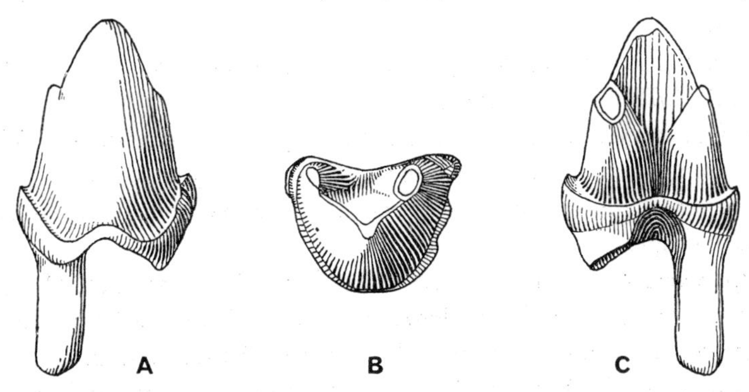

FIGURE 1. **Spalacotherium taylori**, sp. nov., type M21103 left lower molar (from Clemens, 1963*a*).
A, labial view; **B**, occlusal view; **C**, lingual view. ×30.

Diagnosis. Lower molar with acute-angled trigonid, labial cingulum continuous around protoconid, smaller than lower molars of *Spalacotherium tricuspidens*.

Derivation of name. To honour the late Prof. J. H. Taylor who made significant contributions to understanding the geology of the Weald, discovered the Paddockhurst Park fossil locality and gave us considerable help during the early stages of our work.

The type specimen was described and figured by Clemens (1963*a*). At that time there appeared to be a reasonable chance of obtaining additional material and the tooth was not definitely assigned to a taxon. Additional quarrying has exhausted the fossiliferous lens without producing more specimens of this symmetrodont. Now it appears best to recognize the distinctness of the symmetrodont found at Paddockhurst Park.

The paraconid and metaconid of the type specimen are closely approximated and weakly separated from the protoconid by shallow grooves. In relative height of the protoconid and degree of separation of the other trigonid cusps the tooth most closely resembles the anterior and median molars (M_1 to M_5) of *Spalacotherium tricuspidens.* A basal cingulum encircles the crown, and small mesial and slightly larger, distal cingular cusps are present. Continuity of the basal cingulum and the less acute angle formed by the trigonid cusps distinguish M21103 from lower molars of *Spalacotheroides*. ***Spalacotherium taylori*** is distinguished from *S. tricuspidens* by its much smaller size. Lengths of anterior and median molars of *S. tricuspidens* range from 1·6 to 1·8 mm. The dimensions of M21103 are: length = 0·80 mm, width = 0·55 mm.

ORDER EUPANTOTHERIA

Family Dryolestidae

Melanodon hodsoni **sp. nov.**

(Plates 2**F**, **G**, **H**; 3 and 5**A**, **C**)

Type. Department of Zoology, University College, London, W-14 a right upper molar.

Horizon. A pebble bed stratigraphically probably slightly higher than the Telham pebble bed, lower part of the Wadhurst Clay, a member of the Hastings Beds, Neocomian age.

Locality. An abandoned quarry approximately 200 yards south-west of the Tighe farmhouse (93732672), Isle of Oxney, Kent.

Diagnosis. Upper molars smaller than those of *Melanodon oweni* and *M. goodrichi;* mesial crest from paracone extends to mesiolabial, parastylar, cusp; only one cusp at distolabial corner of the crown; mesial and distal crests from paracone do not carry large accessory cusps.

Referred specimen. W-13, a right upper molar from Tighe Farm. M21101, a fragment of an upper molar, described by Clemens (1963*a*), from the Cliff End bone bed. W-5, a fragment of a right lower molar from the Cliff End bone bed.

Derivation of name. To honour Mr James Hodson, the owner of Tighe Farm, and his sister, Miss Janet Hodson. We gratefully acknowledge the opportunity to collect at localities on Tighe Farm and the Hodsons' help and hospitality that added to the rewards of our field work.

The two upper molars from Tighe Farm, W-14 and W-13, differ slightly. They might be from the dentition of the same animal but this appears unlikely. W-13 (Plates 2**F**, **G**, **H** and 5**A**) has undergone somewhat more post-mortem abrasion than W-14 (Plates 3**A**, **B**, **C** and 5**B**). Both teeth are three-rooted. The lingual root is the largest. The two labial roots are smaller and of subequal size. Of the cusps the paracone is the highest. Its lingual base is slightly swollen but lacks a distinct cingulum. Three crests radiate from the apex of the paracone, the mesial extends to the parastylar cusp. This crest lacks any evidence of a distinct cusp or cusplike expansion. The parastylar cusp, which is the lowest of the cusps, forms a large hook-like salient on W-14. Its smaller size on W-13 could reflect either or both a different position of the molar in the dental arcade and post-mortem abrasion. The second ridge from the paracone extends directly to the stylocone, a configuration clearly distinguishing these molars from those of *Herpetairus*, the American Late Jurassic dryolestid that most closely resembles *Melanodon*. The stylocone is slightly lower than the paracone and situated near the edge of the concave labial margin of the crown. A third crest along the distal margin of the crown links the paracone and a cusp at the distolabial, metastylar, corner. On W-13 and, particularly, M21101, this crest is slightly expanded just labial to the base of the paracone, but the cusplike expansion is neither as large nor distinct as those found on the molars of the American species of *Melanodon*. The distolabial corner of the crowns of W-14 and W-13 is a large, blunt projection formed by a single cusp. It is higher than the parastylar cusp but lower than the stylocone. The distal edge of the crown is slightly convex. W-14 is slightly larger (length = 1·0 mm; width = 1·3 mm) than W-13 (length = 0·9 mm; width = 1·2 mm).

Only three mammalian teeth have been recovered from the Tighe Farm locality. The two molars of ***Melanodon hodsoni*** just described and an isolated premolar, W-12 (Plate 5**D**,**E**). This tooth might be a premolar of ***M. hodsoni***, but we do not propose this identification. It is mentioned here solely to keep the descriptions of fossils from Tighe Farm together. The two roots of W-12 diverge slightly. Although missing its apex, W-12 appears to have had a high crown modified only by a small, distal projection that did not carry an accessory cusp. Basal cingula are absent.

The Cliff End bone bed has yielded two fossils probably referable to ***Melanodon hodsoni***. One is a fragment of an upper molar, M21101, described elsewhere (Clemens, 1963*a*). It appears to be the lingual half of a molar slightly but not significantly larger than W-13 and W-14. W-5 (Plates 3**D**,**E**,**F** and 5**C**) is part of a therian lower molar from the Cliff End bone bed. Most of the protoconid and adjacent labial surfaces of the crown are lacking. Its talonid is heavily worn and reduced to a sloping shoulder separated by a shallow groove from the distal side of the trigonid. Small size of the trigonid and difference in size of the two roots, the mesial is much smaller than the distal, demonstrate that W-5 is the lower molar of a dryolestid. The paraconid is on the lingual side of the broad mesial end of the trigonid and is the lowest of the cusps. Damage to the protoconid prevents exact determination of its height but this cusp must have been much higher than the paraconid and possibly overtopped the metaconid. It was linked to the paraconid by a crest blocking the labial end of the trigonid basin. The metaconid is a massive, slightly mesiodistally elongated cusp that bifurcates near its apex. Its mesial apex is slightly lower than the distal. This curious structure is not found in any of the European Jurassic dryolestids nor in *Crusafontia*, an Early Cretaceous dryolestid from Spain recently described by Henkel & Krebs (1969). It is duplicated in the American dryolestid *Laolestes*. W-5 (length = 1·1 mm, estimated width = 0·8 mm) falls without the observed range of size of molars of *L. eminens* (1·1–1·4 mm) but is smaller than the molars of *L. grandis* (*c.* 1·6–1·7 mm).

In his comments on *Melanodon* Simpson (1929: 75) noted, 'it is probable that [*Melanodon*] is the upper dentition of *Dryolestes* or *Laolestes*'. Patterson (1956, fig. 13) used the dentitions of *Melanodon* and *Laolestes* to form a composite to illustrate occlusion in dryolestids. The association of the *Melanodon* and *Laolestes* types of dentitions cannot be regarded as demonstrated. In view of the small size of the Wealden sample, the occurrence of just these two types of dryolestid molars is hardly confirmatory evidence. However, because it is of the proper size to be an element of the dentition of ***Melanodon hodsoni*** and in view of the suggestions of Simpson and Patterson we allocate the *Laolestes*-like molar, W-5, to ***M. hodsoni***.

The large, centrally located stylocone and triradiate pattern of crests extending labially from the paracone distinguish ***Melanodon hodsoni*** from the two Purbeckian dryolestids whose upper dentitions are known, *Amblotherium* and *Kurtodon*. Molars of *M. oweni* and *M. goodrichi* are larger than those of ***M. hodsoni***, but one specimen from the Morrison Formation, Y.P.M. 13749, designated *Melanodon* cf. *goodrichi* by Simpson (1929), contains molars of approximately the same size. In general the structure of the molar crowns of ***M. hodsoni*** is simpler than those of the American species and could easily have been derived from one of them. None of the Kimmeridgian eupantotheres from Portugal described by Kühne (1968) appears to be closely related to ***M. hodsoni***. If this remains the case, ***M. hodsoni*** is best regarded as a differentiate of an American lineage that dispersed into England in the Early Cretaceous.

ORDER *INCERTAE SEDIS*

Family Aegialodontidae

Aegialodon dawsoni Kermack, Lees & Mussett

The type specimen, M23345, is an almost complete lower molar and unfortunately remains the only fossil clearly referable to this species and genus. McKenna (1970) speculated that the therian premolar, M21104, found in the Paddockhurst bone bed (Clemens, 1963*a*) might also be referable to *A. dawsoni*. The type of *A. dawsoni* is longer (*c.* 1·4 mm) than M21104 (0·85 mm) but the difference is not great enough to preclude this assignment.

CONCLUSIONS AND SUMMARY

Mammalian fossils have been found at three Early Cretaceous localities in the Weald. All are from strata of the Hastings Beds. The Cliff End bone bed and the fossiliferous strata at Tighe Farm are within the lower part of the Wadhurst Clay and thought to be of Valanginian age. At Paddockhurst Park fossils were obtained from a lens within the Grinstead Clay, which overlies the Wadhurst Clay, and could be either of Valanginian or Hauterivian age.

A search for additional specimens and localities is being continued by the team at University College. At present the faunal lists for the three localities where mammalian remains have been found are as follows:

Cliff End:	*Loxaulax valdensis* and probably other plagiaulacid or paulchoffatiid multituberculates
	Spalacotherium tricuspidens
	Melanodon hodsoni **sp. nov.**
	Aegialodon dawsoni
Tighe Farm:	***Melanodon hodsoni*** **sp. nov.**
Paddockhurst Park:	*Loxaulax valdensis* (?)
	Spalacotherium taylori **sp. nov.**

ACKNOWLEDGEMENTS

We wish to express our gratitude to Dr K. A. Kermack whose continuing interest in and support of this project have made possible the collection of the fossils described here. The assistance and material support provided by Sir Peter Medawar, Prof. M. Abercrombie and Mrs Frances Mussett is also gratefully acknowledged. Discussions with Prof. P. M. Butler added to our understanding of the Wealden mammals. The late Prof. J. H. Taylor, Dr E. R. Shephard-Thorn, Prof. P. Allen and Mr R. W. Gallois have given freely of their knowledge of the geology of the Weald.

The fieldwork was financed by a D.S.I.R. Research Grant awarded to Dr K. A. Kermack and by the Department of Zoology, University College, London. A National Science Foundation Post-doctoral Fellowship made it possible for W.A.C. to spend the 1968–1969 academic year in London. Preparation of the drawings of Wealden mammalian teeth by Mr A. J. Lee was supported by a grant from the Museum of Paleontology, University of California, Berkeley. Figure 1 is copied with the permission of the Palaeontological Association from the original figure used in *Palaeontology*.

REFERENCES

Allen, P., 1949*a*. Notes on Wealden bone beds. *Proc. Geol. Ass.*, **60**: 275–283.

Allen, P., 1949*b*. Wealden petrology: The Top Ashdown Pebble Bed and the Top Ashdown Sandstone. *Q. Jl geol. Soc., Lond.*, **104**: 257–321.

Allen, P., 1960. Geology of the central Weald: a study of the Hastings Beds. *Geol. Ass. Centenary Guide*, no. 24, 28 pp.

Allen, P., 1965. L'age du Purbecko-Wealdien d'Angleterre. *Mém. Bur. Rech. Geol. Miner.*, **34**: 321–326.

Allen, P., 1967. Strand-line pebbles in the mid-Hastings Beds and the geology of the London uplands. Old Red Sandstone, New Red Sandstone and other pebbles. Conclusion. *Proc. Geol. Ass.*, **78**: 241–276.

Anderson, F. W., Bazley, R. A. B. & Shephard-Thorn, E. R., 1967. The sedimentary and faunal sequence of the Wadhurst Clay (Wealden) in boreholes at Wadhurst Park, Sussex. *Bull. geol. Surv. Gt Br.*, **27**: 171–235.

Anderson, F. W. & Hughes, N. F., 1964. The 'Wealden' of north-west Germany and its English equivalents. *Nature, Lond.*, **201**: 907–908.

Clemens, W. A., Jr., 1963*a*. Wealden mammalian fossils. *Palaeontology*, **6**: 55–69.

Clemens, W. A., Jr., 1963*b*. Late Jurassic mammalian fossils in the Sedgwick Museum, Cambridge. *Palaeontology*, **6**: 373–377.

Crompton, A. W. & Jenkins, F. A., Jr., 1967. American Jurassic symmetrodonts and Rhaetic 'pantotheres'. *Science*, N.Y. **155**: 1006–1009.

Crompton, A. W. & Jenkins, F. A., Jr., 1968. Molar occlusion in Late Triassic mammals. *Biol. Rev.*, **43**: 427–458.

Crusafont-Pairó, M. & Adrover, R., 1966. El Primer mamifero del Mesozoico Español. *Publnes Cat. Paleont. Univ. Barcelona*, **13**, 28–33.

Hahn, G., 1969. Beiträge zur Fauna der Grube Guimarota Nr. 3. Die Multituberculata. *Palaeontographica*, (Abt. A), **133**: 1–100.

Henkel, S., 1966. Methoden zur Prospektion und Gewinnung kleiner Wirbeltierfossilien. *Neues Jb. Geol. Paläont. Mh.*, **3**: 178–184.

Henkel, S. & Krebs, B., 1969. Zwei Säugetier-Unterkiefer aus der Unteren Kreide von Uña (Prov. Cuenca, Spanien). *Neues Jb. Geol. Paläont. Mh.*, **8**: 449–463.

Kermack, K. A., Lees, P. M. & Mussett, F., 1965. *Aegialodon dawsoni*, a new trituberculo-sectorial tooth from the Lower Wealden. *Proc. R. Soc.*, (Ser. B), **162**: 535–554.

Kühne, W. G., 1968. Kimeridge mammals and their bearing on the phylogeny of the Mammalia. In E. T. Drake (Ed.), *Evolution and environment*, pp. 109–123. New Haven: Yale University Press.

Kühne, W. G. & Crusafont-Pairo, M., 1968. Mamiferos de Wealdiense de Uña, cerca de Cuenca. *Acta geol. Hispan.*, **3**: 133–134.

Lees, P. M., 1964. The flotation method of obtaining mammal teeth from Mesozoic bone-beds. *Curator*, **7**: 300–306.

Lyddeker, R., 1893. On a mammalian incisor from the Wealden of Hastings. *Q. Jl geol. Soc. Lond.*, **49**: 28.

McKenna, M. C., 1970. The origin and early differentiation of therian mammals. *Ann. N. Y. Acad. Sci.*, **167**: 217–240.

Patterson, B., 1955. A symmetrodont from the early Cretaceous of Northern Texas. *Fieldiana, Zool.*, **37**: 689–693.

Patterson, B., 1956. Early Cretaceous mammals and the evolution of mammalian molar teeth. *Fieldiana, Geol.*, **13**: 1–105.

Patterson, C., 1966. British Wealden sharks. *Bull. Br. Mus. nat. Hist.*, (*Zool.*), **11**: 281–350.

Shephard-Thorn, E. R., Smart, J. G. O., Bisson, G., & Edmonds, E. A., 1966. Geology of the country around Tenterden. *Mem. geol. Surv. U.K.*, no. 304, 123 pp.

Simpson, G. G., 1928, *A catalogue of the Mesozoic Mammalia in the Geological Department of the British Museum* (Natural History). London: Br. Mus. (Nat. Hist.).

Simpson, G. G., 1929. American Mesozoic Mammalia. *Mem. Peabody Mus. Yale*, **3**: 1–235.

Taylor, J. H., 1963. Sedimentary features of an ancient deltaic complex: The Wealden rocks of southeastern England. *Sedimentology*, **2**: 2–28.

Van Valen, L., 1967. The first discovery of a Cretaceous mammal. *Am. Mus. Novit.*, **2285**: 1–4.

Woodward, A. S., 1911. On some mammalian teeth of the Wealden of Hastings. *Q. Jl geol. Soc. Lond.* **67**: 278.

EXPLANATION OF PLATES

Plate 1

Isolated molars of *Loxaulax valdensis* from Cliff End.

A. W-6, Left M_1, occlusal view.
B. W-6, left M_1, labial view.
C. M21098, right M^2, occlusal view.
D. M21098, right M^2, labial view.
E. W-1, left M^2, occlusal view.
F. W–1, left M^2, labial view.
All ×12·5.

Plate 2

A. ? *Loxaulax valdensis*, W-4, upper incisor, lateral view, from Cliff End.
B. W-4, upper incisor, lingual view.
C. *Spalacotherium tricuspidens*, W-3, anterior left upper molar, occlusal view, from Cliff End.
D. W-3, anterior upper molar, labial view.
E. W-3, anterior upper molar, lingual view.
F. ***Melanodon hodsoni***, W-13, anterior right upper molar, labial view, from Tighe Farm.
G. W-13, anterior upper molar, occlusal view.
H. W-13, anterior upper molar, mesial view.
All ×12·5.

Plate 3

A. ***Melanodon hodsoni***, type, W-14, anterior right upper molar, labial view, from Tighe Farm.
B. W-14, type, anterior upper molar, occlusal view.
C. W-14, type, anterior upper molar, mesial view.
D. ***Melanodon hodsoni***, W-5, right lower molar, labial view, from Cliff End.
E. W.5, right lower molar, occlusal view.
F. W-5, right lower molar, distal view.
All ×12·5.

Plate 4

Stereoscopic photographs, all occlusal views.

A. *Loxaulax valdensis*, W-6, left M_1, from Cliff End.
B. *Loxaulax valdensis*, M21098, right M^2, from Cliff End.
C. *Loxaulax valdensis*, W-1, left M^2, from Cliff End.
D. *Spalacotherium tricuspidens*, W-3, anterior left upper molar, from Cliff End.
All *c.* ×12·5.

Plate 5

Stereoscopic photographs (except **D** and **E**).

A. ***Melanodon hodsoni***, W-13, anterior right upper molar, occlusal view, from Tighe Farm.
B. ***Melanodon hodsoni***, type, W-14, anterior right upper molar, occlusal view, from Tighe Farm.
C. ***Melanodon hodsoni***, W-5, right lower molar, occlusal view, mesial end vertical, from Cliff End.
D. Therian, W-12, lateral view, from Tighe Farm.
E. Therian, W.12, lateral view, from Tighe Farm.
All *c.* ×12·5.

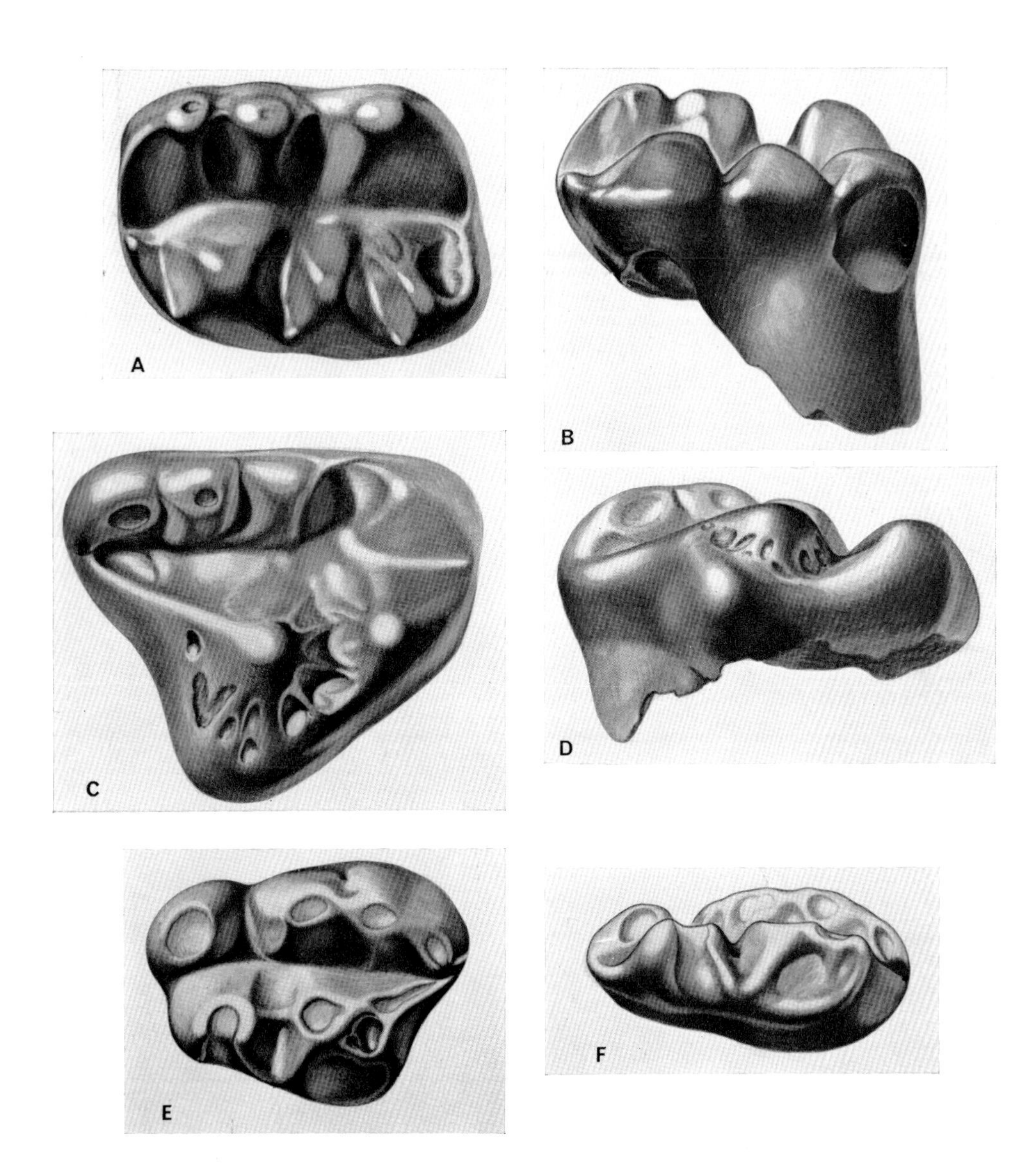
A
B
C
D
E
F

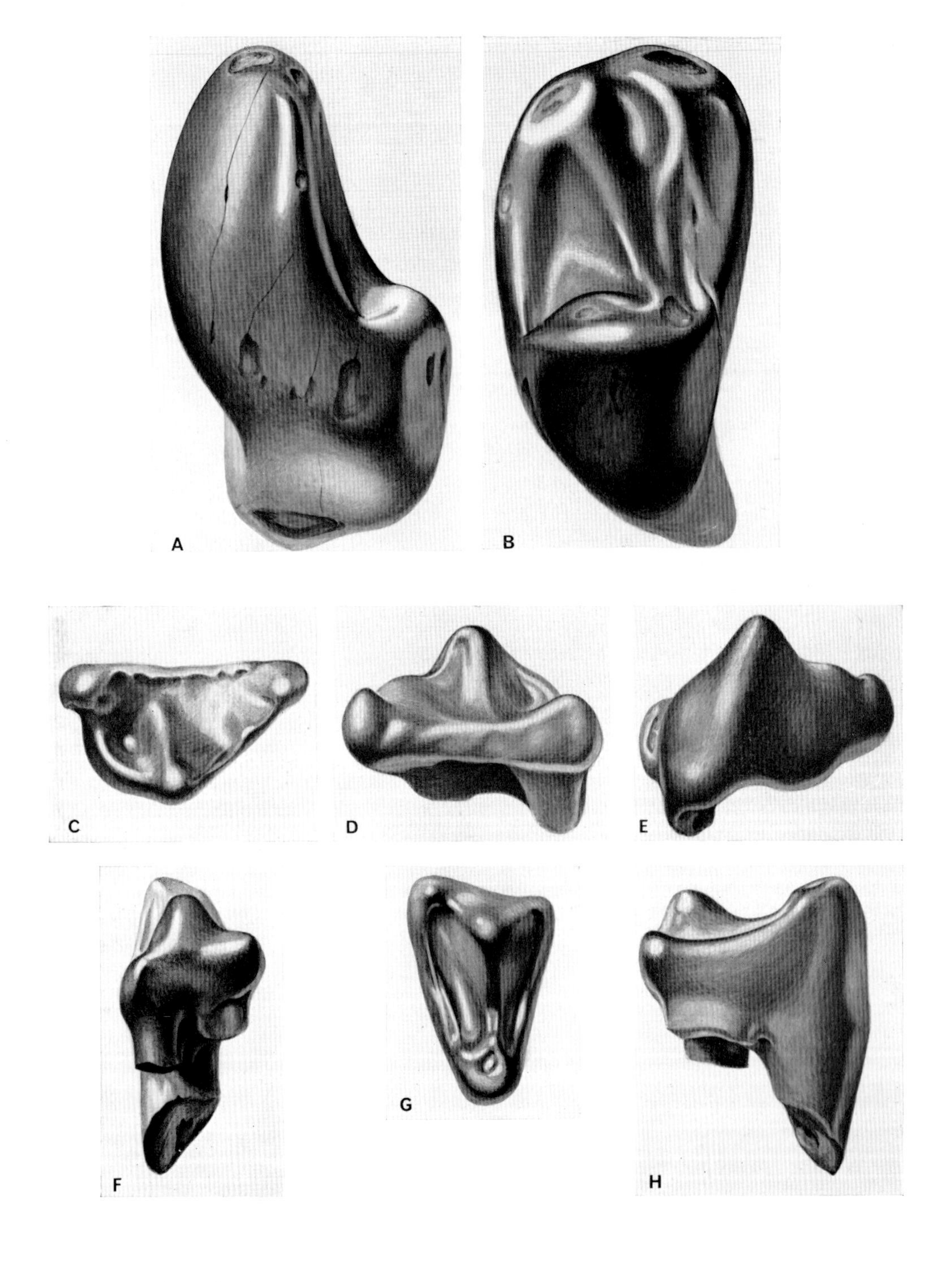

W. A. CLEMENS, JR. AND P. M. LEES

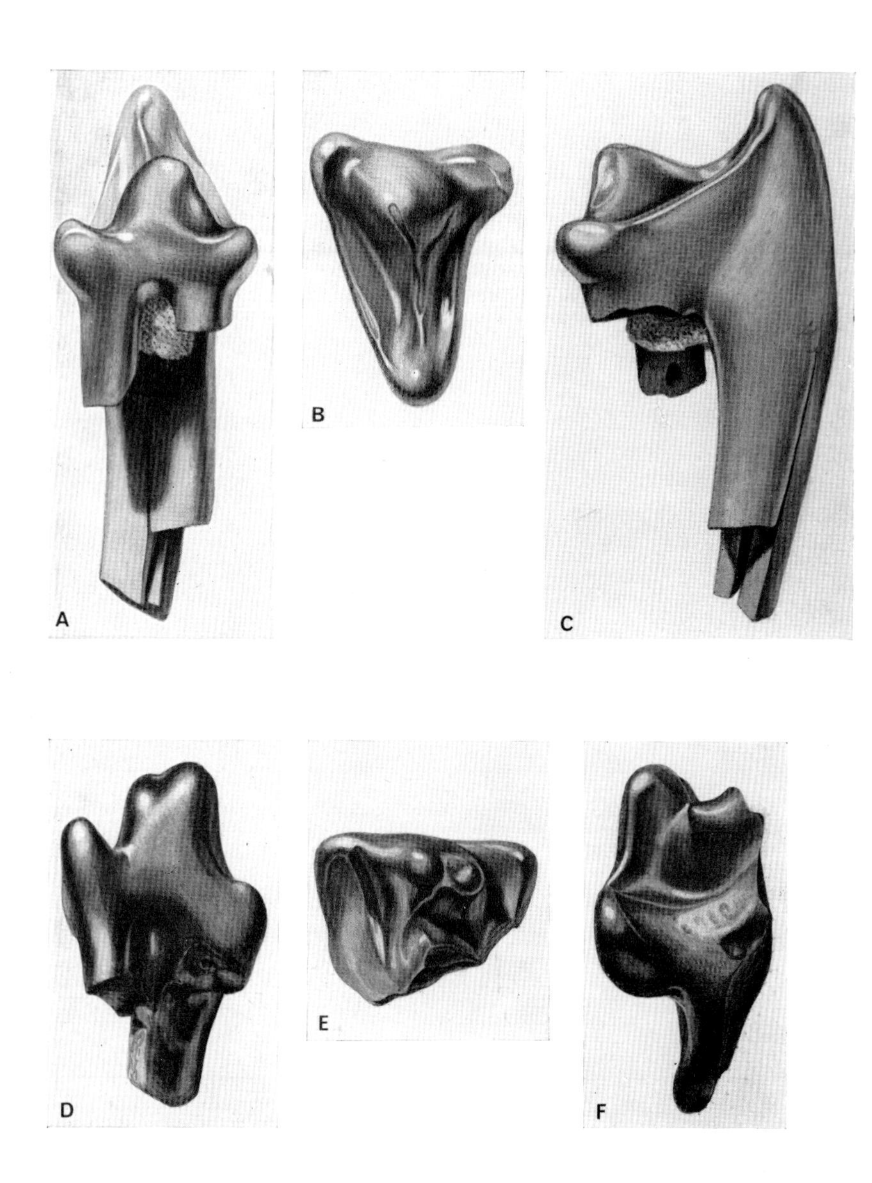

W. A. CLEMENS, JR. AND P. M. LEES

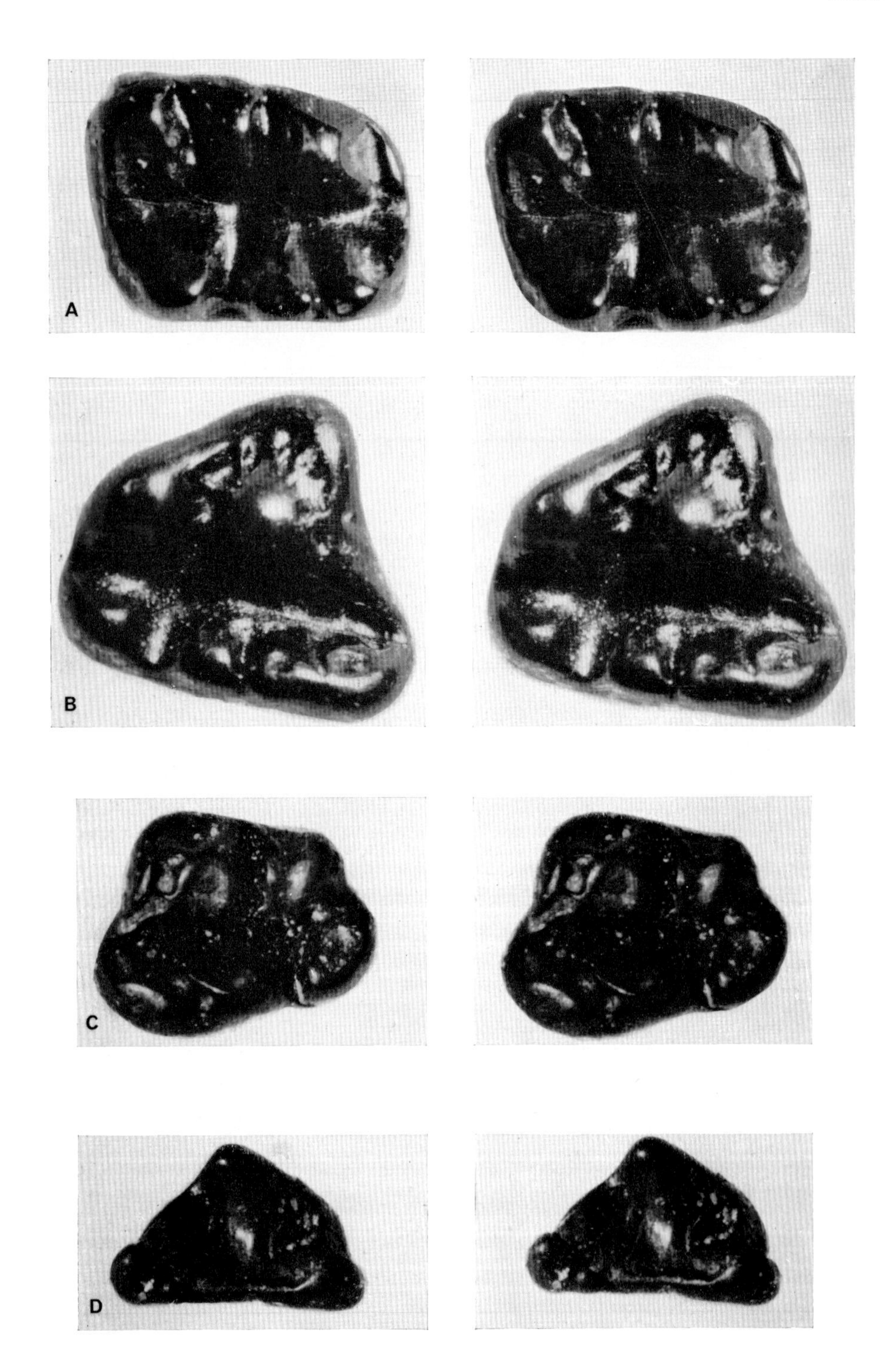

W. A. CLEMENS, JR. AND P. M. LEES

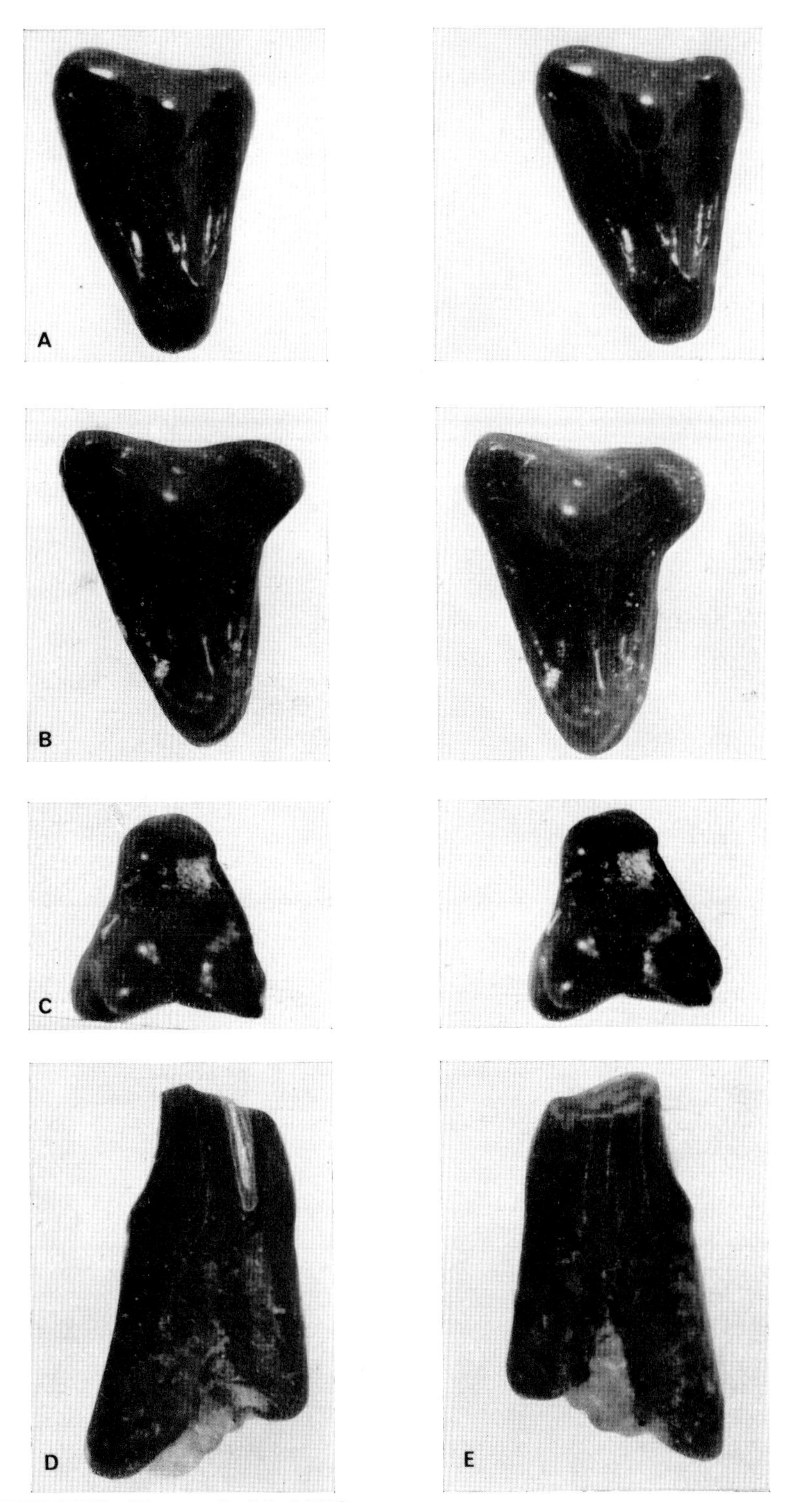

W. A. CLEMENS, JR. and P. M. LEES

Mid-Cretaceous (Albian) therians of the Butler Farm local fauna, Texas

BOB H. SLAUGHTER

Shuler Museum of Paleontology, Southern Methodist University, Dallas, Texas, U.S.A.

The dental features we use to distinguish didelphids from certain insectivores were well established by late Cretaceous, but we may never know which of these were possessed by the stem ancestor and which were modified or developed after divergence. Furthermore, it is not certain that these two surviving groups were the only therians of this grade to take origin from this common ancestor. Even so, it seems better to proceed with the criteria we have developed than to speculate on alternatives that probably can never be dealt with adequately.

An examination of both new (***Kermackia texana,* gen. et sp. nov.**) and described therian specimens from the Albian of Texas demonstrates, to my satisfaction, that representatives of both didelphids and insectivores were associated with triconodonts and symetrodonts in North America by the middle of the Cretaceous. The marsupials are identified on the basis of upper molars containing well developed stylar cusp C (in the mesostylar area), incipient twinning of the entoconid and hypoconulid, and accompanying increase in size of the metacone. That insectivores were also represented is indicated by specimens showing that the penultimate unicuspid premolar was replaced, heel-cusps on incisors, presence of a cusp on the metacrista and three equidistant talonid cusps.

Some specimens, believed to be primitive placentals, have crista obliquas connected to the metaconid rather than joining the trigonid at mid-width. Although this feature is not duplicated in any living form known to me, it does occur in the pantotheres *Peramus* and *Amphitherium*, and therefore is considered a primitive trait.

CONTENTS

INTRODUCTION

Stimulated by Patterson's discovery and description of Albian mammal-teeth (1956), in 1963 the Shuler Museum of Southern Methodist University began systematic prospecting for additional localities of the same approximate age. By 1968, several hundred miles of outcrop had been examined in Bosque, Bowie, Cooke, Erath, Hamilton, Parker, and Somervell counties, Texas. Some 30 vertebrate localities were found, and over 200 tons of the sediments were quarried and returned to the campus for processing. Although interesting specimens of dinosaurs, fishes, small reptiles and amphibians were recovered, only five of these sites produced any mammal material, and only one, the Butler Farm, in any significant amounts. Even this locality was not productive in the usual sense, since only about 50 mammalian specimens were identified. Even so, the material is of considerable interest and importance to our understanding of early mammalian evolution.

Patterson (1956) demonstrated his Albian specimens definitely to be of metatherian-eutherian grade, but was uncertain as to whether they were marsupial, placental, or perhaps of a stem group. In my 1965 paper I, also, hesitated to refer the original Butler Farm material to either of the living infraclasses. Subsequently, however, some specimens have come to light which display characters known only to eutherians and some have typically metatherian features. Unfortunately, dental formulas are still unknown for any of the Butler Farm therians. We may never know which of the characters, we use to distinguish these two major groups, were established after divergence and which were possessed by the stem ancestor. Moreover, it is not certain that the surviving infraclasses were the only groups of this grade. Hence, it seems best to use what we have at this time, rather than to beg the question or conjure up alternatives that cannot be dealt with adequately.

BUTLER FARM LOCAL FAUNA

This assemblage was recovered from the bottom two inches of a 60 foot wide channel-fill. The fill was dissected by an erosional gulley on the farm of Mr Lee Butler, and is 250 yards north-east of U.S. Highway 81, three miles north-west of Decatur, Wise County, Texas. The fresh-water origin of the deposits is indicated by the abundance of frog, salamander and lizard material. The fauna also includes freshwater sharks (*Lonchidion*), lepidonts, amiaids, crocodiles, dinosaurs, multituberculates and triconodonts.

To the south of Wise County, the Trinity Group includes the basal Travis Peak formation (?Aptian), overlain by the shallow marine deposits of the Glen Rose formation (Albian), which in turn is overlain by the terrestrial and brackish water deposits of the Paluxy formation (Albian). The Glen Rose pinches out in southern Wise County; and north and west of this point it is difficult to distinguish the Travis Peak equivalent beds from those of the Paluxy. Thurmond (1970) has recently been able to distinguish the fish faunas from a point 15 feet above the Glen Rose downward from those of the typical Paluxy. Previously, the Butler Farm fauna was tentatively referred as Paluxian (Slaughter, 1965) but now this reference is certain.

BUTLER FARM THERIANS

Trinity molar type 6

(Plate 1)

(Shuler Museum of Paleontology—Southern Methodist University No. 61728)

This molar type, previously described (Slaughter, 1965), has a trenchant protoconid which is much taller than the paraconid or metaconid. The latter are subequal. The protoconid's lingual face is slightly concave. The paraconid and metaconid almost meet at their bases, although they are not connected by a ridge. The cristid obliqua connects to the trigonid near the apex of the metaconid, and there is no entocristid. In these features, the form is more like the Jurassic pantothere, *Peramus*, than like members of the living infraclasses. On the other hand, there is a well-developed hypoconulid, presumably a large hypoconid, and a small protuberance that could be considered an incipient entoconid on the ridge extending lingually from the hypoconulid. This would seem to place the form morphologically intermediate between pantotheres and Cretaceous therians. Also reminiscent of *Peramus*, there is small cusp on the cristid obliqua.

The area of the hypoconid is broken away and I previously interpreted the specimen as lacking a hypoconid, or as having a very small one at best. In the light of a more recently recovered specimen (described below) I now feel I was in error. Not only was there a hypoconid, but it was the largest of the talonid cusps. An artificial ridge created by the break connects the hypoconulid to the cristid obliqua directly anterior to this cusp. This ridge is level with the base of the hypoconulid, thus demonstrating that the upward slope from the lingual side of the talonid toward the cristid obliqua, continued well above the height of the hypoconulid. There must, therefore, have been a hypoconid somewhat taller than the hypoconulid. As there is no entocristid, both labial and lingual slopes from the cristid obliqua continue to the base of the crown. Although this form has certain characters in common with specimens known to be of metatherian-eutherian grade (i.e., three talonid cusps), it also has features that are decidedly more pantotherian, in the lack of entocristid, etc.

***Kermackia texana* gen. et. sp. nov.**
(Plate 2)

Etymology. Generic dedication is made with pleasure to Dr Kenneth A. Kermack, University College of London, in recognition of his contributions to our knowledge of Mesozoic mammals.

Holotype. S.M.P.-S.M.U. No. 62398. Right lower molar complete except for anterior root.

Diagnosis. Cristid obliqua connects with the trigonid near the apex of the metaconid. The entocristid is low, little more than a cingulum at the base of the lingual slope of the cristid obliqua. The hypoconid is the largest talonid cusp, and is placed relatively far forward. The hypoconulid is the next largest, and the entoconid much the smallest.

Description. The enamel has spalled from much of the trigonid, resulting in some rounding of the apices of the trigonid cusps. Even so, the only probable effect is a very slight reduction in trigonid cusp-height and sharpness. The paraconid is slightly smaller than the metaconid, and both are nearly as tall as the protoconid. The most interesting and perhaps most important feature of this tooth is the small size of the basined region of the talonid. The cristid obliqua begins high on the trigonid near the apex of the metaconid, as in *Peramus*, *Aegiolodon* and Trinity molar type 6. It then slopes steeply downward and toward the postero-labial corner of the tooth. There is a large, blade-like hypoconid, a slightly smaller hypoconulid and a much smaller entoconid. The talonid not only slopes labially from the cristid obliqua in the normal fashion, but lingually as well. Only the small cingulum-like entocristid breaks the slope. There are wear-facets on the anterior faces of the hypoconulid and entoconid, demonstrating that there was a protocone on the opposing tooth. However, the height of the anterior half of the cristid obliqua would seem to disallow protoconal development anterior to, or directly lingual to, the paracone. I suspect the upper molars of this form would be similar to McKenna's (1969) hypothetical intermediate between *Kuehneotherium* and something like *Pappotherium.* A protocone lingual or anterior to the paracone would encounter the tall cristid obliqua before the paracone had completed its shear between the protoconid and the hypoconid.

Holoclemensia texana Slaughter, 1968
(Plates 3 and 4)

An upper molar with its protocone missing was described, named (Slaughter, 1968*a*,*b*), and referred to the Didelphidae. A complete ultimate upper molar and a complete lower molar were referred. The metatherian reference was based primarily on the presence of a large stylar cusp C in the middle of the ectoflexus of both the holotype and the paratype. Cusps in this position are common among didelphids, but unknown to me in any Mesozoic eutherian. Certain insectivores (e.g., *Potamogale*) often have well-developed stylar cusps, but only in the A, B, D and E positions. The metacone is smaller and not so lingually placed as the paracone, but is relatively larger than that of *Pappotherium.* Late Cretaceous didelphids (e.g., *Alphadon* and *Pediomys*) have metacones about the same size as the paracones (Lilligraven, 1969); and by the Cenozoic the metacone has surpassed the size of the paracone (Clemens, 1969). I considered the increase in size of the metacone of *Holoclemensia* over that of *Pappotherium* as the beginning of the trend toward the later metatherian condition. Furthermore, I hypothesized that the trend was recognizable in the referred lower molar (S.M.P.-S.M.U. No. 62131; Plate 4) since the hypoconulid was somewhat closer to the entoconid than to the hypoconid. It appears that the distance from the hypoconid to the hypoconulid is related to the length of the antero-lingual face of the metacone of the opposing molar. Therefore, an increase in the size of the metacone is accompanied by an increased crowding of the hypoconulid and the entoconid. The same twinning of the lingual-most talonid cusps is acquired in many placentals without necessarily increasing the size of the metacone in Cenozoic times. The necessary increase in the length of the antero-lingual face of the metacone is accomplished by a deepening of the notch between the paracone and the metacone. This lengthens the antero-lingual face of the metacone and therefore causes twinning of the hypoconulid and entoconid. As this reason for twinning is unknown until the Cenozoic, however, it does not affect the proposition above.

The transverse diameter of the trigonid of the referred *Holoclemensia* lower molar is greater than its antero-posterior diameter. The protoconid is tall, trenchant and concave on its lingual face. The metaconid is two-thirds as tall as the protoconid and has a concave labial face. The paraconid is half as high as the protoconid, and is more conical. The protolophid and paralophid are sharp and deeply notched. As a matter of fact, the sharp crests steepen their angles near the juncture, creating carnassial-like notches. The cristid obliqua joins the trigonid at about the middle of the metaconid. The hypoconid is the tallest talonid cusp. Its lingual face is concave, and is placed slightly farther forward than the entoconid. The hypoconulid is quite far to the rear of the hypoconid.

BUTLER FARM EUTHERIANS

Slaughter (1968*c*) presented two isolated submolariform premolars from Butler Farm as evidence of the presence of eutherians in the fauna. It was pointed out that very early in the known history of placental mammals there was a tendency towards the molarization of the posterior premolars. I do not consider complication-of-premolars not mimicking the molar pattern, as true molarization. If one only considers the presence of a protocone on upper premolars, and a laterally placed metaconid on lower premolars, as evidence of molarization, there is generally one more molariform (or submolariform) premolar above, than below. All known placental groups apparently have passed through a stage in which both P^3 and P^4 have protocones and P_4 has a laterally-placed metaconid (Stage II; Slaughter, 1968*c*). Some have extended molarization forward to P^2/P_3, and in even a few groups to P^1/P_2. Some have demolarized to $P^4/0$, but cases are rare in which demolarization has broken the rule of one more molariform premolar above than below (e.g., *Cimolestes*, $P^3/0$, and *Solenodon*, P^4/P_4). Therefore, the presence of fourth lower molars at Butler Farm having laterally placed metaconids is almost certain evidence that both P^3 and P^4 had protocones.

The protoconid of the referred P_4 (SMP-SMU No. 61947; Plate 5) is flat on its lingual face, has a well-developed metaconid; and a low shelf-like paraconid. The talonid is fairly well developed, with the cristid obliqua extending back from the protoconid. There is also an entocristid extending back parallel with the cristid obliqua from the posterior keel of the metaconid. These enclose a square-basined area that certainly occluded with a well-developed protocone on P^4. The shelf-like paraconid must have received a backwardly-inclined (opisthoclinal) protocone of P^3.

A left upper premolar (SMP-SMU No. 61948; Plate 6), belonging to an animal about the same size, closely resembles P^3 of *Potamogale*. The paracone is rounded anteriorly and sharpened by a posterior keel. This gives the cusp a tear-drop shape in cross-section. There is a crenulated anterior cingulum containing a small cuspule and a posterior cingulum, similarly crenulated, with a larger cusp at its labial border. This posterior cingulum is quite like that of *Potamogale*, even to the labial cusp. The small protocone is broken away but the stump attests to its presence. A scar down the antero-lingual face of the posterior root demonstrates that it was confluent with another root supporting the protocone.

As molar mimicry is unknown among Metatheria, the presence of these teeth demonstrates, I believe, that divergence between Metatheria and Eutheria had already begun.

More recently another sub-molariform lower premolar has been recovered (SMP-SMU No. 62399; Plate 7**D,E,F,G**). This specimen is smaller, the protoconid is taller and more trenchant, and the metaconid is not quite so large. Nevertheless, the metaconid is well developed and there is similar development of the talonid. The crest forming the labial boundary of the talonid basin is an extension of the posterior keel of the protoconid. At its posterior end is a small cuspule (?hypoconulid). An enticristid originates at the small metaconid and a slight bulge at the postero-lingual corner of the tooth could be interpreted as an incipient entoconid.

Pappotherium pattersoni Slaughter (1965)
(SMP-SMU No. 61725; Plate 7**A,B**)

The holotype of *Pappotherium* is a right maxillary with the last two molars preserved. Both have paracones that are much larger and more lingual than the metacones. However, the latter are well developed and separated. Both are tall and piercing, and partially concave on their labial faces. The stylocone is larger than the parastyle and is connected to the paracone by a strong paracrista. There are two tiny cuspules in the metastylar area but there is no stylar cusp C in the mesostylar area. A shelf-like cingulum attached to the parastyle is confluent with the protoconal basin. There are two cusps in the protoconule position and one is the metaconule position, but none is 'winged' (i.e., connected with the paracone or metacone). There is a well-developed cusp mid-length of the metacrista that I have not observed on molars of didelphids, *Holoclemensia* included. The cusp is sometimes present on molars of primitive insectivores, however, (e.g.; *Cimolestes*). The ultimate molar is essentially like the penultimate molar, except for the typically-reduced metastylar area, and the absence of a protoconule.

A smaller pappotherid is represented in the collection by an upper molar (SMP-SMU No. 62402; Plate 7**C**), complete except for the stylocone and parastyle. The cusp on the metacrista is present, and there is a protoconule and a metaconule, both of which are wingless. Perhaps the most interesting aspect of the specimen is the wear. There is wear on the lingual half of the metacrista and the apex of the metacone is worn off on a horizontal plane. There is a continuous facet beginning in the notch between the paracone and the metacone and extending up the posterior side of the paracone. Without changing width appreciably, it crosses the apex and continues down the antero-lingual face. The significance of this is discussed below under the discussion of the *Pappotherium* lower jaw.

There is a left lower jaw fragment (SMP-SMU No. 61992; Plates 8**D,F** and 9) in the Butler Farm collection that I am tentatively referrring to *Pappotherium*, on the basis of size and probable occlusal relationships. The specimen has preserved two simple, unicuspid teeth and two molariform teeth. The specimen could be interpreted in three different ways: (1) That it represents a didelphid and the preserved teeth are P_2-P_3-M_1-M_2; (2) It is a placental that has not begun molarization of the premolar series, and the teeth present are P_3-P_4-M_1-M_2; (3) It is a primitive insectivore at stage II of premolar molarization, and that the preserved teeth are P_2-P_3-P_4-M_1. I subscribe to the third proposition for the following reasons: The second molariform tooth is slightly more worn than the first, particularly on the talonid. If the molariform teeth were M^1 and M^2, the first should show greater wear. On the other hand, if the first molariform tooth were P_4, it was the last of the two to erupt, and should be worn less. In addition, the penultimate unicuspid tooth is just erupting and X-rays show no other unerupted teeth. This would seem to indicate that any milk teeth to the rear have already been replaced. Only the ultimate premolar (P_3) of didelphids (Plate 8**A**) has a predecessor, and therefore the situation seen in the fossil would never happen in marsupials. On the other hand, many insectivores replace their teeth in a wave from back to front (e.g., tupaiids, erinaceids and macroscelids; Plate 8**C**). With such a replacement-sequence, the exact stage present in the fossil can be reached, whether the unicuspid teeth are P_2-P_3 or P_3-P_4. These premolariform teeth have no antero-basal cusps, although there is a slight bulge in that position of the ultimate one. They do have small talonids with a posterior heel-cusp connected to the posterior keel of the protoconid by a short ridge. In silhouette the anterior edges are convex and the posterior edges are straight. This type of premolar is possessed by most Cretaceous didelphids (e.g.; *Alphadon*, *Pediomys*) and many primitive insectivores (e.g. *Cimolestes*, *Gelastops*). It seems probable that it is also the type of premolar possessed by the eupantotherian stem of both infraclasses.

Examination of the wear on these molariform teeth supports the reference of this specimen to *Pappotherium*. Drawing on the wear on the smaller pappotherid described above, the form of the unworn teeth of *Pappotherium pattersoni*, and the wear of the lower molars referred to *P. pattersoni* the following conclusions were reached. Initial contact between upper and lower molars at the labial end of the shear-stroke finds the hypoconid reaching far enough between the paracone and metacone to abraid the post-paracrista and premetacrista. The protoconid contacts the lingual face of the parastyle. As the lower dentition begins to move upward and lingually, it is forced to add a slight forward trend due to the paracrista shearing down the posteriorly sloping metaconid. The consequent actions are: (1) The lingual face of the parastyle is faceted by the protoconid. (2) The metacone participates in the wearing of a notch in the post-cristid between the hypoconid and the hypoconulid. (3) This notch is widened by a sawing action of the post-protocrista which facets the anterior faces of the hypoconulid and entoconid on a single plane. (4) Until the rounded posterior face of the metaconid is worn flush with the protoconid, the bulge wears a broad facet down the anterior face of the paracone. (5) The tip of the paracone first encounters the talonid basin; travels labially and cuts a V-shaped notch in the cristid obliqua before continuing to shear down between the metaconid and the hypoconid and wearing a V-shaped groove in the labial face of the talonid to the crown's base. The horizontal facet on the tip of the paracone is apparently worn during contact with the talonid basin whereas its anterior facet contacts the posterior face of the protoconid, and the posterior one, the hypoconid. The most severe wear on the trigonid is on the apex and posterior slope of the metaconid. The apex is rounded off posteriorly and is confluent with a broad facet that extends to the talonid basin.

Trinity molar type 4
(SMP-SMU No. 61726; Plate 10**C**,**D**)

This specimen is tentatively referred to Eutheria on the basis of its equidistant talonid cusps. The trigonid is extremely compressed antero-posteriorly, and the protoconid is but slightly taller than the paraconid and metaconid (which are about equal). Although the notches in the paralophid and protolophid are not deep, they are aligned in such a way that there is a V-shaped valley that joins the notches across the trigonid. The labial faces of the metaconid and paraconid slope toward this valley on a single plane. The drop at each end of this trigonid valley is abrupt, and the posterior face of the trigonid is square in the lack of a metaconid bulge. Such a trigonid's occluding tightly between its two upper partners would provide an efficient but curious 'clipping' mechanism. The talonid is long and curiously hooked. Neither entocristid nor cristid obliqua is notched by occlusion with the protocone or paracone, as is the case in *Pappotherium*. This might suggest: (1) less labial initial contact with the upper dentition; or (2) a more nearly vertical occlusal movement and a long cingulum-like protocone. This form clearly could not belong to either of the proposed taxa based on upper teeth.

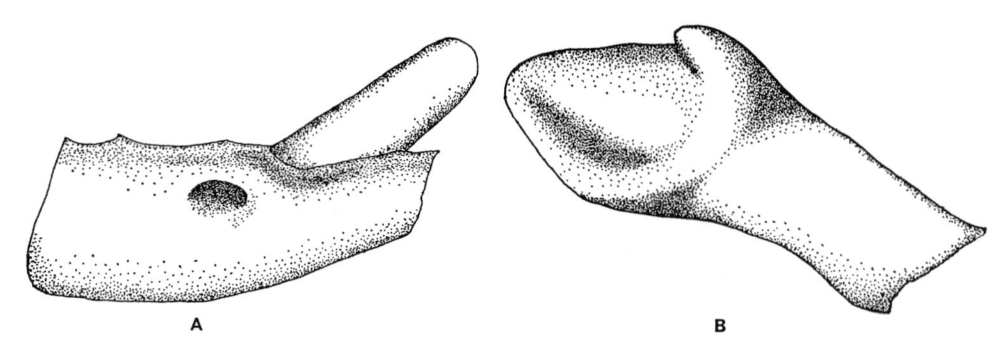

FIGURE 1. **A**. Labial view of lower jaw fragment SMP-SMU No. 62400. **B**. Lingual view of upper incisors SMP-SMU No. 62401.

The collection includes an anterior fragment of a right lower jaw (SMP-SMU No. 62400; Plate 10**B**) referable to Eutheria. There are alveoli of a double-rooted P_1, in front of which there is a single-rooted tooth which I believe to be the canine. The tooth is rather blade-like but without sharpened edges. Its apex is rounded rather than trenchant. In fact the tooth is very similar to the large anterior incisor of *Erinaceus*. The root is preserved of the outermost incisor; and it appears nearly to have equalled the size of the canine and to also be procumbent. Part of the jaw is broken away between this tooth and the symphysis, but there appears to have been one more incisor, nearly as large as the canine. It seems probable that there were but two incisors; but certainly there were no more than three. There is a mental foramen directly beneath the mid-length of P_1. An upper incisor (SMP-SMU No. 62401; Fig. 1**B**, Plate 10**A**) probably I^3, is bifid, slightly basined lingually, and resembles I^3 of some tenrics and erinaceids. I have never observed such incisors in any marsupial, but they are not uncommon in primitive insectivores.

EVOLUTIONARY SIGNIFICANCE OF THE BUTLER FARM THERIANS

The talonid

The homologies of the single talonid cusp of Jurassic pantotheres has been the subject of considerable discussion. Each of the talonid cusps has been proposed as the primary one; the hypoconid (Osborn, 1888); the entoconid (Gregory, 1934); the hypoconulid (Butler, 1939*b*). Simpson (1928) was undecided between the entoconid and the hypoconulid; Patterson (1956) considered both the hypoconid and hypoconulid possibilities. My earlier interpretation of the Trinity molar type 6, which has the hypoconid area broken away, led me to conclude that the hypoconulid was indeed the primary talonid cusp (1966). This was based on my belief that the hypoconid was either absent or at best was poorly developed. Now, however, after comparing the type 6 molar with that of ***Kermackia***, I believe there not only was a hypoconid but that it was the largest talonid cusp. Although I still consider the hypoconulid a good candidate as the primary talonid cusp, the rationale I used earlier is invalid.

Mills (1964, 1967) insists that the single talonid cusp of Jurassic pantotheres is the hypoconid. And drawing on Butler's observations (1939*a*) concerning unchanging occlusal relationships between upper and lower cusps during evolutionary molar pattern-changes, he speculated as to the form of the unknown upper molars of *Amphitherium* and *Peramus*. In the hypothetical upper molar, the metacone is near the tooth's labial border. This would require the labial movement of the metacone after the *Kuehneotherium*-stage only to migrate back to the original position between the *Peramus*-stage and the Albian. If the single talonid cusp of *Amphitherium* is indeed the hypoconid Mills reconstruction seems reasonable. Another possibility, however, seems equally probable. The hypoconid is the largest talonid cusp on all Butler Farm therians, but it is positioned relatively far forward. Perhaps, a kink first formed in the cristid obliqua and later the hypoconid arose in connection with the increased accentuation of the groove between the paracone and metacone.

Vandebroek (1961) offered evidence for the hypoconulid's being primary, which requires that the cristid obliqua originated from the posterior crest of the protoconid. He pointed out that the posterior crest of the protoconids of unicuspid didelphid premolars have a labial flexure near their posterior ends. This appeared to Vandebroek to correspond with the looping of the cristid obliqua around the postero-labial corner of the molar talonid. He then would suggest that the hypoconid arose from this flexure and that the cusp at the end of the posterior crest (cristid obliqua) is the hypoconulid. As I examine the submolariform premolars from Butler Farm, I feel that this is the true origin of the cristid obliqua on their talonids. In both of these specimens the crest forming the labial border of the talonid basin clearly takes origin from the crest descending from the apex of the protoconid. The crest analogous to the entocristid of molars is connected with a crest that ascend the posterior slope of the metaconid. As Patterson (1956) has pointed out, the molarization of the premolar series does not necessarily recapitulate the origin of the molar pattern. This, I think, is another example of a similar pattern's being reached through mimickry, but not following the same developmental path. The cristid obliqua of *Amphitherium*, *Peramus*, *Aegialodon*, Trinity molar type 6, and ***Kermackia*** take origin from the apex of the metaconid. Even the more advanced Butler Farm therians, *Pappotherium* and *Holoclemensia*, which are more typical of later tribosphenic dentitions, have their cristid obliqua's connecting lingual to mid-width. It seems unlikely that the connection of the cristid obliqua to the trigonid would have originated on the protoconid, shifted to the metaconid before the *Amphitherium* stage and then shifted back after the middle Cretaceous. Therefore, although I am inclined to agree with Butler and Vandebroek as to the probability of the hypoconulid's being the primary talonid cusp, Vandebroek's reasoning is, in my view, incorrect.

The trigonid

The homologies of the trigonid cusps are not really in debate. It does seem worthy of mention, however, that one feature in common to all Butler Farm therians, regardless of infraclass affinities, is the rather extreme antero-posterior compression of the trigonid. Surely this reflects the condition of the stem ancestor. The angle of the trigonid of *Kuehneotherium* is upwards of 80°. It is considerably reduced by the time we reach the *Peramus*-grade. Reduction in the trigonid angle is even greater among the Butler Farm therians (i.e., 40°–45°). Curiously, the trigonid of the only known tooth of *Aegiolodon*, which is intermediate in age between *Peramus* and the Butler Farm therians, has an angle of about 60°. This is greater than that of either *Peramus* or any of the Butler Farm molars. One possibility is that the *Aegialodon* specimen is a molariform premolar. It is not uncommon for a molarized premolar to have a greater trigonid angle than the molars. For example, the referred *Pappotherium* p4 has an angle of 50°+, while the angle of ml is 45°–. Such a reference would suggest that the cristid obliqua of some molariform premolars had the same origin as that of true molars (i.e., metaconid crest), and some had an entirely different origin (i.e., protoconid crest). This is possible; but I prefer to identify the *Aegialodon* specimen as an anterior molar of a series with no molariform premolars or a typical molar of a form of similar grade to that of the Butler Farm therians, but not morphologically intermediate between them and their Jurassic ancestors.

Upper molar

Butler's (1941) and Patterson's (1956) contention that the paracone is the primary cusp of the upper mammalian molar is now well established. All known Albian upper molars have wide stylar shelves, regardless of infraclass affinities. Furthermore, the paracone is usually more lingual than the metacone. The similarity between forms presumed to represent different infraclasses surely is a reflection of the close proximity to the point of divergence. Even though the protocone of *Pappotherium* is rather small relative to later forms, it is large enough to reach across the talonid basin and notch the entocristid. There are forms in the Butler Farm fauna that I feel had even smaller and more restricted protocones. The anterior portion of the cristid obliqua of the Trinity molar type 6, and to a lesser extent, ***Kermackia***, is so tall as to make it impossible for the paracone to complete its shear labial to the crest of the protocone were developed antero-lingually. The cingulum-type protocone proposed by Mills (1964, 1967) meets the pre-requisites for an upper molar partner of this tooth, except that the forwardly-placed hypoconid requires a more lingual metacone. I prefer the hypothetical upper molar proposed by McKenna (1969) for earliest Cretaceous therians.

SUMMARY AND CONCLUSIONS

The Butler Farm local fauna was recovered from a small channel-fill about 100 feet below the top of the Trinity Group (Albian) and is considered as belonging to the

Paluxy formation. It is slightly lower in the section than the Greenwood Canyon local fauna of Patterson (1956). That the deposits are of fresh-water origin is indicated by the abundance of frog, salamander and lizard material, and by the absence of rays and/or galeoid sharks.

The therian represented by Trinity molar type 6 shares such characters as the cristid obliqua connecting to the metaconid and lack of entocristid, with the Jurassic pantotheres, *Amphitherium* and *Peramus*. It appears to have had a well-developed hypoconid, hypoconulid, and an incipient entoconid. The form therefore seems to be morphologically intermediate between certain pantotheres and primitive mammals of metatherian-eutherian grade.

The newly described form, ***Kermackia***, also has its cristid obliqua connected to the metaconid, but has added a low cingulum-like entocristid. This molar makes a proper intermediate between type 6 and animals such as *Holoclemensia* and *Pappotherium.*

The presence of submolariform premolars in the collection indicates the divergence of placentals and marsupials had begun. As a matter of fact, these teeth can almost be duplicated in living insectivores such as *Potamogale. Pappotherium* is referred to Eutheria on the basis of the presence of a well-developed cusp on the metacrista which I have not observed in marsupials but which occurs in certain primitive insectivores. Through the referred lower jaw, a sequence of eruption of permanent premolars from back to front is indicated. This is the sequence of tupaiids, erinaceids and macroscelids; but the reverse of that of tenrics (Plate 8**B**). No significance of this is offered other than proof that the form is indeed eutherian.

Holoclemensia is referred to the Metatheria on the basis of the large cusp in the middle of the ectoflexus (stylar cusp C) adjacent to the notch between the metacone and paracone. Although many primitive insectivores have well-developed stylar cusps, I know of none with stylar cusp C. The cusp is common among didelphids. The trend of twinning of the hypoconulid and entoconid so prevalent among late and post Cretaceous marcupials is incipient in the lower molars referred to *Holoclemensia.*

Other forms referred to Eutheria demonstrate that considerable specialization had occurred in spite of the close proximity to the point of divergence of Metatheria and Eutheria.

ACKNOWLEDGEMENTS

Special thanks are due to Mr Lee Butler for his kind persmission to make large excavations on his property, and other courtesies.

I am also indebted to Drs W. A. Clemens, Jr., University of California, Berkley; C. L. Gazin, U.S. National Museum; M. C. McKenna, American Museum of Natural History, and A. J. Sutcliffe, British Museum (Natural History); and K. A. Kermack, University College of London, for permission to examine specimens under their care.

The recovery and study of the Butler Farm therians was carried on by NSF Grants GB2092, 3805, and 6102. Ronald Ritchie and Roy Pickerrell were of special aid throughout the fossil recovery. Discussions with Drs J. R. E. Mills, Leigh Van Valen, Fred Szalay, S. W. Geiser, as well as those listed above, were most helpful.

REFERENCES

BUTLER, P. M., 1939*a*. Studies of the mammalian dentition. I. Differentiation of the post-canine dentition. *Proc. zool. Soc. Lond.*, **109**: 1–36.

BUTLER, P. M., 1939*b*. The teeth of Jurassic mammals. *Proc. zool. Soc. Lond.*, **109**: 329–356.

BUTLER, P. M., 1941. A theory of the evolution of mammalian molar teeth. *Am. J. Sci.*, **239**: 421–450.

CLEMENS, W. A., JR., 1968. Origin and early evolution of marsupials. *Evolution, Lancaster, Pa.*, **22** (1): 1–18.

GREGORY, W. K., 1934. A half century of trituberculy. The Cope-Osborn theory of dental evolution, with a revised summary of molar evolution from fish to man. *Proc. Am. phil. Soc.*, **73**: 169–317.

KERMACK, K. A., LEES, P. M. & MUSSETT, F., 1965. *Aegialodon dawsoni*, a new trituberculosectorial tooth from the lower Wealden. *Proc. R. Soc.* (Ser. B), **162**: 535–554.

LILLIGRAVEN, J. A., 1969. Latest Cretaceous mammals of the upper part of Edmonton Formation of Alberta, Canada, and review of marsupial-placental dichotomy in mammalian evolution. *Paleont. Contr. Univ. Kans.*, Art. 50, 1–122.

McKENNA, M. C., 1969. The origin and early differentiation of therian mammals. *Ann. N. Y. Acad. Sci.*, **167**: 1: 217–240.

MILLS, J. R. E., 1964. The dentitions of *Peramus* and *Amphitherium*. *Proc. Linn. Soc. Lond.*, **175**: 117–133.

MILLS, J. R. E., 1967. Development of the protocone during the Mesozoic. *J. dent. Res.*, Suppl. 5, 787–791.

OSBORN, H. F., 1888. The evolution of the mammalian molar to and from the triangular type. *Am. Nat.* **22**: 1067–1079.

PATTERSON, B., 1956. Early Cretaceous mammals and the evolution of mammalian molar teeth. *Fieldiana Geol.*, **13** (1): 1–105.

SLAUGHTER, B. H., 1965. A therian from the Lower Cretaceous (Albian) of Texas. *Postilla*, **93**: 1–18.

SLAUGHTER, B. H., 1966. New fossil evidence concerning which is the primary cusp of the talonid on mammalian molars. *Tex. Acad. Sci.* (Abstract), **18** (1): 132

SLAUGHTER, B. H., 1968*a*. Earliest known Eutherian Mammals and the evolution of premolar occlusion. *Tex. J. Sci.*, **20** (1): 3–12.

SLAUGHTER, B. H., 1968*b*. Earliest known marsupials. *Science, N.Y.* **62**: 254–255.

SLAUGHTER, B. H., 1968*c*. *Holoclemensia* instead of *Clemensia*. *Science, N.Y.*, **162**: 1306.

THURMOND, J. T., 1970. Lower vertebrates and paleoecology of the Trinity Group (Lower Cretaceous) in north-central Texas, Southern Methodist University, 127 pp., 34 figs. Unpubl. Ph.D. dissertation.

VANDEBROEK, G., 1961. The comparative anatomy of the teeth of lower and non-specialized mammals. International colloquium on the evolution of lower and non-specialized mammals. *Kon. Vlaamse Acad. Wetensch. Lett. Sch. Kunsten België*, Part I, pp. 215–320, Part II, 44 pls. Brussels.

EXPLANATION OF PLATES

PLATE 1

Trinity molar type 6.

A. Lingual view of SMP-SMU No. 61728, ×25. **B**. Occlusal view of same. **C**. Posterior view of same. **D**. Labial view of same.

PLATE 2

***Kermackia texana* gen. et sp. nov.**

A. Lingual view of holotype, ×30. **B**. Occlusal view of same. **C**. Posterior view of same.
D. Lingual view of holotype SMP-SMU No. 62398. **E**. Posterior view of same. **F**. Occlusal view of same.

PLATE 3

Holoclemensia texana Slaughter.

A. Labial view of ultimate upper molar (SMP-SMU No. 62009), ×20. **AA**. Occlusal view of same.
B. Labial view of holotype (SMP-SMU No. 61497), ×20. **BB**. Occlusal view of same.
C. Occlusal view of ultimate upper molar (SMP-SMU No. 62009), ×10.
D. Occlusal view of holotype (SMP-SMU No. 61497), ×10.

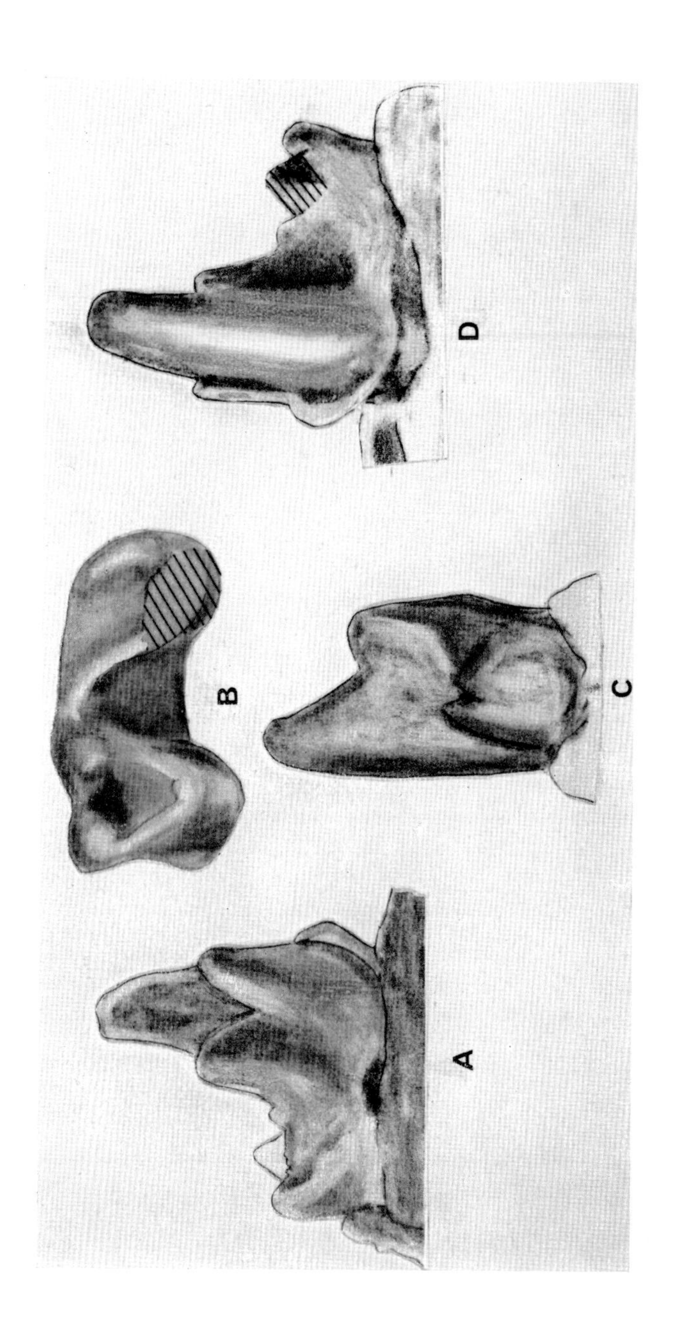
A
B
C
D

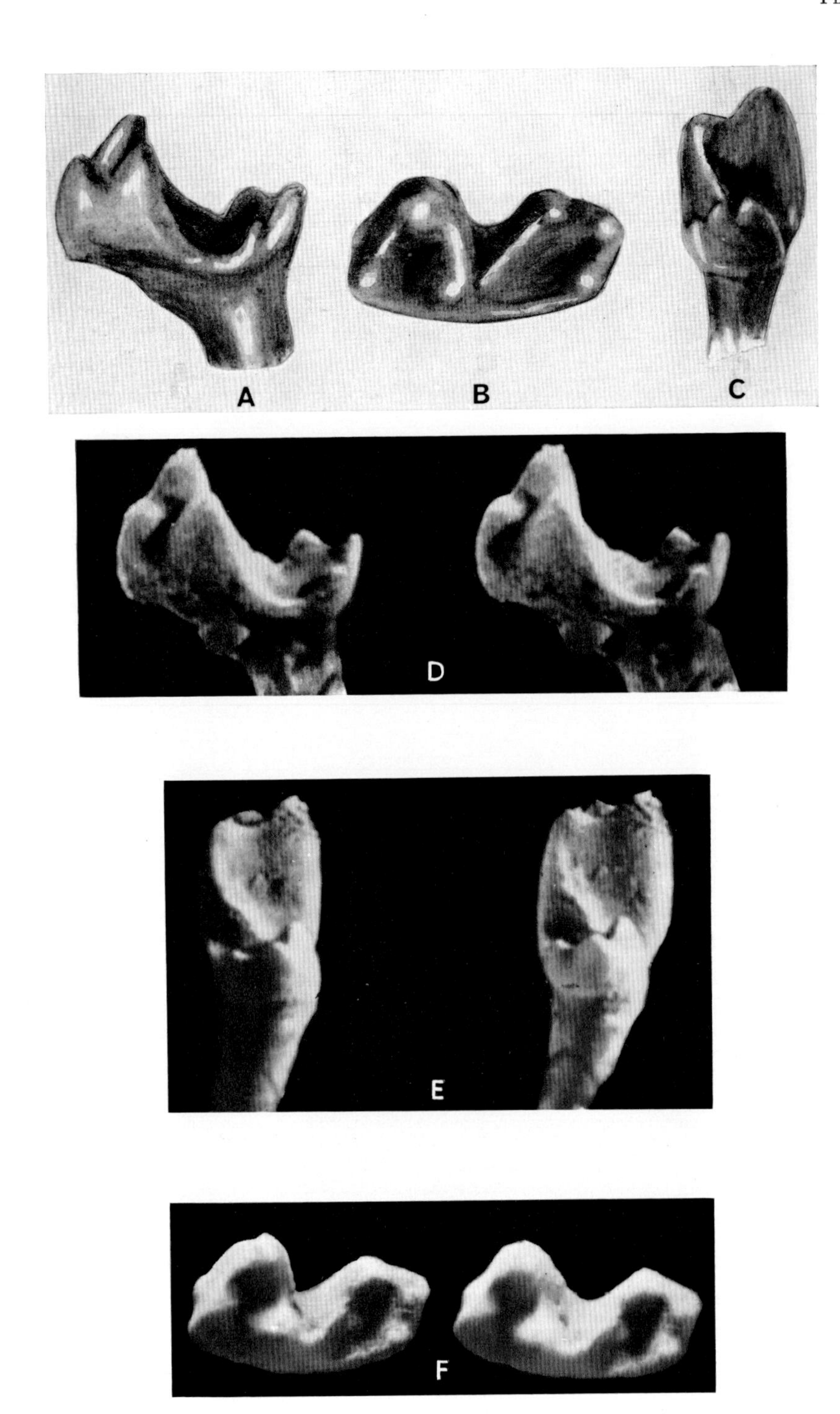

B. H. SLAUGHTER

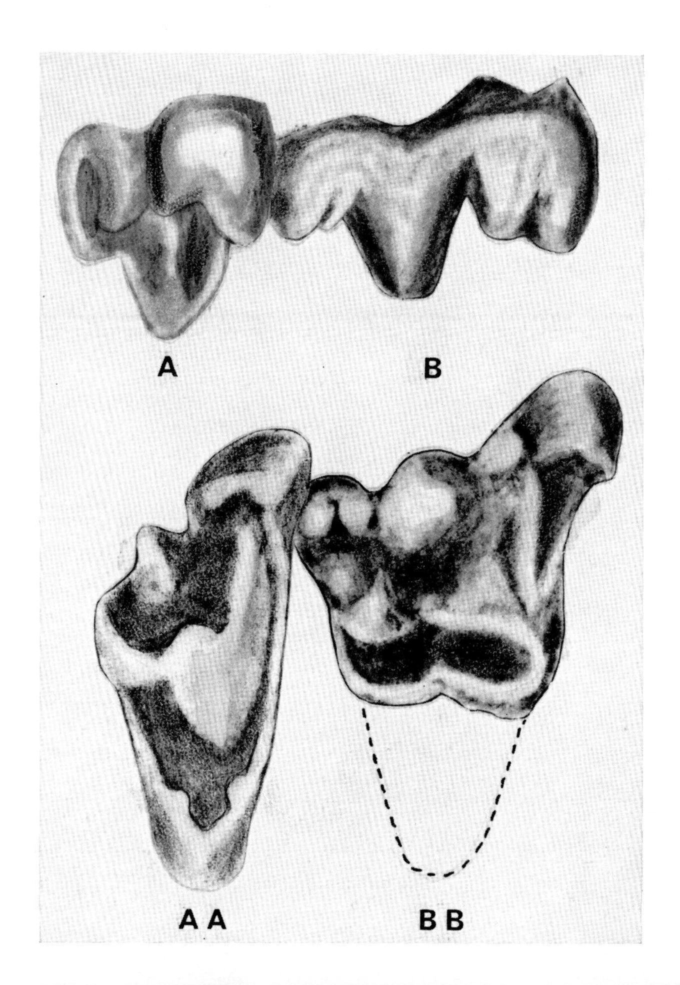

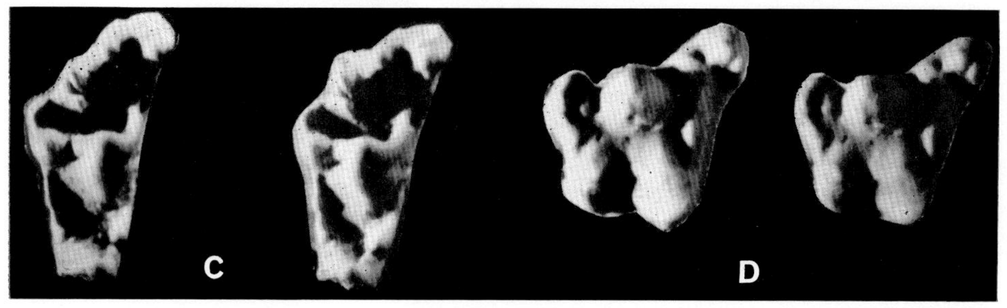

B. H. SLAUGHTER

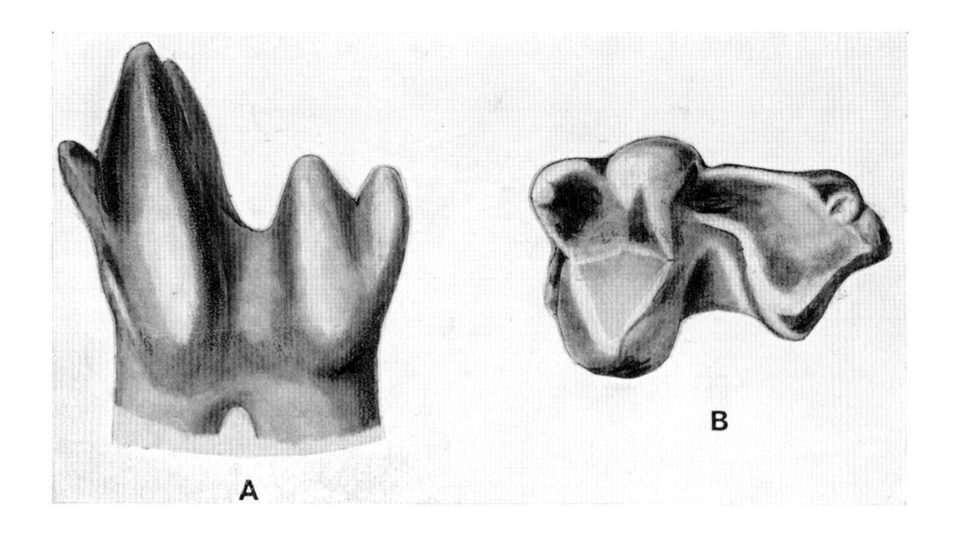

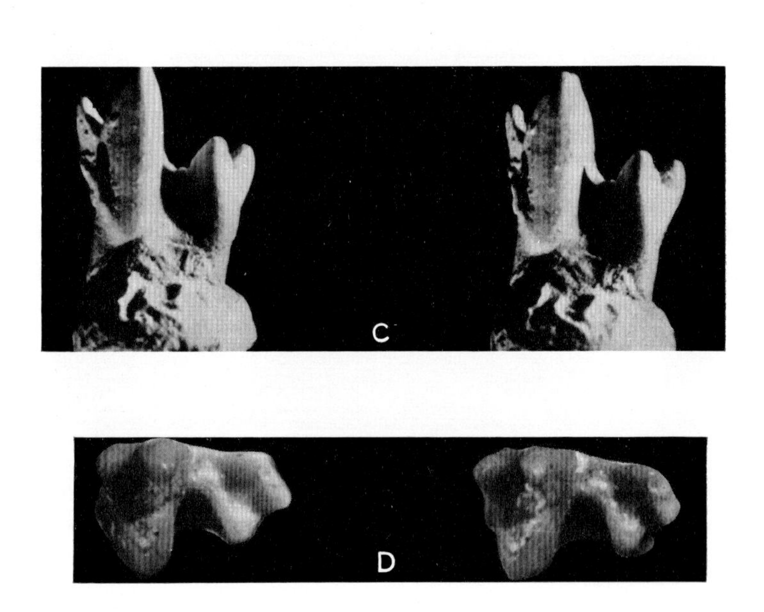

B. H. SLAUGHTER

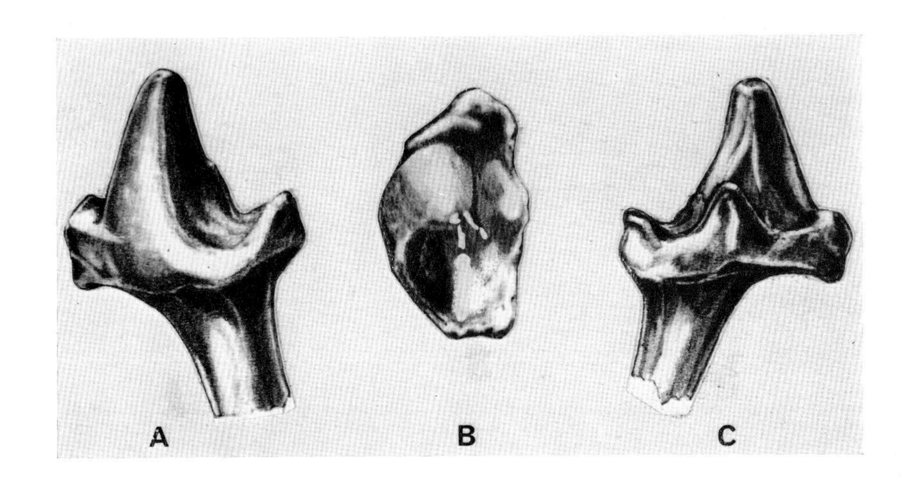

B. H. SLAUGHTER

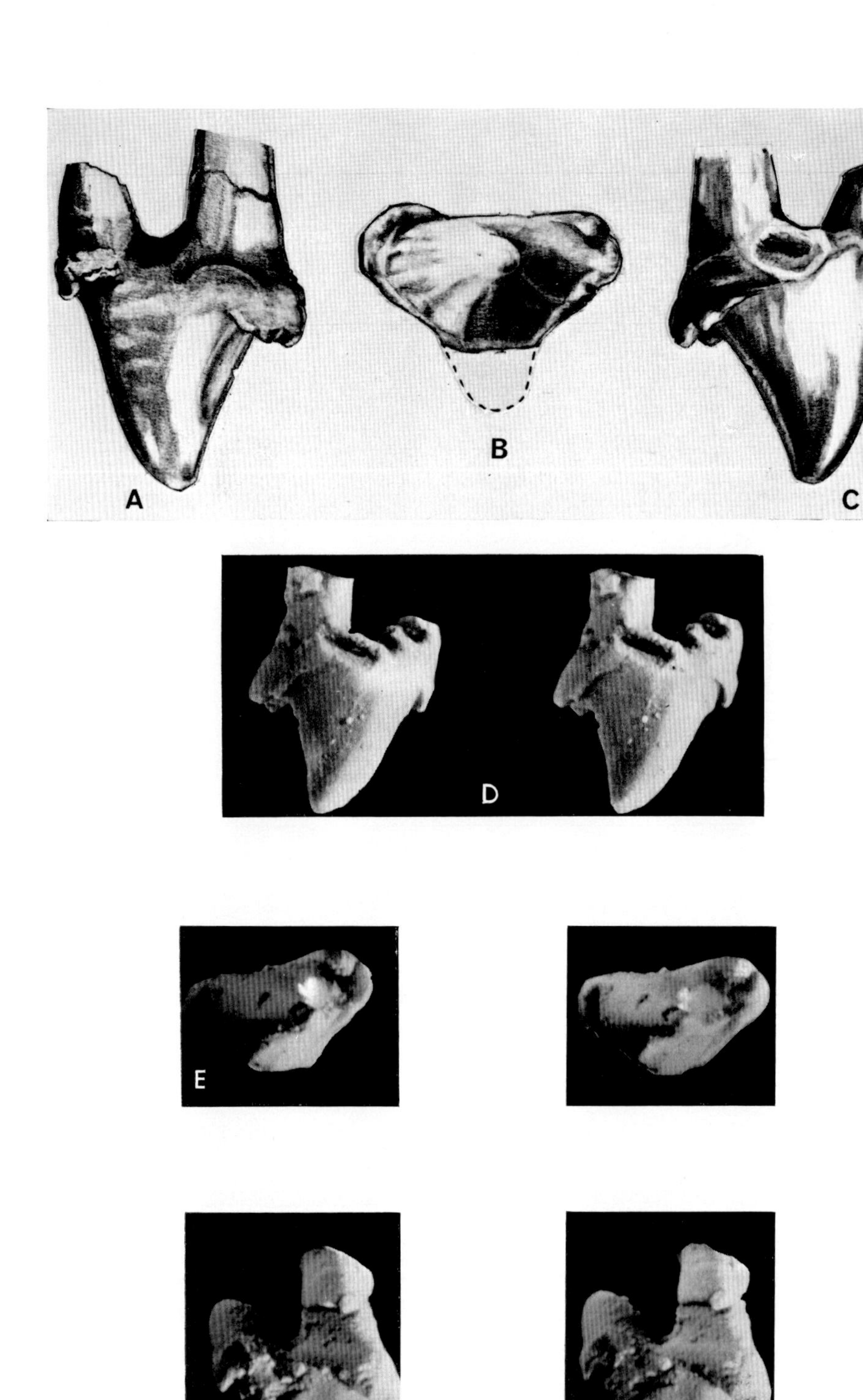

B. H. SLAUGHTER

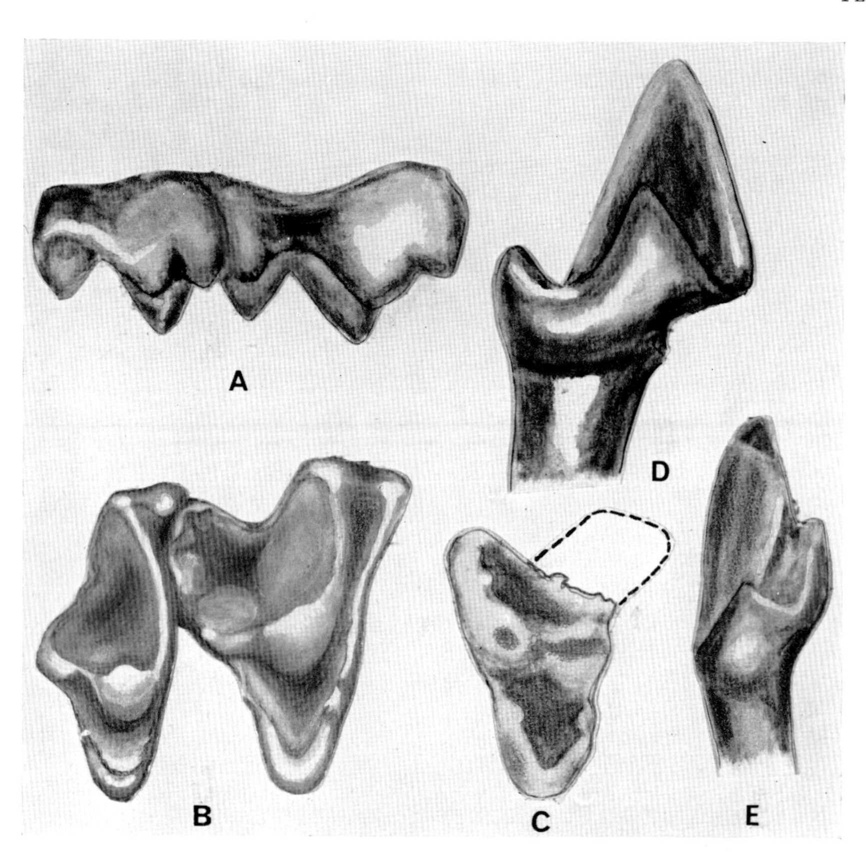

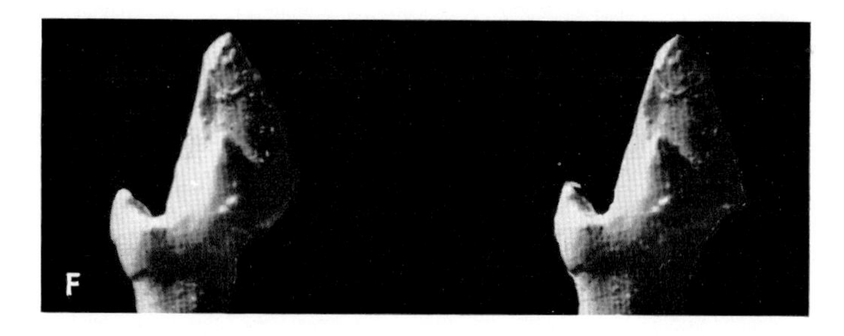

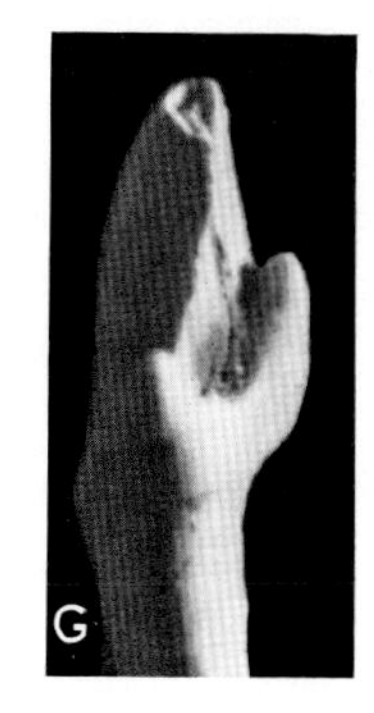

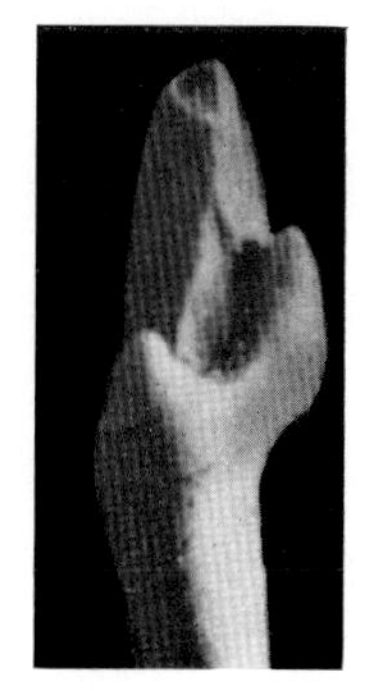

B. H. SLAUGHTER

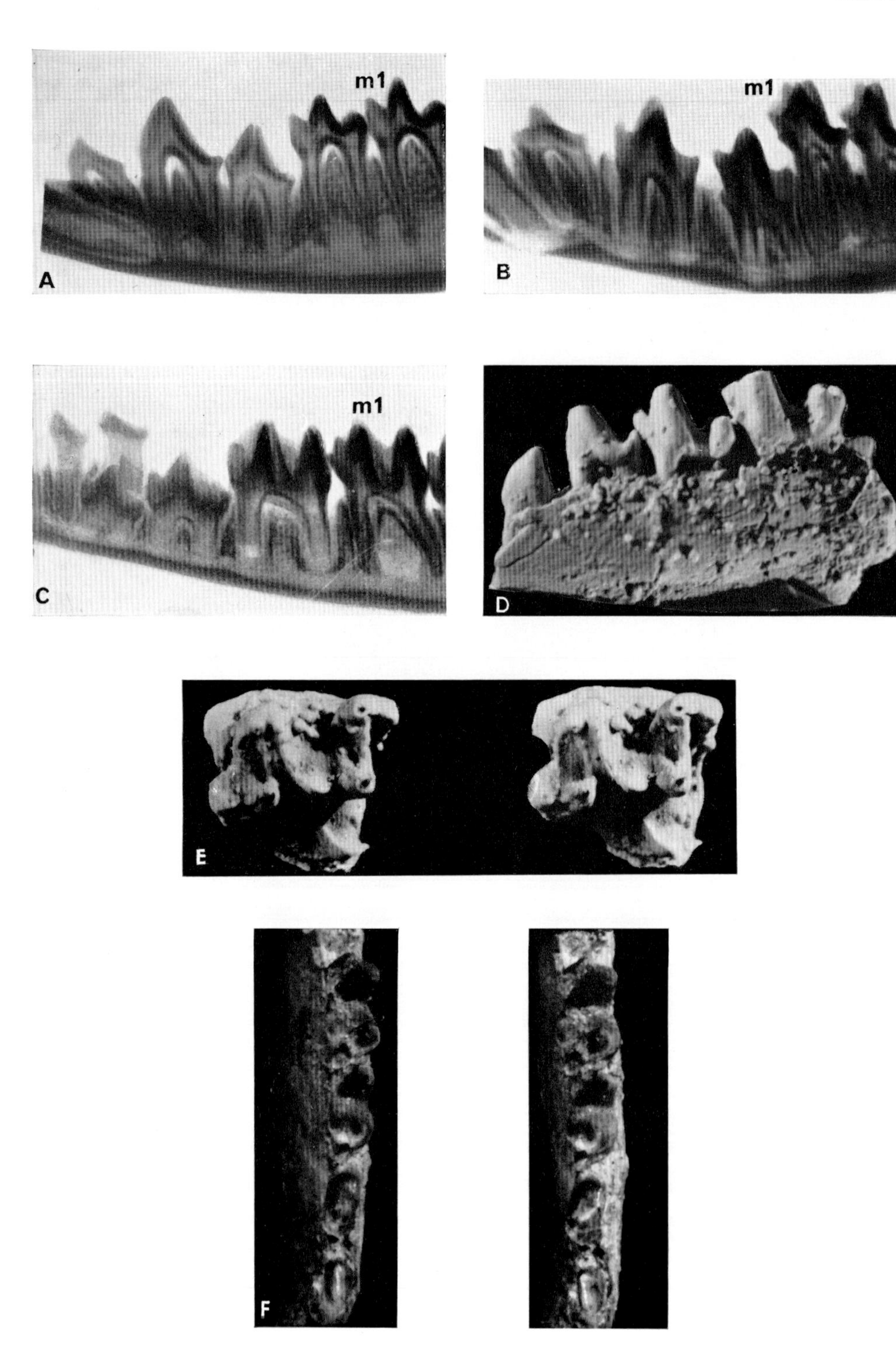

B. H. SLAUGHTER

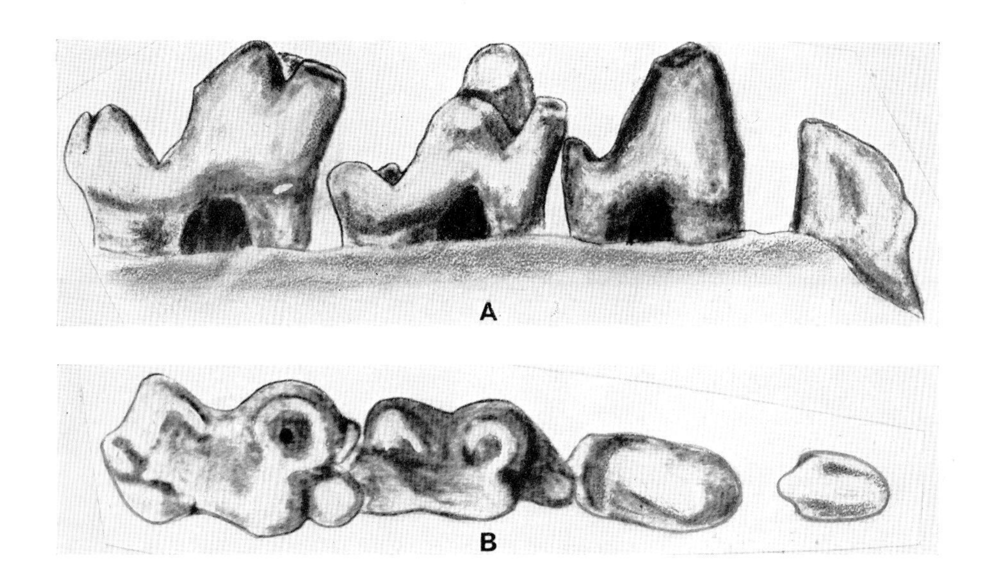
A
B

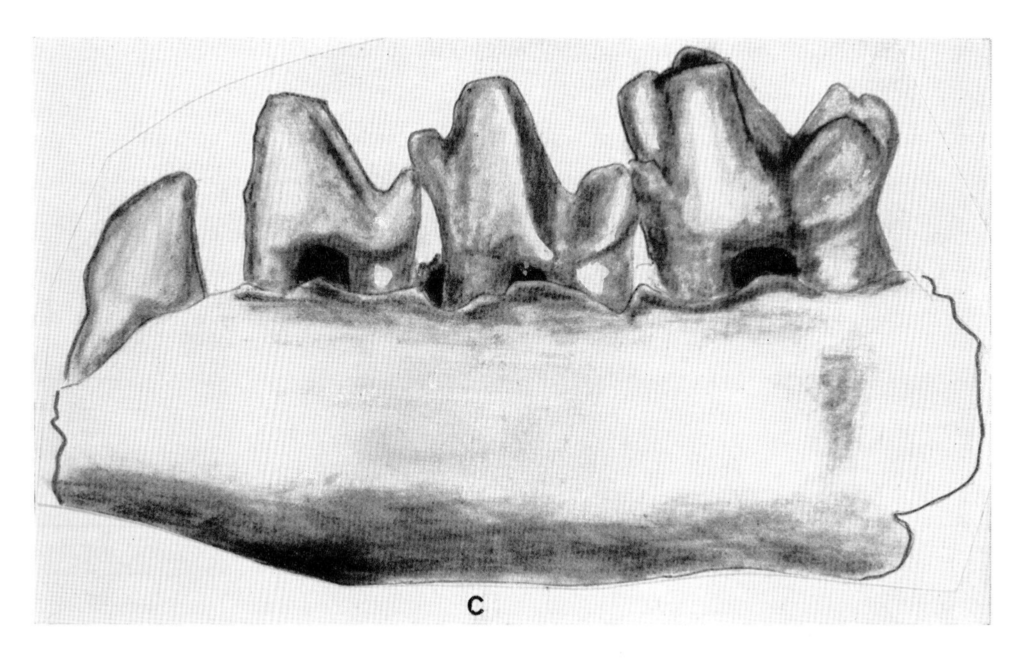
C

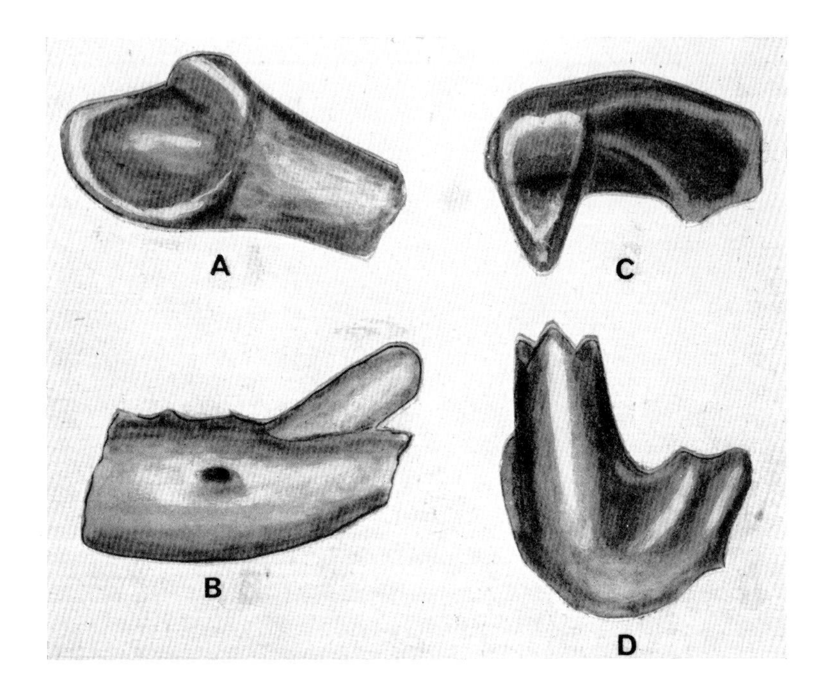

B. H. SLAUGHTER

PLATE 4

Holoclemensia texana Slaughter.

A. Labial view of referred lower molar (SMP-SMU No. 62131), ×25. **B**. Occlusal view of same.
C. Labial view of lower molar (SMP-SMU No. 62131), ×10. **D**. Occlusal view of same.

PLATE 5

A. Labial view of referred P_4 (SMP-SMU No. 61947), ×25. **B**. Occlusal view of same. **C**. Lingual view of same.

PLATE 6

Referred eutherian P^3 (SMP-SMU No. 61948).

A. Labial view, ×25. **B**. Occlusal view of same. **C**. Lingual view of same.
D. Lingual view, ×13. **E**. Occlusal view. **F**. Labial view.

PLATE 7

A. Labial view of ultimate and penultimate molars of ***Pappotherium patersoni*** Slaughter (SMP-SMU No. 61725), holotype, ×25. **B**. Occlusal view of same.
C. Upper molar of un-named pappotherid (SMP-SMU No. 62402), ×30.
D. Lingual view of referred eutherian P_4 (SMP-SMU No. 62399), ×25. **E**. Posterior view of same.
F. Lingual view of referred lower P_4 (SMP-SMU No. 62399), ×23. **G**. Posterior view.

PLATE 8

X-rays and photographs showing eruption of permanent premolars.

A. *Metachirus*, a didelphid. **B**. *Setifer*, a tenric.
C. *Elephantulus*, a macroscelid. **D**. *Pappotherium pattersoni* (SMP-SMU No. 61992).

Pappotherium pattersoni Slaughter.

E. Occlusal view of holotype (SMP-SMU No. 61725), ×14·5.
F. Occlusal view of referred lower jaw (SMP-SMU No. 61992).

PLATE 9

A. Lingual view of lower jaw referred to *Pappotherium pattersoni*, Slaughter, SMP-SMU No. 61992 × 28. **B**. Occlusal view of same. **C**. Labial view of same.

PLATE 10

A. Lingual view of upper incisor (SMP-SMU No. 62401), ×25.
B. Labial view of lower jaw fragment (SMP-SMU No. 62400), ×25.
C. Occlusal view of Trinity molar type 4. (SMP-SMU No. 61726) ×25. **D**. Labial view of same.

Marsupial mammals from the early Campanian Milk River Formation, Alberta, Canada

RICHARD C. FOX

Departments of Geology and Zoology, The University of Alberta, Edmonton, Alberta, Canada

---

The discovery of mammals in the Milk River Formation, Alberta, early Campanian in age, has revealed a diverse assemblage of didelphoid marsupials represented by isolated teeth and fragmentary bone. The Milk River marsupials are the oldest Late Cretaceous marsupials known, and include: the didelphids ***Alphadon creber* sp. nov.**; *Alphadon* sp.; ***Albertatherium primus* gen. et sp. nov.**; the pediomyids ***Pediomys exiguus* sp. nov.**; ***Aquiladelphis incus* gen. et sp. nov.**; ***Aquiladelphis minor* sp. nov.**; and a stagodontid, *Eodelphis* sp. With the exception of *Alphadon* sp. and *Eodelphis* sp. none of the new marsupials seems antecedent to later Cretaceous species; *Alphadon* sp. and *Eodelphis* sp. may be parts of lineages leading to later species, but confirmation of these lineages must await collection of more complete dentitions than presently available. The roles of facies differences, migration, extinction, evolutionary radiation, and sampling error in producing taxonomic differences between the Milk River local fauna and local faunas of marsupials from later Cretaceous horizons are discussed.

CONTENTS

INTRODUCTION

In May 1968, the author discovered the first mammalian fossil to be found in the Late Cretaceous Milk River Formation, southern Alberta, Canada (Fox, 1968).

After the discovery was made, a collecting programme that employs underwater washing and screening of large quantities of fossiliferous rock matrix (Clemens, 1965; McKenna, 1965) was developed, and about 30 tons of rock from eight sites were processed during the next three months. During 1969, a like amount was processed during two months. At present, the collecting programme has yielded over 300 teeth and rare fragmentary jaws of small mammals, including remains of the only known Late Cretaceous triconodont (Fox, 1969), diverse species of multituberculates (Allotheria) (Fox, 1971), a few placentals (Eutheria) (Fox, 1970), and relatively abundant marsupials (Metatheria). The fossils are from the continental sandstones and shales that comprise the upper part of the Milk River Formation (Russell & Landes, 1940) and are early Campanian in age (Williams & Burk, 1964). The Milk River local fauna constitutes the geologically oldest assemblage of mammals now known from the North American Upper Cretaceous and provides the first significant evidence concerning the directions of mammalian evolution in North America between the Albian (Patterson, 1956; Slaughter, 1965, 1968*a,b*) and the late Campanian (Sahni, 1968; Fox, in progress).

Previously, remains of Late Cretaceous marsupials had been recorded in North America only from rocks of late Campanian and Maestrichtian age (Sloan & Van Valen, 1965; Clemens, 1966; Sahni, 1968; Lillegraven, 1969; Fox, in progress). These fossils are thought to pertain to the Superfamily Didelphoidea because of their possession of:

(1) three premolars and four molars in each jaw;
(2) upper molars of tribosphenic type;
(3) numerous stylar cusps on the stylar shelves;
(4) closely approximated or 'twinned' entoconid and hypoconulid cusps on the lower molars.

Among living didelphoid marsupials, this suite of features is seen in the opossum *Didelphis*, for example; however, it is not seen in the dentitions of any known placental mammals, living or extinct. The taxa described here are as yet known only from isolated teeth; consequently, no statement can now be made as to the numbers of premolars and molars living individuals might have possessed. The teeth are, however, didelphoid-like in the morphology of their crowns (see, for example, Clemens, 1968:8), and are assigned to the Order Marsupialia on that account.

In this paper, terminology of stylar cusps and methods of measurement of molars follow Clemens (1966), the most recent reviewer of Late Cretaceous marsupials. All measurements are in millimetres (mm.); additional abbreviations used here are:

L—length of molar (anteroposteriorly),
W—width of molar (transversely),
Loc.—locality.

All specimens are catalogued in the collections of the Department of Geology, The University of Alberta, Edmonton, Canada.

SYSTEMATICS

Order Marsupialia
Superfamily Didelphoidea Illiger, 1841
Family Didelphidae Gray, 1821
Alphadon Simpson, 1927
***Alphadon creber* sp. nov.**
(Plates 1 and 2; Table 1)

Etymology. creber, L., numerous, in reference to the abundance of teeth of the species relative to those known of other therians from the Upper Milk River Formation.

Holotype. 5545, an isolated RM^3.

Horizon and type locality. Upper Milk River Formation (Russell & Landes, 1940), exposed in Verdigris Coulee, approximately 18 miles east of the village of Milk River, Alberta, at Loc. UA-MR-6, with co-ordinates on file in the Department of Geology, The University of Alberta.

Referred specimens. 5533, LM^1; 5554, 5555, RM^1; 5541, 5558, RM^2; 5556, RM^3; 5529, 5542, 5543, 5544, 5557, 5583, LM^3. From Loc. UA-MR-4, 6, and 8.

Diagnosis. A small species of *Alphadon*, resembling *A. wilsoni* Lillegraven, 1969; upper molars with large stylar cusp B and smaller but well developed cusp D; stylar cusp C mostly absent.

Description. Upper molars of ***Alphadon creber*** resemble those of *A. wilsoni*, Late Maestrichtian Upper Edmonton Formation, Alberta, with the following differences displayed by the Milk River teeth: (1) cusp C is usually absent, but on some specimens a small cuspule is present on the anterior margin of the middle cleft, behind cusp B, or on the posterior margin, in front of cusp D; (2) the paracone and metacone are separated by a more narrow valley; (3) anteroposterior lengths are usually less, especially for M^1; (4) the protocone is usually less broad, particularly on M^1; (5) the posterolabial lobe of the stylar shelf can be larger and the labial border of the crown more oblique in relation to a line through the paracone and metacone, especially on M^3. However, some individual molars of ***A. creber*** (5544, 5545, M^3; 5541, M^2) are virtually identical in size, occlusal outline and coronal features, other than (1) and (2) above, with individuals referred to *A. wilsoni* (2567, 2921, 3156, M^3; 3440, M^2).

Lower molars probably pertaining to ***Alphadon creber***. Numerous lower molars that are virtually identical with those of *A. wilsoni* occur in the Milk River Formation. They are probably referable to ***A. creber***, and appear to differ from lower molars of *A. wilsoni* only by displaying less inflated coronal walls than do molars of the latter species.

Discussion. Current authority (Clemens, 1966, 1968) holds that upper molars of primitive didelphoid marsupials had wide stylar shelves, large stylocones, and numerous stylar cusps in addition to the stylocones. Reduction and loss of stylar cusps (as in pediomyids and stagodontids among Late Cretaceous didelphoids) would be secondary, according to this scheme. If this interpretation is correct, and it is fully supported by the upper molar morphology of known Cretaceous therians, upper molars of ***Alphadon creber*** must have been derived from a morphology of upper molars in still earlier

Table 1. ***Alphadon creber* sp. nov.,** Milk River Formation, Late Cretaceous, Alberta, Canada; dental dimensions

Specimen no.	L	W
5533 (M^1)	1·7	1·7
5554 (M^1)	1·7	1·9
5555 (M^1)	1·8	1·8
5541 (M^2)	2·0	2·2
5558 (M^2)	1·9	—
5529 (M^3)	1·9	2·3
5542 (M^3)	1·9	[2·2]
5543 (M^3)	2·1	2·5
5544 (M^3)	2·1	2·4
5545 (M^3) type	2·2	2·4
5556 (M^3)	[2·0]*	2·4
5557 (M^3)	—	[2·3]
5583 (M^3)	—	[2·4]
5535 (M_x)	2·0	1·2
5564 (M_x)	—	1·1
5565 (M_x)	1·9	1·0
5567 (M_x)	1·9	1·2
5568 (M_x)	1·9	1·1
5569 (M_x)	2·0	1·2
5580 (M_x)	1·9	1·2
5581 (M_x)	2·0	1·1

* Dimension incomplete owing to breakage [ ].

marsupials in which stylar cusp C was present and well developed. ***Alphadon creber***, from the oldest horizons in which *Alphadon* is known to occur, is the most advanced species of *Alphadon* in morphology of the stylar shelves, for cusp C is present and well developed in all later species of *Alphadon* except the late Maestrichtian *A. lulli* Clemens, 1966, and in that species it is only occasionally lacking (Clemens, 1966: 10). If it were not for this apparent progressiveness of the stylar shelves, ***A. creber*** would be an excellent, indeed irresistible, candidate for being the direct ancestor of *A. wilsoni*, and perhaps for *A. lulli*, as well; as it is, however, no known species of *Alphadon* could have descended from ***A. creber***.

The upper molars referred here to ***A. creber*** exhibit a variability that may be the result of inclusion of a taxonomically heterogeneous sample within this single species. Most available third upper molars of ***A. creber*** can be assigned to either one of two groups that differ from one another in several minor features. The first group, which includes these molars more closely resembling third upper molars of *A. wilsoni*, is composed of teeth that are slightly smaller, with (1) the labial margin of the crown more nearly parallel to an imaginary line through the tips of the paracone and metacone; (2) the stylocone relatively smaller and farther from the paracone; (3) the crest between the paracone and stylocone failing to meet the stylocone or interrupted along its extent toward the stylocone by a notch; (4) a relatively smaller D cusp. The other group, resembling less closely the third upper molars of *A. wilsoni*, includes those teeth that (1) are slightly larger; (2) are more oblique labially; (3) have a relatively larger stylocone, closer to the paracone; (4) have a continuous crest between the paracone and stylocone; (5) exhibit a relatively larger cusp D, than those of the first group. In the samples of second upper molars and the first upper molars of ***A. creber***, no comparable bimodality has been observed, but only one complete M^2 is known at present, and but three first upper molars. Perhaps lending strength to the likelihood of heterogeneity of the sample is the fact that one upper molar (5556, M^3), from which the enamel is mostly eroded, is nearly as large as the largest individual of the second group, but resembles teeth in the first group in all features other than size. This suggests that if, in fact, two species are included in what is now recognized as ***A. creber***, the total variation in dimensions of the molars of each is similar, but that their qualitative morphological differences are nevertheless maintained. Among the lower molars probably referable to ***A. creber*** it is not possible as yet to distinguish among differences owing to (1) position of teeth in the jaw; (2) individual variation among teeth at the same position; (3) difference in species, if, in fact, teeth from more than one species are included in the sample.

Alphadon sp.
(Plate 2; Table 2)

Referred specimens. 5560, 5561, 5579, LM_1; 5527, LM_2; 5559, RM_3. From Loc. UA-MR-6, only.

Description. The presence in the Upper Milk River Formation of a second, larger species of *Alphadon* is indicated by five moderately large isolated lower molars, virtually identical to lower molars in jaws in as yet undescribed species of *Alphadon* from the late Campanian Oldman Formation, Alberta. The crista obliqua of each of the molars is clearly directed to the notch between the protoconid and metaconid, not to the posterior wall of the protoconid, as on lower molars of pediomyids. No upper molars that might pertain to this species have yet been discovered.

Table 2. *Alphadon* sp., Milk River Formation, Late Cretaceous, Alberta, Canada; dental dimensions

Specimen no.	L	W
5560 (M_1)	3·0	1·7
5561 (M_1)	—	1·5
5579 (M_1)	—	1·7
5527 (M_2)	3·3	1·7
5559 (M_3)	3·6	2·0

***Albertatherium* gen. nov.**

Etymology. after the Province of Alberta, Canada; *therium*, G., mammal.

Type species. ***Albertatherium primus***, new species.

Diagnosis. The same as that of the type and only speci̇cs.

Albertatherium primus sp. nov.

(Plate 3; Table 3)

Etymology. primus, L., first.

Holotype. 5528, an isolated RM^3.

Horizon and type locality. Upper Milk River Formation (Russell & Landes, 1940), exposed in Verdigris Coulee, approximately 18 miles east of the village of Milk River, Alberta, at Loc. UA-MR-6, with co-ordinates on file in the Department of Geology, The University of Alberta.

Referred specimens. 5532, 5547, broken RM^3; 5547, LM^4. From Loc. UA-MR-4 and 6 only.

Diagnosis. Small North American Late Cretaceous didelphoid marsupials; upper molars short anteroposteriorly relative to transverse width in jaw; paracone and metacone subequal in size, separate, but close together; metaconule slightly lower and smaller than protoconule; principal cusps and conules define a deep, narrow protoconal basin; stylar shelf with deep middle cleft and five marginal stylar cusps (A–E), of which A and E are the smallest and C the largest; ridge from paracone anteriorly concave and directed toward cusp A or notch between cusps A and B; cusps, conules, and ridges relatively high in comparison to other Late Cretaceous marsupials.

Description. In occlusal outline M^3 resembles a slender isosceles triangle, with a short labial side, nearly equal anterior and posterior sides, and a narrow apical protocone. The paracone and metacone are placed more labially than lingually on the crown; the paracone is slightly the taller and its base extends slightly more lingually than does that of the metacone. The principal cusps are tall and slender; the conules are tall, narrow, well spaced lingually from the paracone and metacone, and bear medial wings. The protocone is high, but lower than the metacone, is narrow in coronal aspect, and lacks inflation at base and sides common on upper molars of later Cretaceous marsupials. The protoconal basin is anteroposteriorly narrow and is deep.

The stylar shelf is less than one-third the transverse width of the crown, and consists of two rounded lobes separated by a deep middle cleft. Five well-defined stylar cusps (A–E) are present on the labial margin of the stylar shelf; these are widely spaced from and smaller than the paracone and metacone. Cusp A is an erect, labiolingually compressed pyramid on the anterolabial salient of the crown. Cusp B is adjacent to and larger than A; it sweeps upward from the stylar shelf to form an erect, compressed blade; a conical rather than blade-like B cusp is usual among Late Cretaceous marsupials. The B cusp is well anterior of the middle cleft and anterolabial to, but relatively far from, the paracone. Cusp C, largest and tallest of the stylar cusps, is centrally located at the apex of the middle cleft, equidistant from the paracone and metacone, but widely spaced from B and D. Cusp C is nearly conical but somewhat flattened on its labial side; it is sharply pointed and leans anterolabially to a minor degree. Stylar cusp D is an erect blade, lower than the stylar cusps anterior to it and located on the posterolabial margin of the middle cleft, labial to and slightly posterior to the metacone. Cusp E is the smallest stylar cusp, placed at the junction of the metastylar ridge with the labial margin of the stylar shelf, close to D. The long axis of

E parallels the metastylar ridge and is continuous with it. The long axes of the other stylar cusps parallel the labial margin of the stylar shelf; the cusps are connected to one another by a narrow marginal crest.

A high, anteriorly concave ridge extends from the paracone towards stylar cusp A, joining with A on 5528. In other Late Cretaceous marsupials in which stylar cusp B is present and well developed on M^3, the ridge extends to that cusp (although variably in the bunodont *Glasbius*, and variably in some modern didelphines, as well (Clemens, 1966)), and is generally convex anteriorly, as in *Alphadon*. A second ridge, from cusp D, extends for a short distance towards but fails to meet the metacone. The metastylar ridge is high and narrow, and runs outward to the posterolabial corner of the crown, where it supports cusp E. A moderately wide cingulum extends from the protoconule to cusp A; a more narrow cingulum extends from the metaconule labially, beyond the base of the metacone, at its side. Accessory cuspules are present on both cingula; protoconal cingula are lacking.

On LM^4 (5547) the crown labial to the paracone and metacone is less extensive transversely than on M^3. The protoconule is stronger than the metaconule, and protoconule, metaconule and protocone are weaker than on M^3, although they nevertheless define a narrow protoconal basin. The metacone, although shorter than the paracone, is strong. A high crest extends towards but fails to meet the small stylar cusp A. Cusp B is broken on the only specimen now available; the base of B, however, is larger than the base of cusp A. Cusp C is the largest stylar cusp, more compressed labiolingually than on M^3 and slightly closer to the metacone than to the paracone. Posterior to cusp C, the stylar shelf is lacking; the posterolabial border of the crown posterior to the C cusp is provided by the base of the metacone; no shelf extends beyond it.

Lower molars probably pertaining to ***Albertatherium primus***. Five isolated lower molars occlude tightly with upper molars of ***Albertatherium primus***, but with no other upper molars of marsupials from the Milk River Formation and, hence, are likely referable to ***Albertatherium primus***. Coronal morphology of these teeth resembles that of Campanian and Maestrichtian species of *Alphadon;* the Milk River molars differ from those of *Alphadon* in possessing (1) a higher protoconid relative to the metaconid and higher metaconid relative to the paraconid; (2) a paraconid more lingually placed; (3) a relatively higher entoconid; (4) a hypoconulid more lingual in position and more nearly horizontal in orientation; (5) generally more slender and more nearly erect cusps (excepting the hypoconulid), with straighter, less inflated sides.

Discussion. Among the three recognized families of Late Cretaceous marsupials, ***Albertatherium*** seems to resemble explicitly only the Didelphidae in upper molar morphology. Upper molars of Pediomyidae display a narrow stylar shelf labial to the paracone, or, in some species, the shelf may be entirely lacking there; stylar cusp B is reduced or lost altogether. Upper molars of Stagodontidae are generally larger than those of other Cretaceous marsupials; the paracone is small relative to the metacone, B is the largest stylar cusp, C is lacking or exceedingly diminished, and the wide stylar shelf is deeply but asymmetrically bilobed (Clemens, 1966). Upper molars of Cretaceous Didelphidae display subequal paracone and metacone, a stylar shelf with numerous stylar cusps and a large stylocone; these features are seen on molars

Table 3. ***Albertatherium primus* gen. et sp. nov.,** Milk River Formation, Late Cretaceous, Alberta, Canada; dental dimensions

Specimen no.	L	W
5528 (M^3) type	2·2	2·7
5547 (M^3)	2·2	—
5546 (M^4)	1·8	2·7
5526 (M_x)	2·4	1·3
5550 (M_x)	2·4	1·5
5551 (M_x)	[2·1]	1·3
5566 (M_x)	2·3	1·3
5585 (M_x)	2·2	1·3

of ***Albertatherium***, and may be taken as sufficient for referring ***Albertatherium*** to the Didelphidae.

Holoclemensia texana Slaughter, 1968*b*,*c*, from the Albian of Texas, is the oldest known marsupial, and is the only marsupial known from pre-Campanian strata. It has been referred to the Didelphidae. The known upper dentition of *Holoclemensia* consists of a penultimate and ultimate molar only; both teeth display a strong, hooked anterolabial salient of the stylar area that supports a large cusp A and a smaller cusp B, closely posterior to A. Cusp C is the largest stylar cusp on the penultimate molar, and is large on the ultimate molar, as well. The penultimate molar bears a closely approximated pair of cusps at the posterolabial corner of the stylar shelf; the smaller and more anterior of these Slaughter (1968*b*) equates with stylar cusp D. The stylar shelf on each molar is wide; the crowns are anteroposteriorly short and transversely long. The base of the metacone does not extend as far lingually as does the base of the paracone.

Similarities in upper molars of *Holoclemensia* and ***Albertatherium*** are in the relatively short crowns and enlarged C cusps that, on preultimate molars at least, can be larger than the stylocones. ***Albertatherium*** displays smaller, more marginal stylar cusps, a weaker anterolabial salient of the stylar shelf, a deeper middle cleft, and paracone and metacone more nearly equal in the lingual extent of their bases. In ***Albertatherium***, the stylocone, although smaller than cusp C, as in *Holoclemensia*, is larger than cusp A.

Holoclemensia is the only marsupial known to be geologically older than ***Albertatherium***; this fact, and the possession by both of relatively short upper molar crowns and C cusps that are distinctly larger than the stylocones, suggest that ***Albertatherium*** might have been a descendant of *Holoclemensia*. If this were the case, changes in upper molar morphology leading to an increase in symmetry of the lobes of the stylar shelf, a reduction in the size of the stylar cusps, a reduction in the size of stylar cusp A relative to stylar cusp B, a reduction in the width of the stylar shelf, a deepening of the middle cleft, and an increase in the size of the metacone relative to the paracone would have had to occur.

Slaughter (1968*b*) suggests that *Holoclemensia* may have been ultimately ancestral to *Alphadon*, represented by numerous species from late Campanian and Maestrichtian strata (Sloan & Van Valen, 1965; Clemens, 1966, 1968; Lillegraven, 1969; Sahni, 1968; Fox, in progress), and thought to be the stem radicle or close to the stem radicle of Late Cretaceous didelphoid evolution and, ultimately, of all Cenozoic marsupials (Clemens, 1966, 1968). Comparison between ***Albertatherium*** and *Alphadon* suggests that upper molars of *Alphadon* are anteroposteriorly longer, with larger, more nearly conical, and less marginally placed stylar cusps. The stylocone (cusp B) is the largest stylar cusp in *Alphadon;* the anterolabial salient of the stylar shelf is stronger and middle cleft less deep in *Alphadon* than in ***Albertatherium***. Derivation of upper molars of *Alphadon* from ***Albertatherium*** would require lengthening of the molar crowns, increase in size of the stylar cusps, increase in width of the stylar shelf and strength of the anterolabial salient, increase in size of the stylocone relative to stylar cusp C, and reduction in depth of the middle cleft. Most of the changes required for a lineage stemming from ***Albertatherium*** to *Alphadon* would require trends in upper molar morphology opposite to those required for a lineage between *Holoclemensia* and ***Albertatherium***. Consequently, it seems unlikely that ***Albertatherium*** was included in the ancestry of *Alphadon* if, in fact, the ancestor of both ***Albertatherium*** and *Alphadon* was *Holoclemensia.*

Family Pediomyidae Clemens, 1966

Pediomys Marsh, 1889*a*

Pediomys exiguus **sp. nov.**

(Plate 3; Table 4)

Etymology. exiguus, L., small.

Holotype. 5536, an isolated RM^1.

Horizon and type locality. Upper Milk River Formation (Russell & Landes, 1940), exposed in Verdigris Coulee, approximately 18 miles east of the village of Milk River, Alberta, at Loc. UA-MR-4, with co-ordinates on file in the Department of Geology, The University of Alberta.

Referred specimens. 5537, RM^2; 5548, LM^2; 5538, RM^1. From Loc. UA-MR-4 and 6 only.

Diagnosis. Small pediomyids in which stylar cusp B is present but small, cusp C is mostly absent, and cusp D is a long, low crest.

Description. The occurrence of a small pediomyid in the Milk River Formation is indicated by four upper molars on which the stylar shelf is narrow labial to the paracone. On these teeth, cusp B is reduced, the middle cleft is shallow or lacking altogether, cusp C seems usually absent, and cusp D is a long, low blade parallel to the labial margin of the crown. On the first upper molars the D cusp extends as an elongate crest across the middle cleft and terminates at the base of cusp B. On one specimen (5538), a small cuspule is present on the crest at the C position; on the other (5536), the cuspule is lacking.

The remaining upper molars (5537, 5548) are poorly preserved second upper molars. On these, the stylar shelf is only wide enough labial to the paracone to support a small cusp B; this cusp on the second upper molars is larger than on first upper molars, but much smaller than the paracone, as on first upper molars. On second upper molars, cusp D is a large blade on the posterior margin of the middle cleft; the cleft is shallow. Among marsupials now known from the Milk River Formation, **Pediomys exiguus** is closest to ***Alphadon creber*** in upper molar dimensions, but can be readily distinguished from the didelphid species by possessing more narrow stylar shelves labial to the paracones and distinctly smaller stylar cusps B, the same features that distinguish upper molars of later Cretaceous species of *Alphadon* and *Pediomys* from each other.

Table 4. **Pediomys exiguus sp. nov.,** Milk River Formation, Late Cretaceous, Alberta, Canada; dental dimensions

Specimen no.	L	W
5536 (M^1) type	1·6	1·5
5538 (M^1)	[1·6]	[1·7]
5537 (M^2)	—	1·8
5548 (M^2)	1·6	1·9

Discussion. Upper molars of **Pediomys exiguus** are smaller than those of *Pediomys krejcii* Clemens, 1966, the smallest species of *Pediomys* known from late Maestrichtian horizons (Clemens, 1966; Lillegraven, 1969), but larger than those of the smallest of as yet undescribed species of *Pediomys* from the late Campanian Oldman Formation, Alberta (Fox, in progress). The primitiveness of the stylar shelves of the Milk River molars relative to those of later species of *Pediomys* cannot be now determined with certainty. If, as Clemens (1966, 1968) suggests, species of *Pediomys* are advanced among Cretaceous marsupials in reduction of the stylar shelf labial to the paracone and reduction and loss of stylar cusp B, ***P. exiguus*** is clearly more primitive than Maestrichtian species of *Pediomys* in which stylar cusp B is not known to occur (*P. florencae* Clemens, 1966), or occurs only very rarely (*P. hatcheri* (Osborn), 1898). Among other described species of *Pediomys* from Maestrichtian sites (*P. elegans*, *P. cooki*, *P. krejcii*), a small cusp B occurs variably on preultimate molars (Clemens, 1966); a meaningful statement of frequency of the cusp in ***P. exiguus*** is hardly possible here, owing to the small sample of teeth now available, but the cusp does occur on each of the four teeth known, and is larger on them than on upper molars of Maestrichtian *Pediomys*. A stylar cusp B has not been observed on any of the upper molars of species of *Pediomys* from the Oldman Formation. Available evidence suggests, then, that the early appearance of the Milk River species is paralleled by a morphology of stylar cusp B that is relatively primitive in comparison to that in later species of *Pediomys*.

The posterior part of the stylar shelf on molars of ***P. exiguus*** seems not primitive, however. Upper molars of late Campanian and Maestrichtian species of *Pediomys* usually show a small to medium sized cusp D posterior to the middle cleft, and often a small separate cusp is present in the C position, as well (Clemens, 1966; Fox, in progress). The D cusp on these molars usually is somewhat elongate parallel to the labial margin of the crown, but does not extend across the middle cleft, as it does on at least some first upper molars of ***P. exiguus***. The extent of the cusp in ***P. exiguus*** would likely be the result either of an increase in its own length anteriorly or fusion with cusp C. In either case, the peculiarities of the cusp in the Milk River species are not, as far as is known to me, seen on upper molars of other species of *Pediomys*, and are not intermediate between those species and didelphids ancestral to pediomyids, in which the stylar cusps were presumably separate, discrete cones. Perhaps further study enabled by larger dental samples will demonstrate the suggestion made here that ***P. exiguus*** was not a part of the ancestry of late Campanian and Maestrichtian species of *Pediomys* now known.

Aquiladelphis **gen. nov.**

Etymology. aquila, L., eagle, in reference to the Aquilan Stage, proposed (Russell, 1964), to include the time of deposition of the Milk River Formation.

Type species. ***Aquiladelphis incus*** **sp. nov.**

Diagnosis. Medium-sized to large North American pediomyids. Upper molars with two small stylar cusps B, stylar cusp C largest stylar cusp, stylar cusp D larger than B cusps and smaller than cusp C; paracone and metacone labiolingually compressed, low, massive, amd divergent; conules strong; protoconal basin broad. Lower molars with low and heavy trigonid cusps; talonid basins broad and shallow.

Aquiladelphis incus **sp. nov.**

(Plates 4 and 5; Table 5)

Etymology. incus, L., anvil, in reference to the broad occlusal surfaces of the molars.

Holotype. 5522, an isolated, broken LM^3.

Horizon and type locality. Upper part of the Milk River Formation (Russell & Landes, 1940), exposed in Verdigris Coulee, approximately 18 miles east of the village of Milk River, Alberta, at Loc. UA-MR-8, with co-ordinates on file in the Department of Geology, The University of Alberta.

Referred specimens. 5523, LM^4; 5524, LP_x; 5525, 5587, RM_x. From Loc. UA-MR-4, 6, and 8.

Diagnosis. The larger species of ***Aquiladelphis***. Known premolars strong but not bulbous; pre-ultimate upper molars longer anteroposteriorly than wide transversely; lower molars broad.

Description. Dimensions of the teeth referred to ***Aquiladelphis incus*** indicate that it is the largest mammal now known from the Milk River Formation.

On 5522, an M^3, the paracone and metacone are relatively low cusps that are subequal in size, labiolingually compressed, widely divergent from one another, and labially placed on the crown. The conules are set well apart from the paracone and metacone, and are long and wide, with strong internal wings, which extend labially to meet short descending ridges on the paracone and metacone. The protocone is low, long, and wide; its tip is in a relatively anterior position and a patch of short, vertical, crenulated ridges occurs on its posterolingual wall. The principal cusps and conules define a shallow and unusually broad protoconal basin, nearly equal in size to that on upper molars of the largest Cretaceous marsupial known, the Maestrichtian stagodontid *Didelphodon* Marsh, 1889*a*. On 5522, only the anterior parts of the stylar shelf are preserved; the shelf is narrow and lacks a middle cleft. Stylar cusp A is mostly worn away from the strong anterolabial salient of the stylar shelf; two small, closely approximated blade-like cusps occupy the B position. A short, low ridge from the paracone extends towards but fails to meet the anterior B cusp. Stylar cusp C is nearly conical, slightly taller than and well apart from cusps B; it is accompanied by a small anterior accessory cusp. Posterior to cusp C, the stylar shelf is broken away. A partly damaged, heavily worn cingulum extends from the protoconule to cusp A and the parastyle. A wide unworn cingulum runs from the metaconule towards the missing posterolabial corner of the crown.

LM^4, 5523, agrees with the holotype in distinctive features. The metacone and enamel adjacent to it are broken from the crown, but the tooth is clearly anteroposteriorly long. The paracone is less compressed labiolingually and the protocone and protoconal basin are slightly smaller and conules slightly weaker than on the more anterior molar. On 5523 the stylar area is bilobed; the anterior lobe is a strong salient separated from the weak posterior lobe by a shallow middle cleft. Stylar cusp A rises as a pyramid at the terminus of the salient; two blade-like B cusps, higher than A and separated from it by a vertically deep notch, follow. A heavy, low ridge runs from the paracone to the anterior B cusp. A swelling of the labial edge of the crown at the position of cusp C suggests the presence of a cusp there, but damage to the occlusal surface of the crown from the middle cleft posteriorly has destroyed it, if present. A wide, heavily worn cingulum extends from the protoconule to cusp A. The presence or absence of a posterior cingulum cannot be determined on 5523 owing to damage.

The lower premolar (5224), referred to ***Aquiladelphis*** because of its large size, is a tall, heavily constructed tooth; the crown is not inflated and is covered with subtley wrinkled enamel. The anterior edge of the crown is trenchant and faintly serrated; on either side of the edge are deep wear facets. A small posterior heel is present. A deeply hollowed, posterolabial wear facet curves from the apex of the crown to its base and occupies more than half of the labial side of the tooth. No roots are preserved on the specimen.

The lower molars are poorly preserved, but clearly are large, broad, heavy teeth. The paraconid is labially placed and smaller than the metaconid; the crista obliqua meets the posterior wall of the protoconid and the hypoconulid and entoconid are 'twinned' at the posterolingual corner of the talonid. Heavy wear facets extend across the metaconid, are deeply incised into the protoconid basin, and run posteriorly along the hypoconid to terminate on the hypoconulid.

Discussion. The isolated premolar and upper and lower molars referred to ***Aquiladelphis incus*** are associated because of their size and general adaptations for powerful dental function. In the upper molars, the narrowness of the stylar shelves and the size, number, and distribution of the stylar cusps indicate that ***Aquiladelphis*** is a pediomyid and not a didelphid or stagodontid; this estimate is supported by the position of the crista obliqua on the lower molars. The upper molars of ***A. incus*** differ

Table 5. ***Aquiladelphis incus* gen. et sp. nov.,** Milk River Formation, Late Cretaceous, Alberta, Canada; dental dimensions

Specimen no.	L	W
5522 (M^3) type	[5·6]	[5·5]
5523 (M^4)	[4·5]	4·4
5524 (P_x)	4·7	3·1
5525 (M_x)	5·6	3·0

from those of the late Campanian and Maestrichtian species of *Pediomys* in their larger size, length-width ratios, and in their possession of two blade-like, moderately strong cusps at the B position. When present in *Pediomys*, stylar cusp B is single, small, and conical, labial or posterolabial to the paracone. In ***A. incus*** the stylar shelf is continuous labial to the paracone; in species of *Pediomys*, it may be continuous labial to the paracone or entirely absent (Clemens, 1966). Stylar cusp C in ***A. incus*** is somewhat larger than either of the B cusps; in species of *Pediomys* stylar cusp C may be enlarged (as in *P. elegans* Marsh, 1889*a*), or small and only variably present (in *P. hatcheri* (Osborn), 1898) or absent altogether (*P. krejcii*) (Clemens, 1966; Lillegraven, 1969). In ***A. incus*** the paracone and metacone are more highly compressed labiolingually, more massive, more divergent from one another, and are higher than on molars of large species of *Pediomys*. Upper molars of ***A. incus*** also differ from those of large *Pediomys* by displaying a stronger anterolabial salient of the stylar shelf, wider anterior and posterior singula, a lower, broader protocone, and a shallower, broader protoconal basin.

The lower molars are similar to those of *Pediomys* other than in their large size and the probability that the crista obliqua meets the posterior wall of the protoconid more internally in ***Aquiladelphis***; wear makes determination of the exact position of the crista impossible.

Aquiladelphis minor sp. nov.

(Plate 6; Table 6)

Etymology. *minor*, L., smaller.

Holotype. 5539, an isolated, broken RM^2.

Horizon and type locality. Upper part of the Milk River Formation (Russell & Landes, 1940), exposed in Verdigris Coulee, approximately 18 miles east of the village of Milk River, at Loc. UA-MR-8, with co-ordinates on file in the Department of Geology, The University of Alberta.

Referred specimens. 5532, LM^2; 5553, fragmentary $RM^{2?}$; 5540, fragmentary RM^4; 5534 $LM_{2?}$; 5531, 5586, $RM_{3?}$. From Loc. UA-MR-4, 6, and 8.

Diagnosis. A species of ***Aquiladelphis*** smaller in dental dimensions than ***A. incus***. Upper molars with two stylar cusps B; stylar cusp C very large and the largest stylar cusp; stylar cusp D large, cusp E small.

Description. None of the upper molars available is complete, but in combination they probably reveal most of the upper molar morphology of the species. The lingual parts of the crowns resemble those of ***Aquiladelphis incus***, on a smaller scale. The paracone and metacone are low, labiolingually compressed, and divergent. The conules are strong, and their internal wings meet descending ridges on the paracone and metacone. The protocone is broad and its basin shallow. Proportions of the teeth are much like those of ***A. incus***, although the anteroposterior lengths of the crowns seem somewhat less in proportion to their widths than in the larger species. Two cusps, blade-like in shape, are present in the B position, as on molars of ***A. incus***; stylar cusp C in ***A. minor*** is large, tall, and conical. The lingual part of its base extends between and is closely adjacent to the labial third of the bases of the paracone and metacone, and the base of C extends outward to the labial margin of the stylar shelf. Posteriorly adjacent to cusp C is cusp D, also large; the cusp is nearly conical, but somewhat compressed labiolingually. At the posterolabial corner of the crown is a small but distinct stylar cusp E.

The lower molars assigned to ***A. minor*** can best be described by noting that they show close resemblance to lower molars of both *Alphadon* and *Pediomys*; however, the crista obliqua in ***A. minor*** meets the back wall of the protoconid, resembling its position in *Pediomys*.

Table 6. ***Aquiladelphis minor* sp. nov.,** Milk River Formation, Late Cretaceous, Alberta, Canada; dental dimensions

Specimen no.	L	W
5539 (M^2) type	—	3·9
5552 (M^2)	3·7	3·7
5531 (M_x)	4·3	3·6
5534 (M_x)	[3·6]	2·1

Discussion. Clear differences other than in size are evident from comparison of molars assigned to ***Aquiladelphis incus*** and ***A. minor***; when bettter preserved dentitions are available, discovery of additional differences may indicate that the species should be included in separate genera.

Comparison of the stylar areas on the upper molars of ***A. incus*** and ***A. minor*** at hand is hampered by the loss from breakage of the parastyle in teeth of ***A. minor***, and damage to or loss of the stylar shelf posterior to cusp C in specimens referred to ***A. incus***. Nevertheless, it seems clear that stylar cusp C in preultimate molars of ***A. minor*** is enormous relative to that in preultimate molars of ***A. incus,*** even with consideration for possible differences in sizes of the cusp owing to different positions of the teeth in the jaw.

Comparison of the lower molars is hampered by the poor preservation and extensive wear on molars of ***A. incus***. On these teeth, however, the paraconid is more labial and the talonid relatively more narrow transversely and longer anteroposteriorly than on molars of ***A. minor***, with the latter features presumably in correlation with the great anteroposterior dimensions of upper molars of ***A. incus***. These differences seem not great enough, however, to outweigh the observable resemblances in known teeth of the two species, and, although comparison is difficult, the differences among these teeth seem no greater than variation among dentitions of other marsupial species included, for example, within the single genera *Alphadon* and *Pediomys*.

The characters of the upper molars of ***Aquiladelphis***—most particularly the relatively great anteroposterior lengths and low, compressed, and divergent paracone and metacone—make ***A. incus*** and ***A. minor*** unlikely candidates for ancestors of Campanian and Maestrichtian species of *Pediomys*. More probably, species of ***Aquiladelphis*** were part of a lineage of pediomyids in which crushing specializations of the molar dentition were emphasized, to an unusual degree—in ***A. incus***, at least—among Cretaceous therians.

Family Stagodontidae Marsh, 1889*b*
Eodelphis Matthew, 1916
Eodelphis sp.
(Plate 6; Table 7)

Referred specimens. 5530, RM_x; 5562, LM_x (trigonid only). From Loc. UA-MR-6 and 8 only.

Description. In the collections of the mammalian teeth from the Upper Milk River Formation are one lower molar and a broken trigonid, closely similar to lower molars of *Eodelphis*, hitherto known only from rocks of late Campanian age.

Table 7. *Eodelphis* sp., Milk River Formation, Late Cretaceous, Alberta, Canada; dental dimensions

Specimen no.	L	W
5530 (M_x)	4·4	2·5

As in geologically younger stagodontids (*Eodelphis*, *Boreodon?*, *Didelphodon*), the trigonid in the Milk River specimens is short anteroposteriorly and the metaconid is the smallest and lowest trigonid cusp. The paraconid is smaller and lower than the protoconid and is compressed into a blade-like structure. The talonid is wider than the trigonid, and carries 'twinned' entoconid and hypoconulid cusps. A well-defined cingulum obliquely crosses the anterior face of the trigonid; a wider, posterior cingulum extends from the hypoconulid to the labial side of the base of the crown, beneath the hypoconid.

The new teeth differ from those of the Maestrichtian stagodontid *Didelphodon* Marsh, 1889, in their smaller size, less compressed and proportionally higher trigonid cusps, and weaker anterior and posterior cingula.

Lesser differences are seen in comparison with the lower molars of *Eodelphis* (=*Boreodon ?*) *cutleri* Woodward, 1916. The specimens from the Milk River Formation are slightly smaller, with less bulbous walls, more trenchant cusps, more narrow anterior cingula, and proportionally higher trigonids. Compression of the trigonids anteroposteriorly and relative heights of the trigonid cusps are approximately the same in both.

A close resemblance, including in size, is seen in comparison with the lower molars of *Eodelphis browni* Matthew, 1916. The Milk River teeth have slightly higher trigonids, taller protoconids, and wider talonids. The observed differences are small enough to indicate that the Milk River molars pertain to a species of *Eodelphis;* however, the identity of that species must await collection of larger samples, more adequate for comparison.

CONCLUSIONS

Preliminary knowledge of the mammals of the Upper Milk River Formation, Alberta, indicates that by middle Late Cretaceous time, a minimum of seven species of didelphoid marsupials were extant in North America. In comparison to this number, at least 11 species of didelphoids are known from late Campanian sites (Sahni, 1968; Fox, in progress), and 13 species have been recorded from Maestrichtian horizons (Sloan & Van Valen, 1965; Clemens, 1966; Lillegraven, 1969, for example) in the Western Interior. The number of known species of didelphoids nearly doubles, therefore, in the interval early Campanian–late Maestrichtian.

One might be tempted to ascribe this increase in species to a dramatic evolutionary radiation of didelphoids during the final stages of the Cretaceous. A rapid and, hence, similar diversification beginning in the late Campanian has been suggested for North American Late Cretaceous eutherian and multituberculate mammals (Van Valen & Sloan, 1966; Lillegraven, 1969). It might be supposed that those broad forces of natural selection that caused an evolutionary 'explosion' in the non-marsupial mammals, would have also affected marsupial species contemporary with them, and in the same manner. However, the reality of a marked latest Cretaceous radiation of multituberculates is dubious, and has been questioned elsewhere (Fox, 1968, 1971), and the nature of the apparent diversification affecting Late Cretaceous eutherians is not entirely clear; the former seems wholly or nearly wholly a function of limited sampling of multituberculates from the older horizons, while the latter may be owing to the combined effects, in as yet undertermined proportions, of limited sampling from the older horizons, local facies control, and migration from East Asia near the close of the period, with subsequent speciation (Lillegraven, 1969). At the present state of knowledge, it seems most informative to ascribe the relatively small numbers of marsupial species from early Campanian horizons solely to sampling error; the Milk River marsupials are known only from three sites within an area of about three square miles in southern Alberta. In contrast, late Campanian didelphoids are known from numerous sites in Wyoming, Montana, and Alberta (nearly 20 localities have yielded didelphoid teeth in Alberta), and Maestrichtian species have been described from a comparable number of localities in Alberta, Montana, Wyoming, Utah, and New Mexico. Plainly, the difference in number of marsupial species between the early

Campanian and later Cretaceous horizons must reflect to a significant degree nothing more important than numbers of samples taken from exposures available at each stratigraphic level.

Three genera of didelphoids (*Alphadon*, *Pediomys*, *Eodelphis*) have been described from late Campanian localities, and four (*Alphadon*, *Glasbius*, *Pediomys*, *Didelphodon*) are known to occur in the Maestrichtian. The greater generic diversity (*Alphadon*, ***Albertatherium***, *Pediomys*, ***Aquiladelphis***, *Eodelphis*) is unexpected, particularly on account of the few samples collected, and it seems highly probable that the number of genera will increase as continued collecting reveals the rarer elements of the fauna. The five Milk River genera already known, and the fact that they can be readily included within the already known families of Late Cretaceous marsupials (Didelphidae, Pediomyidae, Stagodontidae) indicate that the major evolutionary radiation that resulted in these families was a pre-Campanian event; further, it would appear probable, although by no means certain, that this radiation was post-Albian in occurrence (Albian beds in Texas having yielded the oldest known marsupial, *Holoclemensia*, a didelphid) and North American *in locale* (see Clemens, 1968).

A striking aspect of the Milk River marsupials is that none of the species described from upper molars in this paper is a likely candidate for the ancestry of the known late Campanian and Maestrichtian didelphoids. They are barred from this role either by the configuration of their stylar shelves (***Alphadon creber, Albertatherium primus***, ***Pediomys exiguus***) or by other molar features (***Aquiladelphis***) that seem to represent trends in dentition directed away from those seen in later Cretaceous species. Of the the Milk River didelphoids, only *Alphadon* sp. and *Eodelphis* sp. seem likely to include descendants among late Campanian assemblages, but both are known only from exceedingly fragmentary lower dentitions and the evidence that these dentitions provide is not sufficient to demonstrate ancestral-descendant lineages to particular late Campanian species. Whether the taxonomic discontinuities between the Milk River marsupials and those that occur at younger horizons in the North American Late Cretaceous are owing to extinction and speciation, preservation of faunal elements from different habitats, or migration is, of course, not known, nor is it likely that future sampling and study will permit a positive choice among these alternatives or determine the relative effect of each, if all, in fact, combined to cause the faunal differences observed. Nevertheless, work in progress suggests that the marsupial assemblages from the late Campanian Oldman Formation, Alberta, exhibit closer taxonomic relationships to those of standard facies from Maestrichtian sites in Alberta, Montana, Wyoming, and New Mexico than to the Milk River local fauna, and that the Oldman species are mostly without antecedents among the Milk River forms. It is hoped that continued sampling from Milk River sites will show whether these differences are a simple function of sampling error, or whether other factors—migration or preservation of different facies at the two horizons—can best account for the differences observed.

ACKNOWLEDGEMENTS

The help of others has been essential both to the initiation and completion of this project. The credit for development and success of the field programme for collection of the Milk River mammals largely belong to L. A. Lindoe, and he is responsible as well for the stereo-photographs in this paper, using, in part, equipment and techniques of his own design. D. Krause, L. Krishtalka, D. Schowalter, and M. Skinner provided willing assistance in the field and/or laboratory, in the collection, screening, and sorting of fossiliferous matrix. K. Forsen, Warden, Writing-on-Stone Provincial Park, was generous in his aid to the 1968 collecting party, and I thank W. A. Clemens for his continued encouragement of my study of the Cretaceous mammals of Alberta. Financial assistance has been provided by the National Research Council of Canada, the Geological Survey of Canada, and the General Research Fund, and Department of Zoology, The University of Alberta.

REFERENCES

CLEMENS, W. A., JR., 1965. Collecting Late Cretaceous mammals in Alberta. *15th Ann. Field Conf. Guidebook of Alberta Soc. Petroleum Geologists:* 137–141.

CLEMENS, W. A., JR., 1966. Fossil mammals of the type Lance Formation, Wyoming: Part. II. Marsupialia. *Univ. Calif. Publs. geol. Sci.*, **66**: 1–122.

CLEMENS, W. A. JR., 1968. Origin and early evolution of marsupials. *Evolution, Lancaster, Pa.*, **22**: 1–18.

FOX, R. C., 1968. Studies of Late Cretaceous vertebrates. II. Generic diversity among multituberculates. *Syst. Zool.*, **17**: 339–342.

FOX, R. C., 1968. Early Campanian (Late Cretaceous) mammals from Alberta, Canada. *Nature, Lond.*, **220**: 1046.

FOX, R. C., 1969. Studies of Late Cretaceous vertebrates. III. A triconodont mammal from Alberta. *Can. J. Zool.*, **47**: 1253–1256.

FOX, R. C., 1970. Eutherian mammal from the early Campanian (Late Cretaceous) of Alberta, Canada. *Nature, Lond.*, **227**: 630–631.

FOX, R. C., 1971. Early Campanian multituberculates (Mammalia: Allotheria) from the Milk River Formation, Alberta. *Can. J. Earth Sci.*, in press.

LILLEGRAVEN, J. A., 1969. Latest Cretaceous mammals of upper part of Edmonton Formation of Alberta, Canada, and review of marsupial-placental dichotomy in mammalian evolution. *Paleont. Contr., Univ. Kans.*, **50**: 1–122.

McKENNA, M. C., 1965. Collecting microvertebrate fossils by washing and screening. In B. Kummel & D. Raup (Eds), *Handbook of paleontological techniques*. W. H. Freeman: San Francisco.

MARSH, O. C., 1889*a*. Discovery of Cretaceous Mammalia. Part I. *Am. J. Sci.*, **38**: (3) 81–92.

MARSH, O. C., 1889*b*. Discovery of Cretaceous Mammalia. Part II. *Am. J. Sci.*, **38**: (3) 177–180.

MATTHEW, W. D., 1916. A marsupial from the Belly River Cretaceous. With critical observations upon the affinities of the Cretaceous mammals. *Bull. Am. Mus. nat. Hist.*, **35**: 477–500.

OSBORN, H. F., 1898. Evolution of the *Amblypoda*. Part 1. *Taligrada* and *Pantolambda*. *Bull. Am. Mus. nat. Hist.*, **10**: 169–218.

PATTERSON, B., 1956. Early Cretaceous mammals and the evolution of mammalian molar teeth. *Fieldiana, Geol.*, **13**: 1–105.

RUSSELL, L. S., 1964. Cretaceous non-marine faunas of northwestern North America. *Contrib. R. Ont. Mus. Geol.*, **61**: 1–24.

RUSSELL, L. S. & LANDES, R. W., 1940. Geology of the southern Alberta plains. *Mem. geol. Surv. Brch. Can.*, **221**: 1–223.

SAHNI, A., 1968. The vertebrate fauna of the Judith River Formation, Montana, 241 pp. Unpubl. Ph.D. dissertation, Univ. Minnesota.

SIMPSON, G. G., 1927. Mesozoic Mammalia. VIII. Genera of Lance mammals other than multituberculates. *Am. J. Sci.*, **14** (5): 121–130.

SLAUGHTER, B. H., 1965. A therian from the Lower Cretaceous (Albian) of Texas. *Postilla*, **93**: 1–18.

SLAUGHTER, B. H., 1968*a*. Earliest known eutherian mammals and the evolution of premolar occlusion. *Tex. J. Sci.*, **20**: 1–12.

SLAUGHTER, B. H., 1968*b*. Earliest known marsupials. *Science, N.Y.*, **162**: 254–255.

SLAUGHTER, B. H., 1968*c*. *Holoclemensia* instead of *Clemensia*. *Science, N.Y.*, **162**: 1306.

SLOAN, R. E. & VAN VALEN, L., 1965. Cretaceous mammals from Montana. *Science, N.Y.*, **148**: 220–227.

VAN VALEN, L. & SLOAN, R. E., 1966. The extinction of the multituberculates. *Syst. Zool.*, **15**: 261–278.

WILLIAMS, G. D. & BURK, C. F., JR., 1964. Upper Cretaceous. In R. McCrossan & R. Glaister (Eds), *Geological history of Western Canada*. Calgary: Alberta Soc. Petroleum Geologists.
WOODWARD, A. S., 1916. On a mammalian mandible (*Cimolestes cutleri*) from an Upper Cretaceous formation in Alberta, Canada. *Proc. zool. Soc. Lond.*, **1916**: 525–528.

EXPLANATION OF PLATES

PLATE 1

Molars of early Campanian and late Maestrichtian species of *Alphadon*, Alberta, Canada.

A. *Alphadon wilsoni* Lillegraven, 3740, RM^2, occlusal view, Loc. KUA 1, Upper Edmonton Formation. *c.* ×8·5.
B. ***Alphadon creber* sp. nov.,** 5533, LM^1, occlusal view, Loc. UA-MR-6, Milk River Formation. *c.* ×9·4.
C. ***Alphadon creber* sp. nov.,** 5541, RM^2, occlusal view, Loc. UA-MR-6, Milk River Formation. *c.* ×8·5.
D. ***Alphadon creber* sp. nov.,** 5535, LM_x, occlusal view, Loc. UA-MR-4, Milk River Formation. *c.* ×8·3.
E. *Alphadon* sp., 5527, LM_2, occlusal view, Loc. UA-MR-6, Milk River Formation. *c.* ×8·5.

PLATE 2

Upper third molars (M^3's) of early Campanian and late Maestrichtian species of *Alphadon*, Alberta, Canada.

A. ***Alphadon creber* sp. nov.,** 5542, LM^3, occlusal view, Loc. UA-MR-6, Milk River Formation. *c.* ×7·9.
B. ***Alphadon creber* sp. nov.,** 5545, type, RM^3, occlusal view, Loc. UA-MR-6, Milk River Formation. *c.* ×8·2.
C. ***Alphadon creber* sp. nov.,** 5543, LM^3, occlusal view, Loc. UA-MR-4, Milk River Formation. *c.* ×8·1.
D. ***Alphadon creber* sp. nov.,** 5529, LM^3, occlusal view, Loc. UA-MR-6, Milk River Formation. *c.* ×8·9.
E. ***Alphadon creber* sp. nov.,** 5544, LM^3, occlusal view, Loc. UA-MR-4, Milk River Formation. *c.* ×8·5.
F. *Alphadon wilsoni* Lillegraven, 2921, LM^3, occlusal view, Loc. KUA 22, Upper Edmonton Formation. *c.* ×8·5.

PLATE 3

Molars of ***Albertatherium primus* gen. et sp. nov.,** and ***Pediomys exiguus* sp. nov.,** early Campanian, Milk River Formation, Alberta, Canada.

A. ***Albertatherium primus* gen. et sp. nov.,** 5528, type, RM^3, occlusal view, Loc. UA-MR-6. *c.* ×8·2.
B. ***Albertatherium primus* gen. et sp. nov.,** 5526, LM_x, labial view, Loc. UA-MR-4. *c.* ×8·3.
C. ***Albertatherium primus* gen. et sp. nov.,** 5526, LM_x, occlusal view, Loc. UA-MR-4. *c.* ×8·3.
D. ***Pediomys exiguus* sp. nov.,** 5537, RM^2, occlusal view, Loc. UA-MR-6. *c.* ×8·3.
E. ***Pediomys exiguus* sp. nov.,** 5538, RM^1, occlusal view, Loc. UA-MR-6. *c.* ×8·3.
F. ***Pediomys exiguus* sp. nov.,** 5536, type, RM^1, occlusal view, Loc. UA-MR-4. *c.* ×8·1.

PLATE 4

Molars of ***Albertatherium incus* gen. et sp. nov.,** early Campanian, Milk River Formation, Alberta, Canada.

A. 5522, LM^3, type, occlusal view, Loc. UA-MR-8. *c.* ×8·2.
B. 5523, LM^4, occlusal view, Loc. UA-MR-8. *c.* ×8·3.
C. 5525, RM_x, occlusal view, Loc. UA-MR-6. *c.* ×8·6.

PLATE 5

***Aquiladelphis incus*, gen. et sp. nov.,** LP_x, 5524, Loc. UA-MR-4, early Campanian, Milk River Formation, Alberta, Canada.

A. Lingual view. *c.* ×8·5. **B.** Labial view. *c.* ×8·5.

PLATE 6

Molars of ***Aquiladelphis minor* sp. nov.,** and *Eodelphis* sp., early Campanian, Milk River Formation, Alberta, Canada.

A. ***Aquiladelphis minor* sp. nov.,** 5539, RM^2, type, occlusal view, Loc. UA-MR-8. *c.* ×7·9.
B. ***Aquiladelphis minor* sp. nov.,** 5534, $LM_{2?}$, occlusal view, Loc. UA-MR-8. *c.* ×8·1.
C. ***Aquiladelphis minor* sp. nov.,** 5531, $RM_{3?}$, occlusal view, Loc. UA-MR-6. ×8·4.
D. *Eodelphis* sp., 5530, RM_x, lingual view, Loc. UA-MR-6. *c.* ×8·0.
E. *Eodelphis* sp., 5530, RM_x, occlusal view, Loc. UA-MR-6. *c.* ×8·0.

PLATE 1

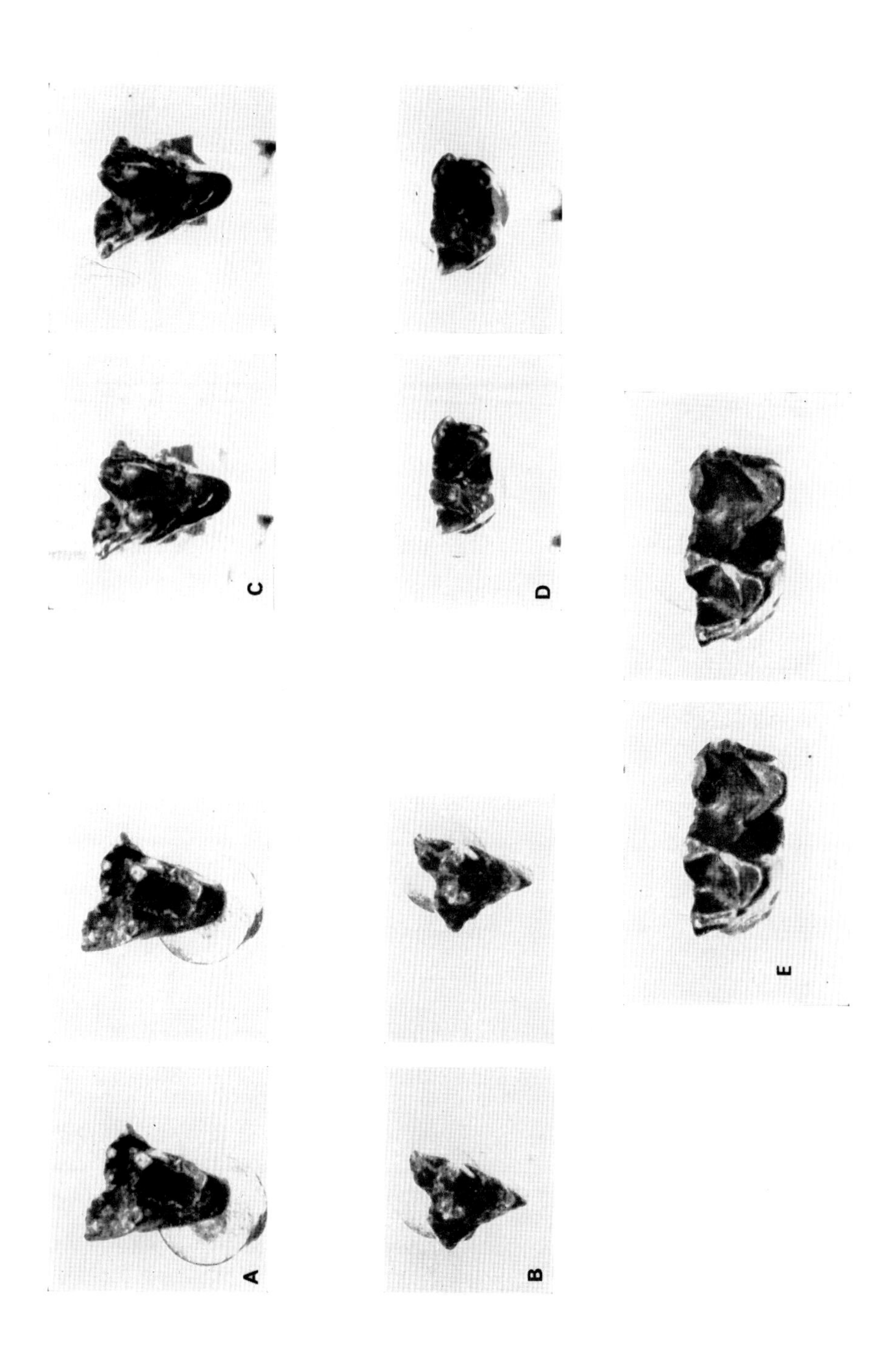

(*Facing p.* 164)

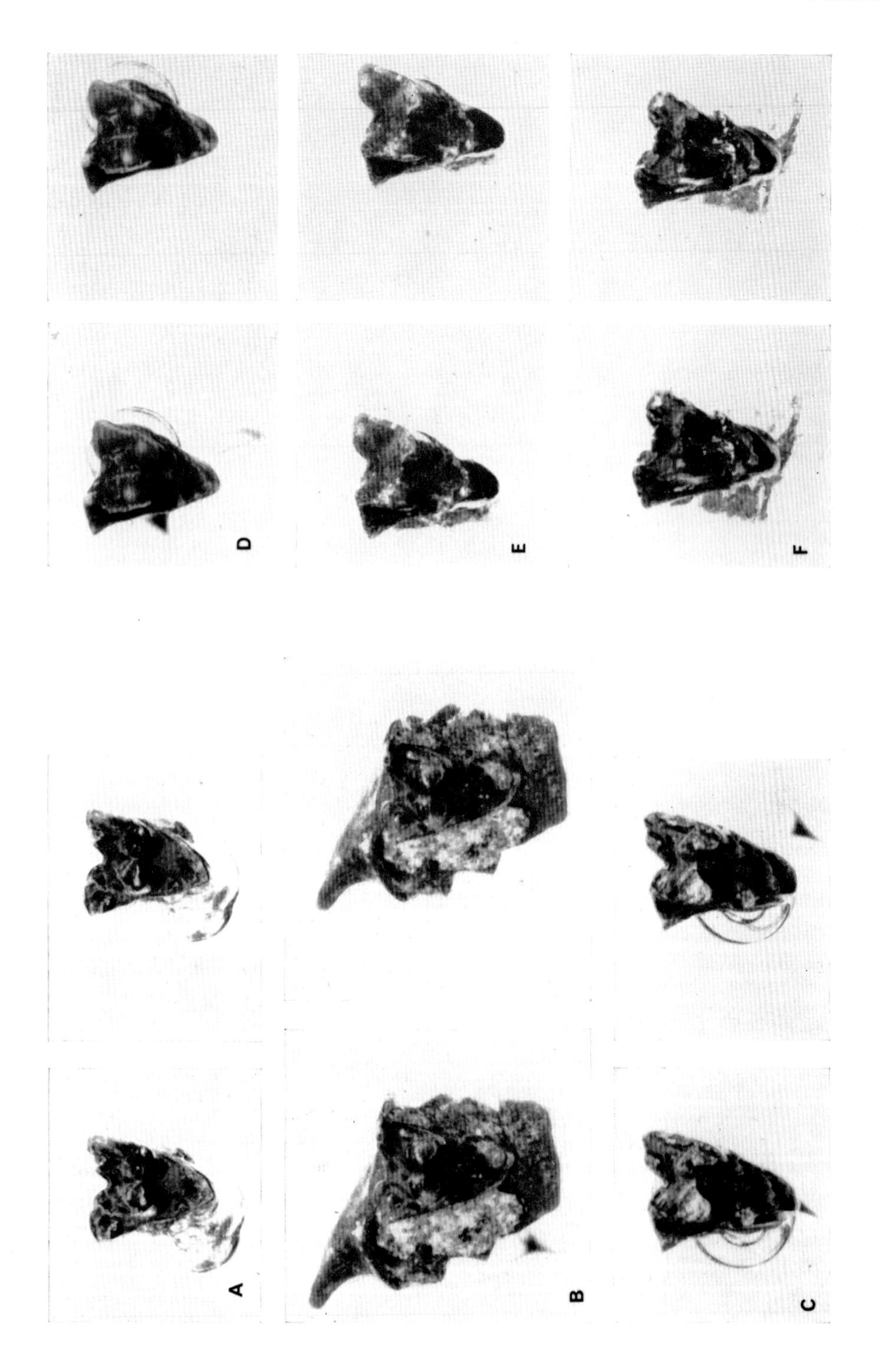
D
E
F
A
B
C

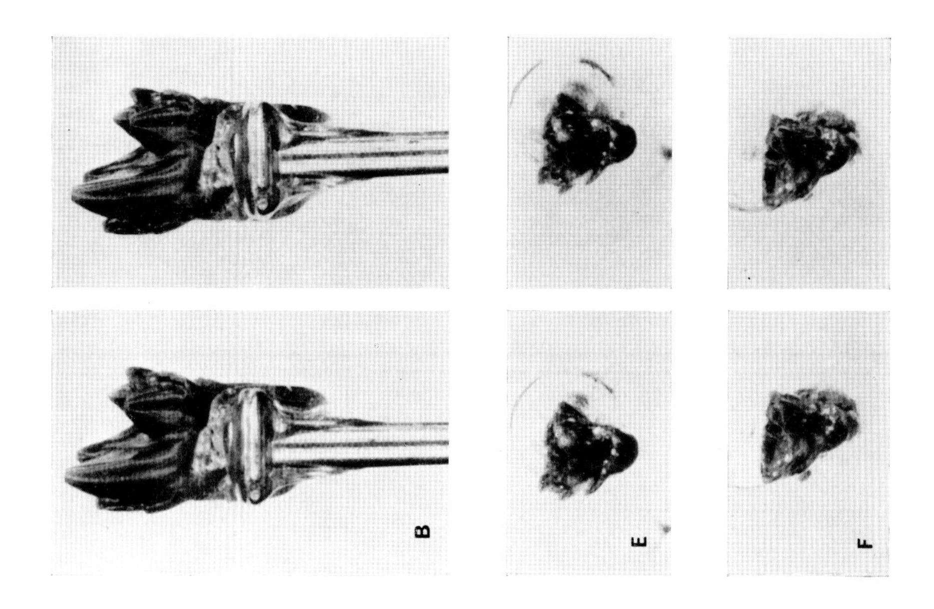

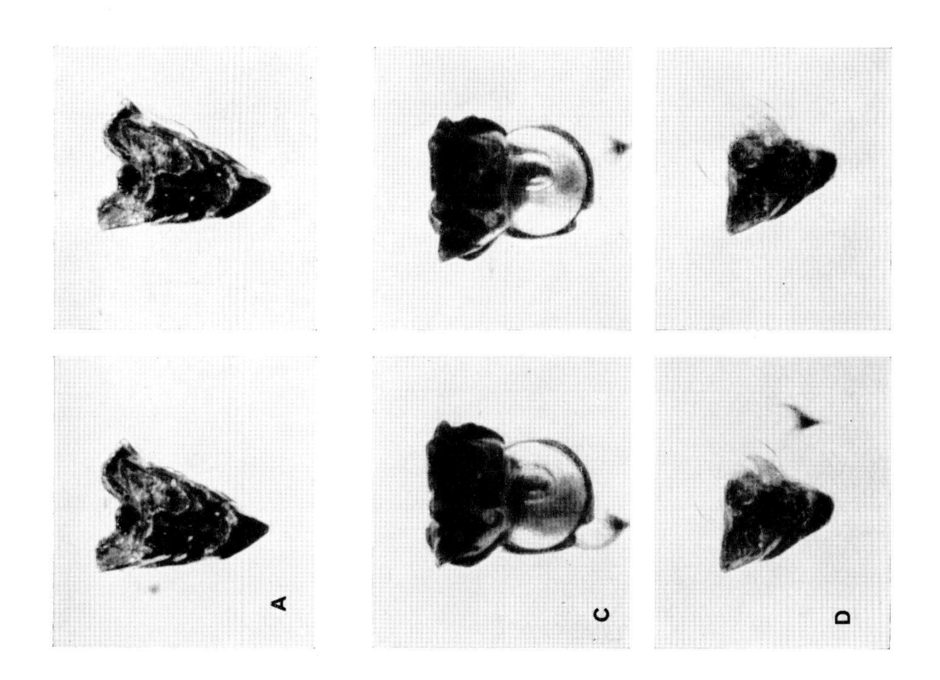

R. C. FOX

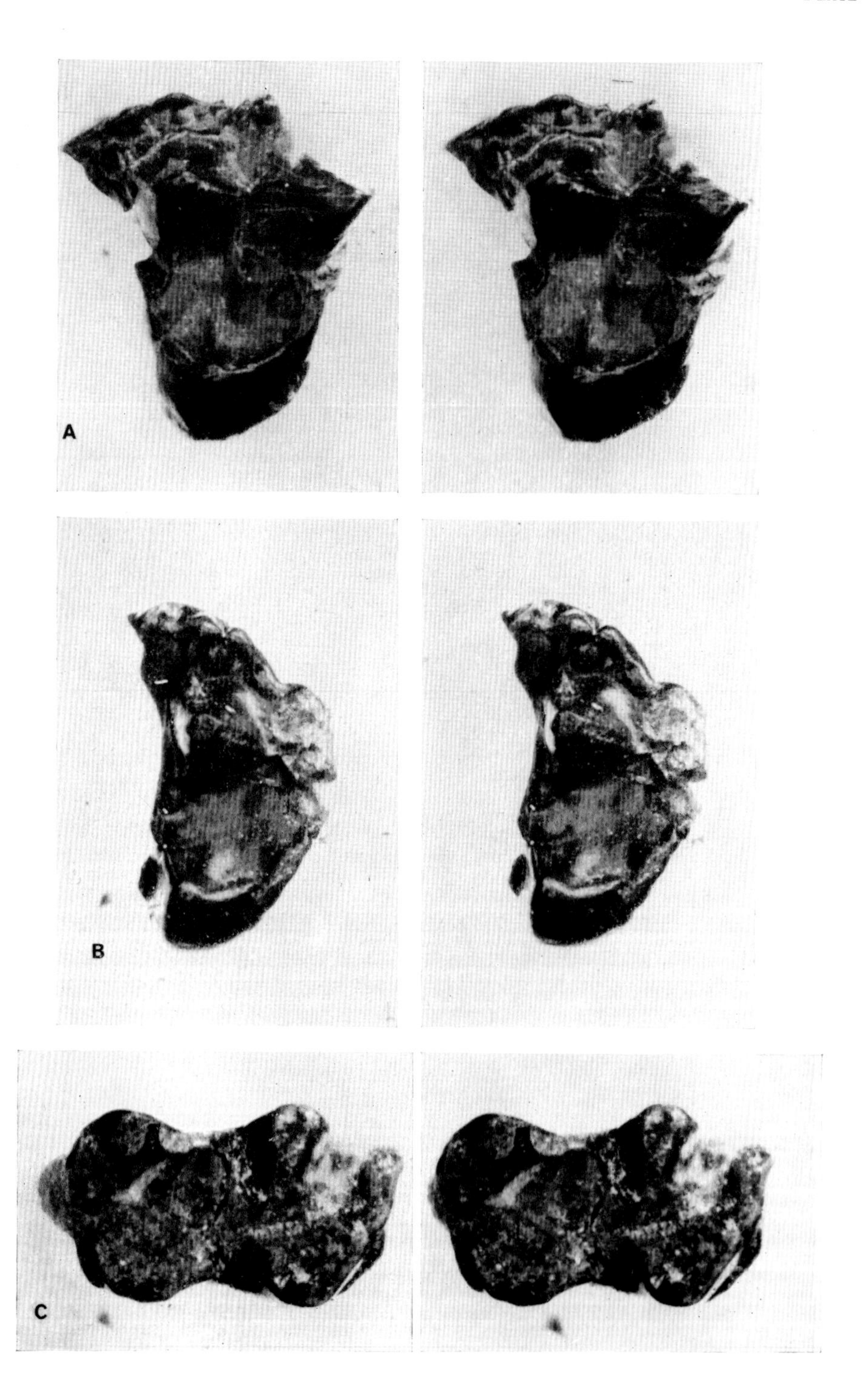

R. C. FOX

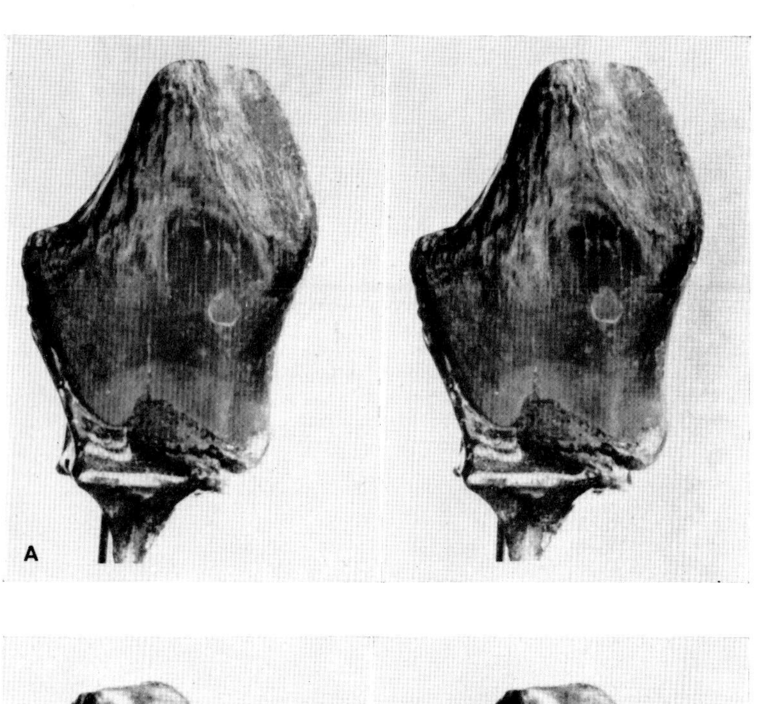

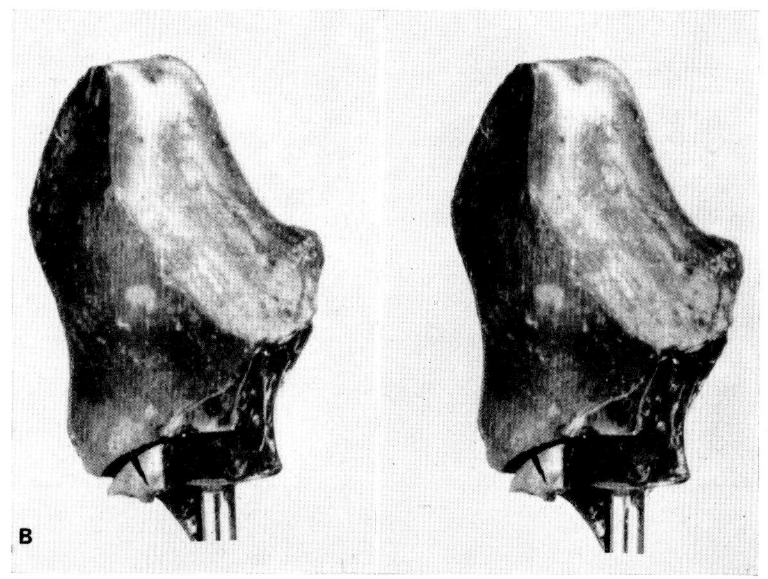

R. C. FOX

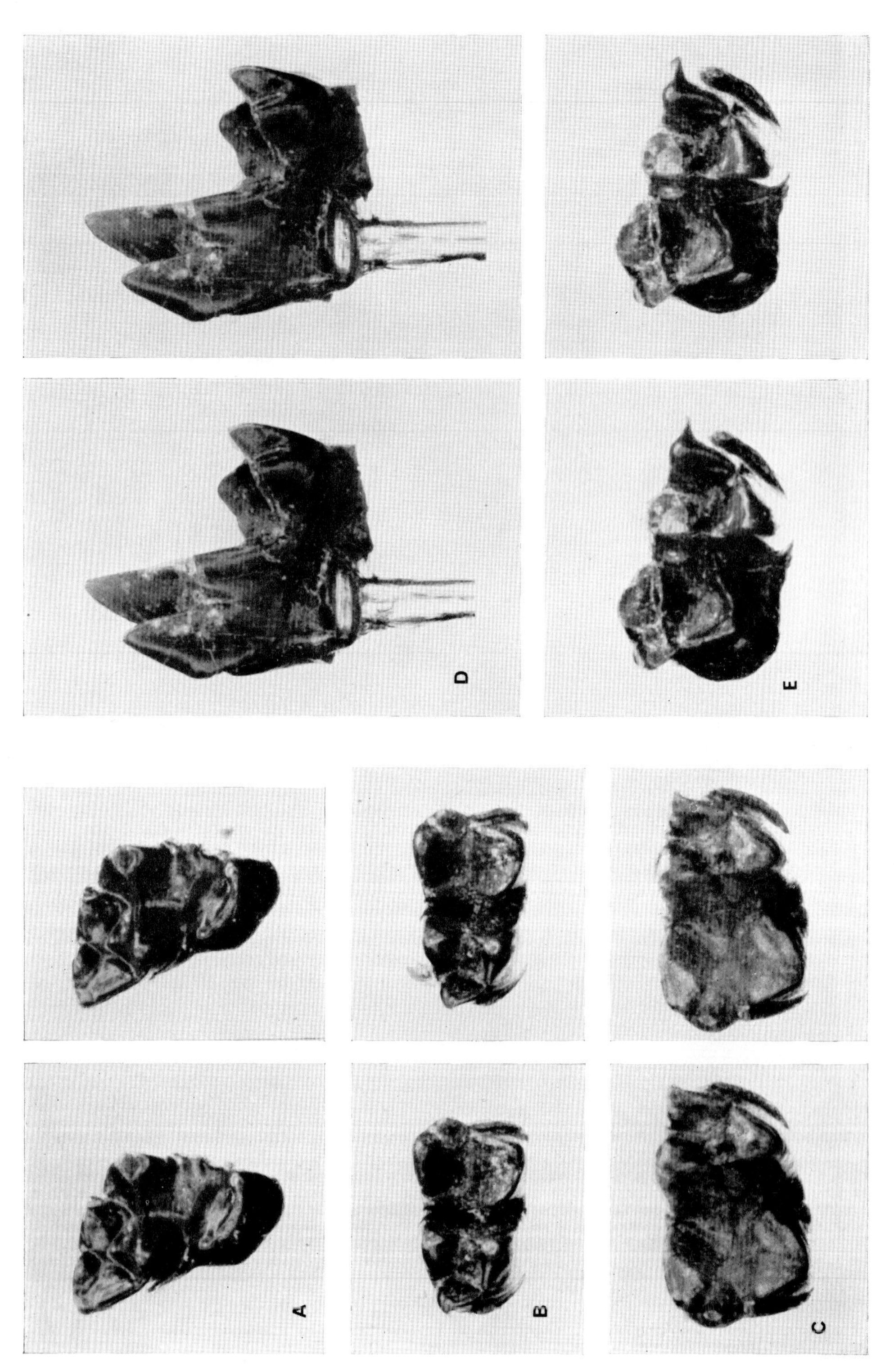

R. C. FOX

Mammalian evolution in the Cretaceous

WILLIAM A. CLEMENS, JR.

Department of Paleontology, University of California, Berkeley, California, U.S.A.

Recent discoveries have made significant additions to the mammalian fossil record of the Early Cretaceous and greatly expanded knowledge of Late Cretaceous mammals. This evolutionary record must be interpreted against a background of fragmentation of terrestrial areas through both continental drift and encroachment of epicontinental seas after their Late Jurassic regression.

During the Early Cretaceous and early Late Cretaceous, significant changes in the mammalian fauna occurred. Some lineages, for example symmetrodonts, became extinct. An adaptive radiation of the herbivorous multituberculates was initiated. Marsupials and placentals differentiated from a eupantotherian stock. These events appear to be correlated with the poleward spread and diversification of angiosperms.

Differences in composition of the Late Cretaceous and Early Tertiary mammalian faunas of Mongolia and North America suggest strong regional differentiation took place during the last half of the Cretaceous. Adaptive radiations of multituberculates and placentals, which characterize Early Tertiary mammalian evolution in the Northern Hemisphere, began in the Late Cretaceous. At the end of the Cretaceous the North American mammalian fauna was modified by extinctions that decimated the marsupials, but had little or no effect on the multituberculates and placentals.

CONTENTS

INTRODUCTION

The Cretaceous is a period of earth history slightly longer than the Cenozoic, beginning approximately 135 million years before the Present and ending about 64 million years ago. It was a time of great change in the terrestrial biota. Earliest Cretaceous floras of low land areas, at least in the Northern Hemisphere, lacked angiosperms but by the close of the period flowering plants were the dominant members of the floras even at high northern latitudes. Reflecting the rise of the angiosperms a pattern of marked evolutionary radiation with the extinction of some older lineages during the Middle Cretaceous characterizes the history of most groups of terrestrial vertebrates. For example, the Early Cretaceous dinosaurs resemble those of the much better known Late Jurassic faunas in which sauropods and camptosaurs were prominent. The dinosaurian faunas that became extinct at the close of the Cretaceous had a quite different composition primarily as a result of evolutionary radiations of the herbivorous ornithischians.

In the last decade knowledge of mammalian evolution during the Cretaceous has increased greatly. In some part this reflects restudy of older collections with improved optical equipment and techniques of preparation. However, contributions from two other areas probably are of greater importance. An increasing spectrum of studies going beyond comparative morphological studies of teeth, the most abundant type of skeletal element in the fossil record, is being focused on problems of Mesozoic mammalian evolution. Important contributions are being made through investigation of the microstructure of dental materials (e.g., Moss & Kermack, 1967) and the functions of the dentition (e.g., Mills, 1964, 1967; Crompton & Hiiemae, 1970). Other parts of the skull and postcranial skeleton of Cretaceous mammals are rare but do provide some valuable information (e.g., Kermack, 1967). Also data from a variety of comparative studies of their living descendants is being incorporated in interpretations of the phylogenetic relationships of Cretaceous mammals (e.g., Butler, 1956; Lillegraven, 1969). Secondly, collections of Cretaceous mammals have been greatly increased through further field work abetted by the underwater screening (McKenna, 1965) and flotation (Lees, 1964; Henkel, 1966) techniques that permit processing of large quantities of rock with an initial concentration of microfossils on the basis of their size and density. These techniques have made it possible to collect at sites where mammalian fossils occur at very low concentrations in the order of one minute tooth in every fifty or more pounds of rock.

Although relatively greatly increased in recent years, the fossil record of Cretaceous mammalian evolution is still plagued by a number of large geographic and stratigraphic gaps. The Early Cretaceous record is disappointingly small for this appears to have been the time of initiation of profound evolutionary changes. During the Neocomian large areas in western France and south-eastern England were covered by shallow, fresh to brackish water. Remains of terrestrial vertebrates including a few mammals (Clemens, 1963*b*; Clemens & Lees, 1971) have been found in bone beds in England that were formed along the strands of this lake. Approximately contemporaneous mammals were recently discovered in Spain (Kühne & Crusafont-Pairó, 1968; Henkel & Krebs, 1969). The Asian record is limited to the remains of two mammals found in what was Manchuria in strata now thought to be of Early Cretaceous age, possibly as recent as Albian (Kermack, Lees & Mussett, 1965). The North American record is restricted to collections from the Trinity Sand and other formations of Albian age in Texas (Patterson, 1956; Slaughter, 1965, 1968). Although there is some reason to suspect that mammals were present in the Southern Hemisphere during the Early Cretaceous, none have as yet been found. Thus studies of mammalian evolution during the Early Cretaceous must use the much better known Late Jurassic mammalian faunas as a base, draw upon the small sample of European mammals for information about evolution during the Neocomian, and then shift to North America for documentation of an Albian mammalian fauna. Because of the uncertainty concerning its age the Asian record is vexing for one of the mammals, *Endotherium*, has a tribosphenic dentition, which is the type of dentition thought to have evolved in the most recent common ancestors of marsupials and placentals. If this mammal is of

Late Jurassic age, as originally proposed (Shikama, 1947), and not Early Cretaceous the chronology of therian evolution currently accepted will have to be significantly modified.

Almost all Late Cretaceous mammalian fossils come from deposits in North America and Asia. An isolated tooth from a mammal of Campanian age found in France (Ledoux, Hartenberger, Michaux, Sudre & Thaler, 1966), which might be a lower molar of a didelphodontine palaeoryctid (*fide* McKenna, 1970), is the only European record. In the Southern Hemisphere the only mammalian fossils possibly of Late Cretaceous age are from a site in Peru (Grambast, Martinez, Mattauer & Thaler, 1967; Sigé, 1968). The problems of correlation of these mammals with the sequence of Northern Hemisphere faunas are considerable and the stage of evolution of the only mammal yet described, *Perutherium*, suggests a Tertiary age.

Discovery of fragmentary mammalian skulls and some postcranial material at Bayn Dzak (Djadokhta), Mongolia, in the later 1920's was of immediate importance for some of these fossils firmly established the existence of pre-Cenozoic placental mammals. Recently the collection has been augmented by fossils obtained by joint Polish-Mongolian expeditions (Kielan-Jaworowska & Dovchin, 1969). These Bayn Dzak mammals are of an early Late Cretaceous age; a Coniacian-Santonian correlation has been suggested (Kielan-Jaworowska, 1969*a*). Although only one mammalian fossil, a fragmentary maxillary of a multituberculate has been found at Khaichin Ula near Bugin Cav, Mongolia, the locality is of considerable importance for it is in younger strata possibly of latest Cretaceous age (Kielan-Jaworowska & Sochava, 1969). This discovery gives promise of resulting both in a sample of a fauna contemporaneous with well-known North American Late Cretaceous faunas and establishing a sequence of Mesozoic mammalian faunas in Mongolia.

Throughout most of the Late Cretaceous North America was bisected by a seaway, the Western Interior Sea, extending across what is now the Great Plains linking the Gulf of Mexico to the Arctic Ocean. The shores of the sea fluctuated greatly and parts of the present Great Plains and Rocky Mountains were intermittently innundated by expansion of this sea. A great volume of sediment derived from source areas in the region of the present Great Basin was carried eastward and the deposits formed from it now record the oscillations of the western coastline (Armstrong, 1968). Isolated teeth from strata of Cenomanian age in the Woodbine Formation of Texas establish the presence of mammals on the eastern North American landmass at the beginning of the Late Cretaceous (McNulty & Slaughter, 1968). Nothing more is known of the North American mammals until the Campanian. The thick section of deposits formed in the lowlands along the western margins of the Western Interior Sea contains many sites yielding the remains of mammals of Campanian and Maestrichtian age. The longest temporal sequence of faunas, ranging from the early Campanian to the end of the Maestrichtian has been found in Alberta (Fox, 1968; Lillegraven, 1969). Contemporaneous faunas are known from localities in the High Plains and Rocky Mountains of the United States as far south as New Mexico (Clemens, 1963*a*, 1966; Sloan & Van Valen, 1965) but in no area is there a sequence as long as that in Alberta.

As is the case with Early Cretaceous mammals, studies of mammalian evolution during the Late Cretaceous must draw information from different continents in order to synthesize a record for this part of the period. However, some of the geographic gaps in the Late Cretaceous record can be filled by inferences based on the much better known Early Tertiary faunas.

EARLY CRETACEOUS MAMMALS

In the Late Jurassic epicontinental seas were greatly restricted (Hallum, 1969). Some geological studies (e.g., Le Pichon, 1968; Allen, 1969; Dewey & Horsfield, 1970) indicate North America was in juxtaposition with Greenland and Europe at this time. South America and Africa also might have been in contact and not separated by an oceanic barrier. Initiation of crustal shifting resulting in formation of first the South Atlantic and later the North Atlantic is thought to be a Cretaceous event. Also it is pertinent to note that although Late Jurassic climates show some evidence of zonation (Vakhrameev, 1965) climatic extremes appear to have been much less severe than they are today.

Large collections of Late Jurassic mammalian fossils have been obtained from two localities in the Northern Hemisphere. One is from Purbeckian strata, part of the Lulworth Beds, in England; the other from approximately contemporaneous strata of the Morrison Formation cropping out near Como Bluff, Wyoming (Simpson, 1928, 1929). The Late Jurassic record from the Southern Hemisphere is limited to a single jaw found at Tendaguru, Tanzania (Dietrich, 1927) and foot prints attributed to mammals discovered in Santa Cruz province, Argentina (Casamiguela, 1961).

Comparisons of the mammals from Purbeck and Como Bluff and the reptiles from these sites and Tendaguru reveal a high degree of faunal similarity (Colbert, 1965).

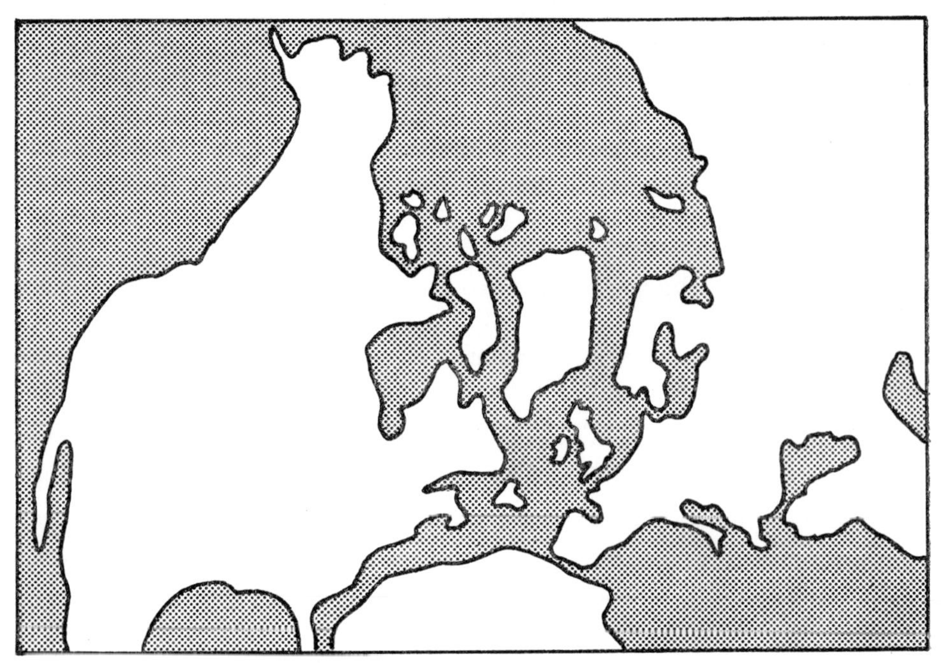

FIGURE 1. Reconstruction of Late Jurassic positions of continents in the Northern Hemisphere (from Hallam, 1969).

It seems reasonable to accept the hypothesis that at the close of the Jurassic, the Americas were in contact with Eurasia and Africa and no marine barrier prevented an east-west dispersal of terrestrial vertebrates. The similarities of Northern and Southern Hemisphere faunas suggest the Tethyian barrier was breached or circumvented in some fashion and climatic zonation did not pose a major obstacle to dispersal of terrestrial vertebrates between the hemispheres. The differences in composition between the Kimmeridgian Guimarota (Kühne, 1968) and Porto Pinheiro (Krusat, 1969) faunas could be an example of local, ecological differentiation. However the orders and major families of Late Jurassic mammals probably had ranges extending over many if not most of the continents.

The collections of Neocomian age from Europe indicate that the close of the Jurassic was not a time of major change in the terrestrial vertebrate fauna. The few mammals known from the English Wealden (Clemens & Lees, 1971) are for the most part closely related to those known in the Purbeckian fauna. Plagiaulacid and, possibly paulchoffatiid (Hahn, 1969, and pers. comm.) multituberculates are present. Specimens of triconodonts have not been discovered in the Weald, but their occurrence in younger North American deposits suggests the lack of a record in Europe at the beginning of the Cretaceous might only be an artifact of sampling. The only group of prototherian mammals known in the Late Jurassic but absent in the Early Cretaceous is the Docodonta. Among the major therian lineages the Symmetrodonta is represented in the Wealden fauna by two species of *Spalacotherium*. The dryolestid eupantotheres are present in Neocomian faunas of both England and Spain (Henkel & Krebs, 1969). The English dryolestid is a species of *Melanodon*, a genus hitherto known only in the Morrison fauna, indicating faunal interchange between Europe and North America continued after the end of the Jurassic. Paurodont eupantotheres have not been discovered at any Early Cretaceous localities but it has been suggested (Kermack, Lees & Mussett 1965) that the Wealden mammal *Aegialodon* is a descendant of *Peramus* which has been classified as a paurodont. *Peramus* has been restudied recently by the author and J. R. E. Mills. The results of our study are being published elsewhere but the major points are summarized below.

In contrast to other Late Jurassic therians, *Peramus*, which is known only from the Purbeckian, has a well differentiated postcanine dentition consisting of four molars and four premolars of which the distal premolar has a high crown towering over those of the adjacent teeth. Both M^1 and M_1 are distinctly different from M^2 and M_2 respectively. The lower molars have long talonids with those of M_2 and M_3 carrying three or four cusps forming the rim of a small talonid basin. The dentition of *Peramus* demonstrates that the evolution of the taolonid basin is not necessarily indicative of the presence of a protocone on the occluding upper molars. A basined talonid, albeit poorly delimited, was formed as a consequence of labial angulation of the principal talonid crest. This change appears to reflect selection favouring increase in length and efficiency of the talonid shearing crest and evolution of a cusp with the functions typical of a metacone of a tribosphenic molar.

Peramus probably is closely related to the ancestors of mammals with tribosphenic dentitions that appear in the Albian, but is not directly ancestral to them. The structure of the talonids of its posterior molars, for example, and other dental characters of

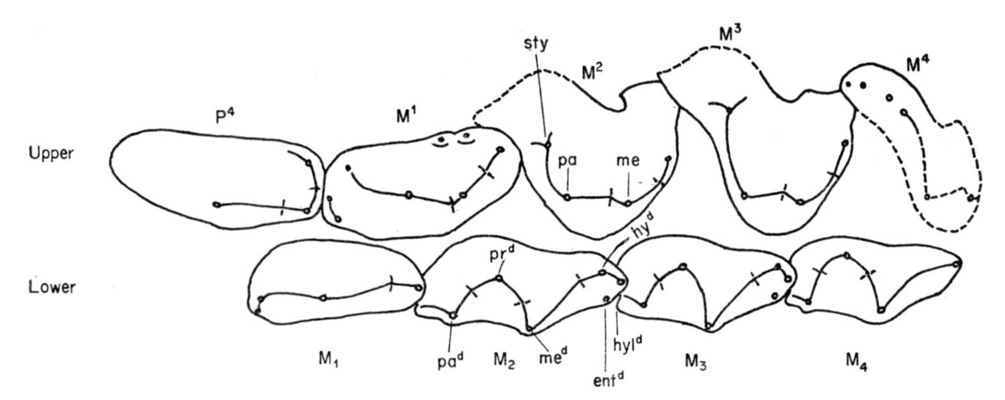

FIGURE 2. Schematic restoration of occlusal views of P^4 through M^4 and M_1 through M_4 of *Peramus tenuirostris*.
sty, Stylocone; pa, paracone; pa^d, paraconid; par^d, protoconid; me, metacone; me^d, metaconid; hy^d, hypoconid; hyl^d, hypoconulid; ent^d, entoconid.

Peramus are of the kinds to be expected in the ancestors of mammals with tribosphenic dentitions. However, great reduction of the stylocone and other features rule *Peramus* out of this lineage and support its allocation to a separate eupantotherian family, Peramuridae. If *Aegialodon* is representative of one of several lineages that reached the tribosphenic grade in dental evolution, *Peramus* could be its ancestor. Until the position in the dental arcade of the type of *Aegialodon* is known and an upper molar discovered, its phylogenetic relationships with *Peramus* and therians with fully tribosphenic dentitions cannot be determined.

In stratigraphic sequence the next sample of Early Cretaceous mammals following the Neocomian collections is from the Albian of Texas. As well as multituberculates, triconodonts, and symmetrodonts this fauna contains several kinds of therians with fully tribosphenic dentitions. The appearance of these mammals and evidence derived from other groups indicates that during the Early Cretaceous the terrestrial biota began to undergo profound changes. A trend toward increasingly warmer, milder climates continued through the Early Cretaceous but was interrupted in the Late Cretaceous (Cousminer, 1961; Lowenstam, 1964; Smiley, 1967; Axelrod & Bailey, 1968). Jurassic lowland floras were dominated by conifers, ferns, cycads, cycadeoides, and ginkgos. Until recently no unequivocal pre-Cretaceous fossil angiosperms had been discovered (Scott, Barghoorn, & Leopold, 1960, but note contrary views of Axelrod, 1961), and they are unknown in the Wealden flora and Neocomian floras of North America (Glaser, 1969). Now Tidwell, Rushforth, Reveal & Behunin, 1970, have described stems of palms from the Middle Jurassic Arapien Shale of Utah. Although the origin of angiosperms can be shown to predate the beginning of the Cretaceous the basic pattern of their evolution remains one of absence from lowland floras of middle and higher latitudes through the Late Jurassic and Neocomian. Through the remainder of the Early Cretaceous they underwent both an evolutionary diversification and a poleward expansion of their range so that by the Turonian floras at high northern latitudes were angiosperm dominated.

Because of the scarcity of samples of Middle Cretaceous terrestrial vertebrates, the course of their adaptation to the change in floral composition is not well documented, but the magnitude of the change can be assessed from comparisons of the better known Late Jurassic and Late Cretaceous faunas. Evolution of the various groups of herbivorous dinosaurs clearly reflects the floral change. Ornithopod dinosaurs (hypsilophodonts and iguanodonts) were moderately abundant in Late Jurassic and Early Cretaceous faunas. As a result of the radiation of hadrosaurians ornithopod abundance and diversity were much greater in Late Cretaceous faunas. The origin and radiation of ceratopsians also appears to reflect adaptations to new sources of food provided by the diversification and increase in geographic range of the angiosperms. This floral change could also have been a causal factor in the diversification of ankylosaurs, extinction of the stegosaurs and limitation of geographic range of the sauropods.

Most Cretaceous mammals, therians and prototherians, have dentitions in which shearing was the primary or only function suggesting they were small carnivores or omnivores. Only multituberculates have dentitions functionally capable of dealing with a herbivorous diet. They include a pair of enlarged, somewhat rodent-like incisors that in most species were suited for puncturing and grasping but not gnawing. The distal premolars were modified to form shearing blades, a specialization independently evolved in kangaroos and a diverse group of other marsupial and placental herbivores (Simpson, 1933). The crowns of the molars are made up of mesiodistally oriented rows of cusps forming a grinding mill capable of dealing with fibrous vegetation. Multituberculata is currently classified in three groups: the Plagiaulacoidea of Late Jurassic age that includes the ancestors of the Late Cretaceous and Early Tertiary Ptilodontoidea and Taeniolabidoidea. All Early Cretaceous multituberculates so far discovered are plagiaulacoids, but they illustrate the beginnings of modifications such as the addition of a third row of cusps to the upper molars that characterize their ptilodontoid and taeniolabidoid descendants. The magnitude of the adaptive radiations of the latter groups during the Late Cretaceous and early part of the Paleocene indicates they were well adapted to foraging in the angiosperm dominated forests (Van Valen & Sloan, 1966).

The Middle Cretaceous was also a time of major evolutionary change of those mammals interpreted as being small carnivores or omnivores. Of the mammalian lineages represented in the Late Jurassic, the symmetrodonts are last recorded in the Albian (Patterson, 1956) while the triconodonts survived until the Campanian (Fox, 1969). The last record of dryolestids is Neocomian, or possibly Aptian (Henkel & Krebs, 1969) and docodonts are unknown after the end of the Jurassic. By the Albian the mammalian fauna of North America contained at least two, *Pappotherium* and *Holoclemensia*, and probably more genera of therian mammals with fully tribosphenic dentitions. *Endotherium*, an Asian mammal with a tribosphenic dentition, might be of Aptian or Albian age (Kermack *et al.*, 1965). Although these mammals were not herbivores the origin and initial radiation of therians with tribosphenic dentitions, as well as the extinction of some of the Jurassic lineages, might indirectly reflect the floral change through the ecological link of the terrestrial invertebrate fauna.

Arachnids, coleopterans, dipterans, hymenopterans and other major groups of terrestrial invertebrates are known to have been in existence at the beginning of the Cretaceous, but the record of their evolution during the period is poor. The Early Tertiary terrestrial invertebrate fauna from European amber deposits contains the

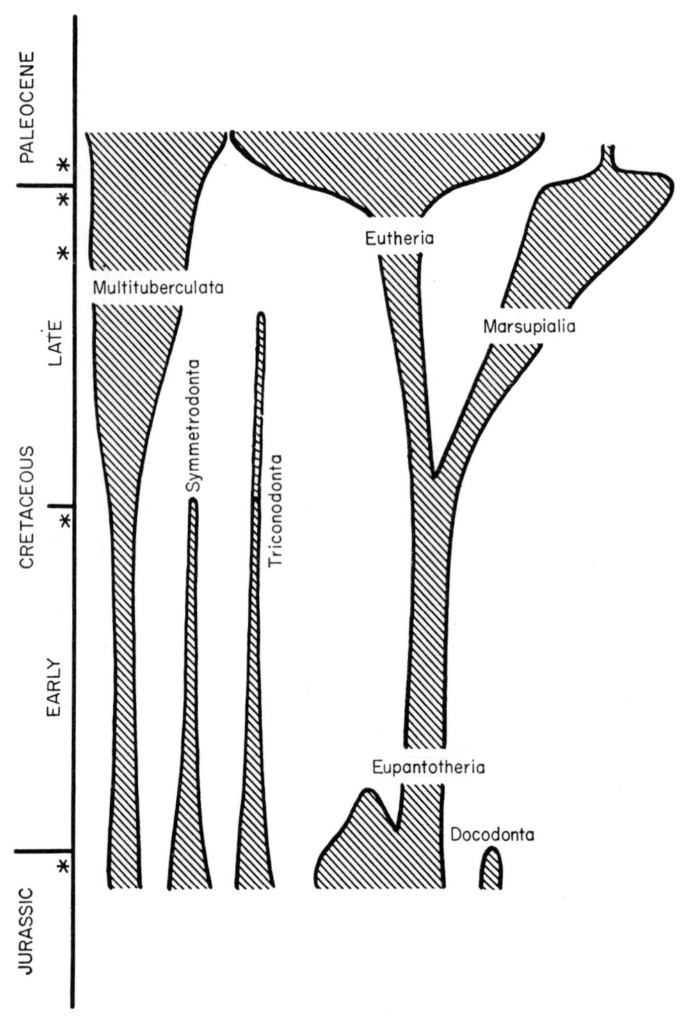

FIGURE 3. Stratigraphic record of North American Cretaceous mammals. Width of columns varied to suggest relative taxonomic diversity. Asterisks mark stratigraphic positions of major fossil localities.

first records of many families (Crowson *et al.*, 1967) indicating a major adaptive radiation took place in the Paleocene or Late Mesozoic. Several lines of evidence suggest this radiation was contemporaneous with and possibly a cause of the initial radiation of angiosperms (note Takhtajan, 1969). Many modern insects and other terrestrial invertebrates feed upon angiosperms and participate in their pollination and seed dispersal. Insects, at least, are not as closely linked to gymnosperms in Recent temperate habitats (Hughes & Smart, 1967). On the basis of ecological relationships of modern terrestrial invertebrates, a hypothesis that they underwent an adaptive radiation in conjunction with the initial radiation of angiosperms appears defensible. Secondly among the few fossil insects from the Late Cretaceous are the first records of the Lepidoptera (MacKay, 1970) and primitive myrmecioid ants not far from tiphiid wasps (Wilson, Carpenter & Brown, 1966); both members of groups now closely related ecologically to angiosperms. Finally, if there was an evolutionary radiation of insects and other terrestrial invertebrates in the Middle Cretaceous these organisms would have provided an expanding source of food for small vertebrate carnivores and omnivores. Hecht (1963) noted the coincidence of the evolutionary radiations of frogs and angiosperms in the Cretaceous and suggested this was related to the increase in abundance and diversity of terrestrial invertebrates.

Evolution of the tribosphenic dentition with its combination of shearing and crushing functions also appears to have been part of the mammalian adaptations to the new source of food made available by the radiation of terrestrial invertebrates. Correlation of the tribosphenic type of dentition with an insectivorous diet can be documented in modern mammals. Some larger insectivorous mammals, anteaters and pangolins for example, have greatly reduced dentitions and swallow their prey with little or no mastication. However, smaller insectivorous mammals that masticate their food, for example shrews, Australian marsupial 'mice', tupaiids, and hedgehogs, have dentitions only slightly modified from the basic tribosphenic pattern and still preserve both shearing and crushing functions. The time of origin of the tribosphenic dentition and the diets of modern mammals with this type of dentition suggest its evolution in the common ancestors of marsupials and placentals was related to the origin and diversification of angiosperm dominated floras and their invertebrate faunas.

At the beginning of the Cretaceous the major groups of mammals appear to have had wide geographic ranges. Continental drift and marine transgression during the Early Cretaceous resulted in subdivision or restriction of their ranges. The evolution of angiosperm dominated floras and their invertebrate faunas provided new food sources for many groups of vertebrates. Herbivorous dinosaurs and multituberculates underwent major adaptive radiations. Changes in the invertebrate faunas were linked with the evolution of the tribosphenic dentition and differentiation of placental mammals from a marsupial or marsupial-like ancestry (Lillegraven, 1969). By the middle of the Cretaceous mammalian faunas were a mixture of groups that were prominent members of Middle and Late Jurassic faunas—symmetrodonts, triconodonts, and plagiaulacoid multituberculates—and mammals that had reached the tribosphenic grade in dental evolution.

LATE CRETACEOUS MAMMALS

During the Late Cretaceous marine transgressions maintained or increased the fragmentation of continental area. Through drift the Americas were being separated from Eurasia and Africa (Le Pichon, 1968). The trend toward warmer, milder climates appears to have been temporarily reversed during the Late Cretaceous. Fragmentation of continental areas is reflected in the development of distinct endemism in Late Cretaceous faunas.

In the Early Cretaceous mammals with fully tribosphenic dentitions were present in Asia (*Endotherium*) and North America (*Pappotherium* and *Holoclemensia*). The phylogenetic relationships of these mammals to the marsupials and placentals that, with the multituberculates, made up the latest Cretaceous faunas of the Northern Hemisphere, are open to question. *Pappotherium* and *Holoclemensia*, studied by

Slaughter (1965, 1968), are known from isolated teeth and small dentulous skull and jaw fragments. Discovery of submolariform premolars, a modification unknown in Late Cretaceous or Cenozoic marsupials, and characters of individual teeth led him (Slaughter, 1968) to suggest that both eutherians, *Pappotherium pattersoni* and probably several other species, and didelphid marsupials, *Holoclemensia*, were present in the Albian Trinity fauna. The evidence certainly warrants this interpretation but does not exclude the possibility that when the tribosphenic grade of dental evolution was attained an adaptive radiation occurred producing lineages in addition to those directly ancestral to modern marsupials and placentals. Also Lillegraven (1969), studying the biogeography of Late Cretaceous mammals, concluded that marsupials were an endemic North American group but the placentals in the Late Cretaceous North American faunas were descendants of stocks that had immigrated from Asia. Because of the paucity of Middle Cretaceous samples the apparent differences between these interpretations cannot be resolved, but it does appear probable that differentiation of marsupials and placentals occurred prior to the Albian.

The evolution of marsupials during the Late Cretaceous was centered in North America. Marsupials are not part of the Mongolian, Late Cretaceous Bayn Dzak fauna (Kielan-Jaworowska, 1969*a*) and they are unknown in all Cenozoic Asian faunas. Only one mammalian fossil, a eutherian lower molar, is known from the Late Cretaceous of Europe (Ledoux *et al.*, 1966) and marsupials have yet to be reported in the well sampled European Late Paleocene faunas (Russell, 1964). *Peratherium*, a didelphid common in European faunas from Eocene to Miocene, was probably one of a host of Early Eocone (Sparnacian) immigrants from North America (note Kurten, 1966). Mammals of Cretaceous age have not been discovered in Africa. The oldest, Late Eocene, Cenozoic mammalian faunas of Africa do not contain marsupials. The endemism of the Early Cenozoic African faunas suggests marsupials did not reach this continent.

During the Late Cretaceous marsupials underwent a marked adaptive radiation in North America where they were abundant and diverse members of most local faunas (Clemens, 1968; Lillegraven, 1969). Several patterns of dental modification are evident in this radiation that were independently paralleled in the Cenozoic evolution of South American and Australian marsupials. For example, in most Late Cretaceous lineages the broad labial shelf or cingulum of the primitive tribosphenic molar was maintained, only in the pediomyids did it undergo reduction. This pattern of reduction of the shelf is also found in the evolution of both the Cenozoic South American marsupial carnivores, borhyaenids, and the ancestors of the Tasmanian wolf, *Thylacinus*.

Several hypotheses concerning the origin of South American and Australian marsupials have been proposed (Clemens, 1968). The differences in composition of their early Cenozoic faunas demonstrate that a strong barrier to dispersal of terrestrial vertebrates separated North and South America during the Late Cretaceous and Early Cenozoic. Studies of the geology of northern South America (Harrington, 1962; Jacobs, 1963; Lloyd, 1963) and composition of its Tertiary faunas (Patterson & Pascual, 1968) indicate the barrier was a marine gulf linking the Caribbean and Pacific that remained uninterrupted, though island dotted, through most of the Tertiary.

Mammalian dispersal was limited to fortuitous crossings presumably by rafting, the sweapstakes route of dispersal. The dental morphology of Early Tertiary South American marsupials suggests they could have been derived from a North American Late Cretaceous species of *Alphadon* or a closely related primitive didelphid (Clemens, 1966). Judging from the diversity of Late Paleocene (Riochican) South American marsupials, it appears most likely that they entered this continent in the Late Cretaceous. If the mammalian fossils found near Laguna Umayo in Peru (Sigé, 1968), which include otherwise undescribed teeth of small marsupials, are of Late Cretaceous age, the dispersal must have been a Mesozoic event.

Although the fossil record of the Australian fauna has been greatly increased in the last two decades the oldest mammalian fauna yet discovered is of Late Oligocene or Early Miocene age (Stirton, Tedford & Woodburne, 1968). The diversity of marsupials in this fauna—dasyurids, peramelids, phascolarctids, thylacoleonids, macropodids, and diprotodontids—suggests their ancestral stock reached Australia well before the beginning of the Oligocene. The time and route by which marsupials reached Australia have been matters of some speculation. Late Cretaceous North American and Paleocene South American faunas contain didelphid marsupials with the relatively primitive type of tribosphenic dentition to be expected in the ancestors of dasyurids, which probably were the ancestors of all other families of Australian marsupials. Marsupials are unknown in Asian and African faunas. On the basis of its dental morphology, *Peratherium*, which reached Europe in the Early Eocene, does not appear to have been ancestral to the dasyurids.

In a study of the early evolution of marsupials I (Clemens, 1968) supported the hypothesis that some time in the Late Cretaceous, or possibly Early Tertiary, *Alphadon* or a closely related didelphid dispersed through the Pacific coastal area of Asia to reach Australia by fortuitous rafting. The absence of marsupials from the Late Cretaceous Bayn Dzak and early Tertiary Mongolian faunas might only reflect their absence from the interior of the continent. Nothing is known of the Late Cretaceous faunas of the Pacific coast of Asia, but the discovery of two common North American perissodactyls, *Homogalax* and *Heptodon*, in Early Eocene deposits of coastal China (Chow & Li, 1965) suggests there was distinct differentiation between coastal and interior faunas at this time and possibly earlier. North American Late Cretaceous marsupials were members of coastal lowland faunas, and might have been ecologically restricted to this environment. If this was the case postulating that their dispersal into Asia was also limited to coastal lowlands and the immigrant marsupials became extinct without penetrating the interior of the continent seems reasonable. The hypothesis of dispersal to Australia via the Asian coastal lowlands is based on two assumptions concerning Late Mesozoic and Cenozoic geography. First, that the Pacific Ocean had been in existence since the beginning of the Jurassic and acted as a barrier to direct dispersal between Australia and the Americas. Also, that the spatial relationships of Australia and Asia have been essentially constant since the Late Triassic.

Some recent research on continental drift (note Le Pichon, 1968; Smith & Hallum, 1970) suggests the second of these assumptions might be erroneous. These studies indicate that until the end of the Eocene Australia, New Zealand, and New Guinea were much closer if not adjacent to eastern Antarctica. According to their reconstructions the wide seaway separating New Guinea and Australia from Asia in the Late Cretaceous and Early Tertiary would have been an effective barrier to the dispersal of terrestrial vertebrates. If this reconstruction is correct, it appears more likely that primitive didelphids reached Australia via a sweapstakes route of dispersal from South America to Antarctica in the Late Cretaceous or Early Tertiary. A choice between these two alternative hypotheses rests on the choice of Late Cretaceous geographic reconstructions. Both hypotheses pose biological problems. The first requires explanations of the presumed restriction of marsupials to Asian coastal areas and their extinction in the Early or Middle Tertiary. Even though Late Cretaceous and Early Tertiary climates were much warmer than the modern climate, the second hypothesis requires the adaptation of primitive marsupials to a polar daylight regime.

Martin (1970) has advanced a third interpretation, 'the Darwin Rise hypothesis'. He discounts the probability of selective dispersal of terrestrial mammals around the periphery of the Pacific Ocean by the sweapstake mechanism. Instead he assumes marsupials originated on a land mass centered over what is now the Darwin Rise in the mid-Pacific. Further it is assumed that this land mass fragmented; one unit carrying its marsupial passengers southward to Australia and the other moving northeastward to North America. Cox (1970) has evaluated the 'Darwin Rise hypothesis', found it wanting, and gone on to propose (p. 769), '. . . it is likely that marsupial mammals were in existence before the placentals, which may have evolved from the latter. It is at least possible that the marsupials were able to migrate throughout Africa, Antarctica, and Australia before the Antarctica-Australian land mass separated from Africa and that this in turn took place before the placentals reached Antarctica-Australia.' As Cox has pointed out the work of Lillegraven (1969) indicates that marsupials were in existence prior to placentals and may have dispersed through the southern continents prior to their fragmentation, which prevented equally wide dispersal of placentals. However, if this hypothesis is accepted the absence of marsupials in the Cenozoic faunas of Africa, Madagascar, and peninsular India must be explained.

A definite choice between these and other hypotheses cannot be made with certainty until more is known of the geography of the southern continents during the Mesozoic and the chronology of therian evolution. At the moment I think that one of the first two alternative hypotheses has the greatest chance of proving to be correct. Choice between these two depends upon choice of reconstruction of Cretaceous geography, in particular the position of Australia relative to Antarctica and Asia. If, as the second hypothesis postulates, Australia and Antarctica were in juxtaposition in the Cretaceous a dispersal route through South America to Antarctica with fortuitous exclusion of placentals could have been operative. The major difference between this and Cox's hypothesis is that it places the time of dispersal of marsupials after the initiation of continental drift in the Southern Hemisphere and suggests the route of dispersal did not traverse the African continent.

The oldest unequivocal records of placentals, based on skulls and partial skeletons, are from deposits in the region of Bayn Dzak, Mongolia. American and, more recently, Polish-Mongolian expeditions have recovered a large sample of these mammals (Kielan-Jaworowska & Dovchin, 1969). The isolation of the collecting localities from areas with well developed marine sections and the endemic nature of the fauna prevent accurate correlation with other Cretaceous faunas. Kielan-Jaworowska (1969*a*) has suggested a Coniacian or Santonian age while McKenna (1970) favoured a slightly older, Cenomanian correlation. Both agree the Bayn Dzak mammals are younger than the Albian mammals from Texas and older than the members of North American faunas usually assigned a Maestrichtian or Campanian age.

The Bayn Dzak eutherians represent three major lineages. *Kennalestes* is '. . . close to the origins of leptictids, but not far from the ancestors of palaeoryctoids' (McKenna, 1970; 228). A second eutherian lineage is represented by *Deltatheridium* and *Deltatheroides*, mammals discovered by field parties from the American Museum of Natural History. Like later zalambdodont insectivores, their teeth were modified to emphasize transverse shear. Finally the curiously long-snouted *Zalambdalestes* is a member of a third lineage possibly most closely related to the ancestors of the Tertiary, Asian Anagalidae and the Lagomorpha.

In tracing the evolution of placental mammals one must shift from this early Late Cretaceous Asian record to faunas of Campanian or Maestrichtian age in North America. Excepting the Late Cretaceous Bug Creek fauna (Sloan & Van Valen, 1965; Van Valen & Sloan, 1965) the abundance of placentals relative to marsupials in the North American faunas is low. At present only two families, Leptictidae and Palaeoryctidae, are represented. The leptictoid *Kennalestes*, although a member of the Bayn Dzak fauna in which one palaeoryctid lineage is represented, could be a little modified descendant of the common ancestry of palaeoryctoids and leptictoids (McKenna, 1970). No eutherians phylogenetically intermediate between palaeoryctoids and leptictoids are known from Late Cretaceous, North American faunas. New collections could drastically alter the fossil record but at present it suggests differentiation of leptictoid and palaeoryctoids occurred in Asia and members of these two lineages then dispersed into North America (Lillegraven, 1969).

During the past decade knowledge of mammalian evolution during the Late Cretaceous has been greatly increased. The picture emerging in North America is one of significant change just prior to and at the close of the Cretaceous. Van Valen & Sloan's (1966) summary of multituberculate evolution shows that some lineages failed to survive beyond the close of the Cretaceous. However, this loss in taxonomic diversity was more than compensated for by radiations stemming from other lineages. Eutherian evolution exhibits a similar pattern involving little or no extinction and an evolutionary radiation beginning prior to the close of the Cretaceous (Lillegraven, 1969). Primates and condylarths appear and there was a great radiation of palaeoryctids. During the Late Cretaceous North American marsupials underwent a major adaptive radiation. In most Late Cretaceous faunas their taxonomic diversity is equal to or greater than that of the eutherians and they are much more abundant. At the close of the Cretaceous most marsupial lineages became extinct; only one genus appears to have had Cenozoic descendants (Clemens, 1968).

SUMMARY

During the past two decades information pertinent to studies of the evolution of Cretaceous mammals has greatly increased. In part this reflects the acquisition of new collections of mammalian fossils but of greater importance has been the development of new areas of research. No longer do studies of the phylogenetic relationships of these mammals focus primarily on comparative studies of dental morphology. Work on dental function and microstructure of teeth, investigations of cranial and post-cranial morphology, and a variety of comparative studies of their modern descendants are making contributions to understanding of the phylogenetic relationships of Cretaceous mammals. Another important contribution comes from the work of geologists and geophysists on the modification of continental areas through marine transgressions and continental drift.

The close of the Jurassic and beginning of the Cretaceous does not appear to have been a time of major change in the evolution of mammals. On the basis of the presumed distribution of continents, evidence suggesting only limited climatic zonation, and similarities of dinosaurian faunas it is suggested that the major groups of mammals may have had nearly world-wide distributions.

Extinction of several mammalian lineages in the mid-Cretaceous, origin of the tribosphenic molar pattern, and differentiation of marsupials and placentals appear to be part of the mammalian adaptation to the diversification and spread of angiosperm dominated floras. The ecological link between changes in the lineages of carnivorous or omnivorous mammals and floral evolution is thought to be the terrestrial invertebrates.

Late Cretaceous mammalian faunas illustrate strong regional differentiation at a time when terrestrial areas are thought to have been fragmented by continental drift and wide marine transgressions. The marsupials, which are first known in the fossil record of North America, probably reached South America and Australia by a circum-Pacific route of dispersal. Prior to the end of the Cretaceous, in North America at least, an adaptive radiation of placental mammals began that was to continue on into the Cenozoic essentially unaffected by the extinctions at the close of the Cretaceous. Similarly no major change in the radiation of multituberculates occurred at the close of the Cretaceous. In contrast all but one lineage of the diverse assemblage of Late Cretaceous marsupials became extinct at the close of the period.

REFERENCES

Allen, P., 1969. Lower Cretaceous sourcelands and the North Atlantic. *Nature, Lond.*, **222**: 657–658.

Armstrong, R. L., 1968. Sevier orogenic belt in Nevada and Utah. *Bull. geol. Soc. Am.*, **79**: 429–458.

Axelrod, D. I., 1961. How old are the angiosperms? *Am. J. Sci.*, **259**: 447–459.

Axelrod, D. I. & Bailey, H. P., 1968. Cretaceous dinosaur extinction. *Evolution, Lancaster, Pa.*, **22**: 595–611.

Butler, P. M., 1956. The ontogeny of molar pattern. *Biol. Rev.*, **31**: 30–70.

Casamiguela, R. M., 1961. Sobre la presencia de un mamifero en el primer elenco (Icnologica) de vertebrados del Jurassico de la Patagonia. *Physis*, **22**: 225–233.

Chow, M. & Li, C., 1965. *Homogalax* and *Heptodon* of Shantung. *Vertebr. palasiat.*, **9**: 15–22.

Clemens, W. A., Jr., 1963*a*. Fossil mammals of the type Lance Formation, Wyoming. Part I. Introduction and Multituberculata. *Univ. Calif. Publs geol. Sci.*, **48**: 1–105.

Clemens, W. A., Jr., 1963*b*. Wealden mammalian fossils. *Palaeontology*, **6**: 55–69.

Clemens, W. A., Jr., 1966. Fossil mammals of the type Lance Formation Wyoming. Part II. Marsupialia. *Univ. Calif. Publs geol. Sci.*, **62**: 1–122.

Clemens, W. A., Jr., 1968. Origin and early evolution of marsupials. *Evolution, Lancaster, Pa.*, **22**: 1–18.

CLEMENS, W. A., JR. & LEES, P. M., 1971. A review of English Early Cretaceous mammals. In D. M. Kermack & K. A. Kermack (Eds), *Early mammals. Zool. J. Linn. Soc.*, **50**, Suppl. 1: 117–130.

COLBERT, E. H., 1965. *The age of reptiles*, 228 pp. New York: W. W. Norton.

COUSMINER, H. L., 1961. Palynology, paleofloras, and paleoenvironments. *Micropaleontology*, **7**: 365–368.

COX, C. B., 1970. Migrating marsupials and drifting continents. *Nature, Lond.*, **226**: 767–770.

CROMPTON, A. W. & HIIEMAE, K., 1970. Molar occlusion and mandibular movements during occlusion in the American opossum, *Didelphis marsupialis* L. *Zool. J. Linn. Soc.*, **49**: 21–47.

CROWSON, R. A., ROLFE, W. D. I., SMART, J., WATERSON, C. D., WILLEY, E. C. & WOOTTON, R. J., 1967. Arthropoda: Chelicerata, Pycnogonida, Palaeoisopus, Myriapoda, and Insecta. In, W. B. Harland *et al.* (Eds), *The fossil record*, pp. 499–534. London: Geol. Soc.

DEWEY, J. F. & HORSFIELD, B., 1970. Plate tectonics, orogeny, and continental growth. *Nature, Lond.*, **225**: 521–525.

DIETRICH, W. O., 1927. *Brancatherulum* n.g., ein Proplacentalier aus dem obersten Jura des Tendaguru in Deutsch-Ostafrika. Centralb [Zentrb.] *Min. Geol. Paläont. Stuttgart* (Abt. B), **1927**: 423–426.

FOX, R. C., 1968. Early Campanian (Late Cretaceous) mammals from Alberta, Canada. *Nature, Lond.*, **220**: 1046.

FOX, R. C., 1969. Studies of Late Cretaceous vertebrates III: A triconodont mammal from Alberta. *Canadian J. Zool.* **47**: 1253–1256.

GLASER, J. D., 1969. Petrology and origin of Potomac and Magothy (Cretaceous) sediments, middle Atlantic Coastal Plain. *Md geol. Surv., Rept. Invest.*, No. 11.

GRAMBAST, L., MARTINEZ, M., MATTAUER, M. & THALER, L., 1967. *Perutherium altiplanense*, nov. gen., nov. sp., premier mammifère mésozoique d'Amerique du Sud. *C. r. Acad. Sci.*, (Sér. D), **5**: 707–710.

HAHN, G., 1969. Beiträge zur Fauna der Grube Guimarota Nr. 3. *Palaeontographica*, (Ser. A), **133**: 1–100.

HALLUM, A., 1969. Faunal realms and facies in the Jurassic. *Palaeontology*, **12**: 1–18.

HARRINGTON, H. J., 1962. Paleogeographic development of South America. *Bull. Am. Ass. Petrol. Geol.*, **46**: 1773–1814.

HECHT, M. K., 1963. A reevaluation of the early history of the frogs, part II. *Syst. Zool.*, **12**: 20–35.

HENKEL, S., 1966. Methoden zur Prospektion und Gewinnung kleiner Wirbeltierfossilen. *Neues Jb. Geol. Paläont. Mh.*, **3**: 178–184.

HENKEL, S. & KREBS, B., 1969. Zwei Säugetier-Unterkiefer aus der Untern Kreide von Uña (Prov. Cuenca, Spanien). *Neues Jb. Geol. Paläont. Mh.*, **8**: 449–463.

HUGHES, N. F. & SMART, J., 1967. Plant-insect relationships in Palaeozoic and later time. In W. G. Harland *et al.* (Eds), *The fossil record*, pp. 107–117. London: Geol. Soc.

JACOBS, C., BURGL, H. & CONLEY, D. L., 1963. Backbone of Columbia. *Mem. Am. Ass. Petrol. Geol.* **2**: 62–72.

KERMACK, K. A., 1967. The interrelations of early mammals. *J. Linn. Soc.* (*Zool.*), **47**: 241–249.

KERMACK, K. A., LEES, P. M. & MUSSETT, F., 1965. *Aegialodon dawsoni*, a new trituberculo-sectorial tooth from the Lower Wealden. *Proc. R. Soc.*, (Ser. B), **162**: 535–554.

KIELAN-JAWOROWSKA, Z., 1969*a*. Preliminary data on the Upper Cretaceous eutherian mammals from Bayn Dzak, Gobi Desert. *Palaeont. pol.*, **19**: 171–191.

KIELAN-JAWOROWSKA, Z., 1969*b*. Discovery of a multituberculate marsupial bone. *Nature, Lond.*, **222**: 1091–1092.

KIELAN-JAWOROWSKA, Z. & DOVCHIN, N., 1969. Narrative of the Polish-Mongolian palaeontological expeditions, 1963–1965. *Palaeont. pol.*, **19**: 7–32.

KIELAN-JAWOROWSKA, Z. & SOCHAVA, A. V., 1969. The first multituberculate from the uppermost Cretaceous of the Gobi Desert (Mongolia). *Acta palaeont. pol.*, **14**: 355–371.

KRUSAT, G., 1969. Ein Pantotheria-Molar mit dreispitzigem Talonid aus dem Kimmeridge con Poartugal. *Paläont. Z.*, **43**: 52–56.

KÜHNE, W. G., 1968. History of discovery, report on the work performed, procedure, technique and generalities. *Mems Servs geol. Port.*, (N.S.), **14**: 7–20.

KÜHNE, W. G. & CRUSAFONT-PAIRÓ, M., 1968. Mamiferos de Wealdiense de Uña, cerca de Cuenca. *Acta geol. Hispanica*, **3**: 133–134.

KURTEN, B., 1966. Holarctic land connexions in the early Tertiary. *Commentat. biol.*, **29**: 1–5.

LEDOUX, J., HARTENBERGER, J., MICHAUX, J., SUDRE, J. & THALER, L., 1966. Découverte d'un Mammifère dans le Cretacé Superieur a Dinosaures de Champ-Garimond pres de Fons (Gard). *C. r. hebd. Séanc. Acad. Sci., Paris*, **262**: 1925–1928.

LEES, P. M., 1964. The flotation method of obtaining mammal teeth from Mesozoic bone-beds. *Curator*, **7**: 300–306.

LE PICHON, X., 1968. Sea-floor spreading and continental drift. *J. geophys. Res.*, **73**: 3661–3697.

LILLEGRAVEN, J. A., 1969. Latest Cretaceous mammals of upper part of Edmonton Formation of Alberta, Canada, and review of marsupial-placental dichotomy in mammalian evolution. *Paleont. Contr. Univ. Kans.*, Art. 50 (Vert. 12): 1–122.

LLOYD, J. J., 1963. Tectonic history of the south Central-American orogen. *Mem. Am. Ass. Petrol. Geol.*, **2**: 88–100.
LOWENSTAM, H., 1964. Paleotemperatures of the Permian and Cretaceous periods. In A. E. M. Narin (Ed.), *Problems in paleoclimatology*. London: Interscience.
MACKAY, M. R., 1970. Lepidoptera in Cretaceous amber. *Science, N.Y.*, **167**: 379–380.
MARTIN, P. G., 1970. The Darwin Rise hypothesis of the biogeographical dispersal of marsupials. *Nature, Lond.*, **225**: 197–198.
MCKENNA, M. C., 1965. Collecting microvertebrate fossils by washing and screening. In B. Kummel & D. Raup (Eds), Handbook of paleontological techniques. San Francisco: W. H. Freeman.
MCKENNA, M. C., 1970. The origin and early differentiation of therian mammals. *Ann. N. Y. Acad. Sci.*, **167**: 217–240.
MCNULTY, C. L., JR. & SLAUGHTER, B. H., 1968. In *Field Trip Guidebook, South Central Section*, pp. 68–72. Geol. Soc. Am.
MILLS, J. R. E., 1964. The dentitions of *Peramus* and *Amphitherium*. *Proc. Linn. Soc., Lond.*, **175**: 117–133.
MILLS, J. R. E., 1967. Development of the protocone during the Mesozoic. *J. dent. Res.*, **46**: 787–791.
MOSS, M. & KERMACK, K. A., 1967. Enamel structure in two Triassic mammals. *J. dent. Res.*, **46**: 745–747.
PATTERSON, B., 1956. Early Cretaceous mammals and the evolution of mammalian molar teeth. *Fieldiana Geol.*, **13**: 1–105.
PATTERSON, B. & PASCUAL, R., 1968. Evolution of mammals on southern continents. V. The fossil mammal fauna of South America. *Q. Rev. Biol.*, **43**: 409–451.
RUSSELL, D. E., 1964. Les Mammifères Paléocènes d'Europe. *Mém. Mus. natn. Hist. nat., Paris*, (Sér. C), **13**: 1–324.
SCOTT, R. A., BARGHOORN, E. S. & LEOPOLD, E. B., 1960. How old are angiosperms? *Am. J. Sci.*, **258** (A): 284–299.
SHIKAMA, T., 1947. *Teilhardosaurus* and *Endotherium*, new Jurassic Reptilia and Mammalia from the Husin coal-field, South Manchuria. *Proc. imp. Acad. Japan*, **23**: 76–84.
SIGÉ, B., 1968. Dents de micromammifères et fragments de coquilles d'oeufs de dinosauriens dans la fauna de vertébraés du Crétacé superieur de Laguna Umayo (Andes peruviennis). *C. r. Acad. Sci., Paris* (Sér. D), **267**: 1495–1498.
SIMPSON, G. G., 1928. *A catalogue of the Mesozoic Mammalia in the Geological Department of the British Museum (Natural History)*. London: Br. Mus. (Nat. Hist.).
SIMPSON, G. G., 1929. American Mesozoic Mammalia. *Mem. Peabody Mus. Yale*, **3**: 1–235.
SIMPSON, G. G., 1933. The 'plagiaulacoid' type of mammalian dentition. *J. Mammal.*, **14**: 97–107.
SLAUGHTER, B. H., 1965. A therian from the Lower Cretaceous (Albian) of Texas. Postilla, **93**: 1–18.
SLAUGHTER, B. H., 1968. Earliest known marsupials. *Science, N.Y.*, **162**: 254–255.
SLAUGHTER, B. H., 1969. *Astroconodon*, the Cretaceous triconodont. *J. Mammal.*, **50**: 102–107.
SLOAN, R. E. & VAN VALEN, L., 1965. Cretaceous mammals from Montana, *Science, N.Y.*, **148**: 220–227.
SMILEY, C. J., 1967. Paleoclimatic interpretations of some Mesozoic floral sequences. *Bull. Am. Ass. Petrol. Geol.*, **51**: 849–863.
SMITH, A. G. & HALLUM, A., 1970. The fit of southern continents. *Nature, Lond.*, **225**: 139–144.
STIRTON, R. A., TEDFORD, R. H. & WOODBURNE, M. O., 1968. Australian Tertiary deposits containing terrestrial mammals. *Univ. Calif. Publs geol. Sci.*, **77**: 1–30.
TAKHTAJAN, A., 1969. *Flowering plants, origin and dispersal*. Edinburgh: Oliver & Boyd.
TIDWELL, W. D., RUSHFORTH, S. R., REVEAL, J. L. & BEHUNIN, H., 1970. *Palmoxylon simperi* and *Palmoxylon pristina*: two pre-Cretaceous angiosperms from Utah. *Science, N.Y.*, **168**: 835–840.
VAKHRAMEEV, A., 1965. Jurassic floras of the Indo-European and Siberian botanical-geographical regions. *Palaeobotanist*, **14**: 118–123.
VAN VALEN, L. & SLOAN, R. E., 1966. The extinction of the multituberculates. *Syst. Zool.*, **15**: 261–278.
WILSON, E. O., CARPENTER, F. M. & BROWN, W. L., JR., 1966. The first Mesozoic ants, with the description of a new subfamily. *Psyche, Camb.*, **74**: 1–19.

Concluding remarks: Mesozoic mammals revisited

GEORGE GAYLORD SIMPSON, F.M.L.S.

Department of Geosciences, University of Arizona, and the Simroe Foundation, Tucson, Arizona, U.S.A.

CONTENTS

INTRODUCTION

The first discovery of a Mesozoic mammal was made in or about 1764, but it went unrecognized. The existence of such ancient mammals was not recognized until further discoveries were made in 1812 and even after that publication was delayed until 1824. That event was shocking, because it was then the firm conviction of most geologists that mammals did not exist in the Secondary. Only Charles Lyell and a few of his followers were pleased, because an early misconception of the doctrine of uniformity required (in their opinion) that Secondary mammals existed and had merely escaped previous discovery. Many following discoveries through the 1860's, almost all in England, left no doubt as to the fact and demonstrated considerable variety among Jurassic mammals, at least.

A first general monograph on Mesozoic Mammalia was published by Richard Owen in 1871. In 1878 to 1892 collectors working for O. C. Marsh found both Jurassic and Cretaceous mammals in North America, summarily described in a series of brief papers by Marsh. In 1888 H. F. Osborn had provided a second general study of the European Jurassic forms, and he considered the American Cretaceous mammals more briefly in 1893. He did not treat the American Jurassic forms because almost all were in Marsh's hands and Osborn was not permitted to examine them. Such animosities did not persist, and in the 1920's I was permitted to study all specimens then known, both European and American. A third review of the whole subject was thus possible in two monographs, on European specimens published in 1928 and on American in 1929. In the meantime, in 1924 and 1925, Mongolian Cretaceous mammals had been found by Walter Granger and his associates and had been promptly described mostly by W. K. Gregory with my collaboration.

There followed a hiatus in significant discovery and in extensive study, but already in a small way in the 1940's both discoveries and studies began to accelerate. In the 1960's they became a flood, which evidently has not yet reached its crest. Change is so rapid at present, and so much new knowledge is still unpublished, that the time has not come for another general review. Indeed it would not be possible now, as it was in 1871, 1888, and 1928–1929, for one person to treat all available material on Mesozoic mammals in anything but a summary and at least partly second-hand way. This symposium offers a sort of interim report and a good sampling of recent discoveries and work now in progress. In these concluding remarks I shall not abstract the papers presented, but rather will give a view of the present status of the subject that is exemplified by the preceding contributions.

MATERIALS AND INFORMATION

The most important phenomenon—and it is indeed phenomenal—is the tremendous increase in information about Mesozoic mammals. That has resulted in part from restudy by new methods or from new points of view of specimens previously known. The important paper by Mills (1964) on some British Jurassic specimens is one of several examples. It has, however, especially involved the discovery of many new specimens and faunas. That rather sudden change, after a long period when only a few specimens dribbled in from time to time, has several causes: the great increase in personnel, especially in England and the United States; the cumulative effect of feed-back from increasing success and heightening interest; the use of new methods. Most important in the last respect is the treatment of large quantities of matrix by washing, screening, and in some circumstances flotation (for example: Henkel, 1966; Kühne, 1947; Lees, 1964).

Materials of Mesozoic mammals now known may be broadly and briefly summarized:

EUROPE

Late Triassic or Rhaeto-Liassic

A most important development is acquisition of large numbers of new specimens of this age, mostly by the Kermacks from fissures in Wales, especially of the genera called *Morganucodon* (possibly a synonym of *Eozostrodon* Parrington) and *Kuehneotherium* (possibly a synonym of *Kuhneon*) by the Kermacks. (See for example Kermack, Kermack, & Mussett, 1968, and references there.) It is disappointing that little progress has been made in understanding of the even more enigmatic and possibly important Haramiyidae (but see Parrington, 1947; Peyer, 1956), known by isolated teeth from both England and the continent.

Jurassic

Only a little new material from the Jurassic of England has turned up, but restudy of Stonesfield and Purbeck specimens has given some new data (e.g., Clemens, 1963*b*). Most important is the discovery of a rich Kimmeridgian fauna in Portugal, under active study but only partly published (Hahn, 1969; Kühne, 1968). (In some recent literature the Purbeck is dated as Berriasian and placed in the earliest Cretaceous; I here follow a consensus in continuing to place the Purbeckian in the Jurassic.)

Cretaceous

The very scanty Wealdan fauna has been reviewed recently and slight additions made (Clemens, 1963; Kermack, Lees and Mussett, 1965).* Of great potential importance are still mostly undescribed discoveries in Spain (Henkel & Krebs, 1969; Kühne & Crusafont, 1969). The only late Cretaceous mammal yet reported from Europe is a single therian (in my opinion eutherian) tooth from France (Ledoux, Hartenberger, Michaux, Sudre & Thaler, 1966).

ASIA

Late Triassic

Two important species are known from Yunnan, China: *Sinoconodon rigneyi* (Patterson & Olson, 1961) and one, still not fully described, belonging or closely related to *Morganucodon* (or *Eozostrodon*) (Rigney, 1963).

Cretaceous

Two jaws, a therian and a somewhat dubious symmetrodont, published as Jurassic but now believed more likely to be of early to middle Cretaceous age, are known from Manchuria (Shikama, 1947; Yabe & Shikama, 1938). The published illustrations are poor, and knowledge of these potentially important specimens is inadequate.

Among the most important discoveries of recent years are a number of skulls and partial skeletons, all but one from Granger's late Cretaceous Mongolian locality Shabarakh Usu (called Bain or Bayn Dzak in recent publications) by Polish-Mongolian expeditions. Some excellent eutherian material has been described (Kielan-Jaworowska, 1968), and further information about multituberculates, partly published or in press, was given by Kielan-Jaworowska in this symposium.

* See Clemens & Lees, pp. 117–130, Plates 1–5, in this volume.

AFRICA

Late Triassic

As is well known, the reptile-mammal sequence is best represented by African specimens, a fact that reflects conditions of preservation and discovery and not (in my opinion) necessarily the place of origin of the Mammalia. Some very mammal-like forms, notably *Diarthrognathus*, are not now generally classified as mammals, but two unquestioned mammals are known: *Erythrotherium*, known from a crushed skull (Crompton, 1964) and *Megazostrodon*, skull and partial skeleton (Crompton & Jenkins, 1968), both from Lesotho. Only very brief and incomplete descriptions have so far been published.

Late Jurassic or Early Cretaceous

The only known specimen is a toothless pantothere jaw from the Tendaguru beds of East Africa, and nothing has been added since the last study of it 42 years ago. No later Mesozoic mammals are known from Africa.

NORTH AMERICA

Jurassic

A few additional late Jurassic (Morrison formation) specimens have lately been found but not yet described, and some specimens have been reexamined (e.g. by Crompton & Jenkins, 1967), but no substantial addition to knowledge of this important fauna has been published for more than 40 years.

Cretaceous

Although scanty, the fauna from the Trinity sands of Texas (Albian, late Lower or Middle Cretaceous) is exceptionally important as giving our only knowledge of the stage when metatherian-eutherian early differentiation was apparently occurring (Patterson, 1956 and citations there; Slaughter, 1968*a*,*b*). A review of the fauna as a whole, including the still undescribed multituberculates, is needed.

Among the most important recent developments is the acquisition of large and varied collections of mammals, mostly by washing and screening, over a considerable span of late Cretaceous time and from a number of North American localities, especially in New Mexico, Wyoming, Montana, and Alberta. Both field work and study are now in full swing, and much remains to be published. Among numerous publications that have appeared are those by Clemens (1964, 1966), Lillegraven (1969), and Sloan & Van Valen (1965)

SOUTH AMERICA

Triassic

Discovery of a possible Triassic mammal in Argentina has recently been stated but not published, and according to definitions some Triassic forms could be classified either as reptiles or as mammals (Romer, 1969).

Jurassic

Jurassic footprints from Argentina have been identified as mammalian (Casamiquela, 1961), but they are doubtful and in any case give no really clear and useful information about the animals that made them. Other claimed Jurassic mammals have been lost, were not described, and may not have been mammals.

Cretaceous

Claims by Ameghino that many mammals described by him were Cretaceous must all be rejected. A fragment of a jaw from Peru, described as Cretaceous, is undoubtedly mammalian (probably a condylarth), but evidence for pre-Cenozoic age is not completely convincing (Grambast, Martinez, Mattauer & Thaler, 1967).

In short, no certainly mammalian and certainly Mesozoic fossil is yet known from South America, and even if all dubious cases be given full credit, the information from them is exiguous.

No Mesozoic mammals are known from Australia, Antarctica, or any island other than Great Britain.

That is an impressive record, and it is conservative to estimate that available information is several times more than when the last general monographs were written more than 40 years ago. Yet this information is still grossly inadequate to answer the questions necessarily put to it. Remember that these few faunal records cover a crucial period of mammalian history some 120 million years in length, hence almost twice the duration of the entire so-called Age of Mammals. Gaps of place and time are still incomparably more extensive in the Mesozoic record than the known data.

BASIC GROUPS

Scientists generally strive to follow the dictum of Occam (or Ockham), the canon of parsimony or of fewest entities or hypotheses, or the principle of simplicity in forming interpretations and theories. But the principle of simplicity is itself extremely complicated, and the attempt to minimize both entities and hypotheses frequently involves diametrically opposed goals. The fact that we now have so much (and yet far too little information about Mesozoic mammals multiplies doubts and complexities. The fewer facts available a generation ago required and permitted fewer hypothetical connections among them. But the more numerous facts now in hand are, as a reasonable estimate of order of magnitude, about a thousandth of those necessary to permit reasonable simplification in the sense of unification of warranted interpretations. A minimum of hypotheses might permit enumeration of 'species,' defined as typologically unified groups of sympatric and synchronous individuals, as the sole entities to be accepted, but that minimization of hypotheses would involve a maximization of disconnected entities to a degree both unscientific and incomprehensible. The least number of hypotheses really tolerable as a basis for discussion may be achieved by gathering species into groups, sets, or taxa, at various presumed hierarchical levels, such that the species within each group may with considerable but never complete confidence be considered more closely related to each other than to members of any other group. The degree of confidence will almost inevitably be inversely related to the level or inclusiveness of the set in question.

The following groups may be taken as basis for further interpretations:

Triassic groups

The known forms can now be placed in four groups that are commonly considered families in taxonomic terms. Their ordinal positions depend on relationships to Middle and Late Jurassic groups and are discussed subsequently.

Haramiyidae (or 'Haramyidae,' an invalid emendation, or 'Microlestidae,' based on an invalid generic name).

Sinoconodontidae (*Sinoconodon* and *Megazostrodon*).

Morganucodontidae or Eozostrodontidae (*Morganucodon* or *Eozostrodon* and *Erythrotherium*). For the sake of simplicity and uniformity the names Morganucodontidae and *Morganucodon* will be used from here on, but this does not involve a personal decision that Eozostrodontidae and *Eozostrodon* are not the correct names for this family and genus.

Kuehneotheriidae (*Kuehneotherium* and *Kuhneon*, if the latter prior name designates a different genus).

Jurassic groups

The following groups are now generally accepted and placed at the ordinal level on the basis of their Middle to Late Jurassic members. Several also survived into the post-Jurassic, as noted. Their possible inclusion of Triassic members is discussed subsequently.

Multituberculata (also in the Cretaceous, Paleocene, and Eocene).

Triconodonta (including Amphilestinae or Amphilestidae; also in the Cretaceous).

Docodonta.

Symmetrodonta (Amphidontidae and Spalacotheriidae; also in the Cretaceous).

Pantotheria or Eupantotheria (Amphitheriidae [with Peramurinae], Paurodontidae, Dryolestidae, and Aegialodontidae; the latter is of Early Cretaceous age, and a dryolestid is also known from the Early Cretaceous).

Cretaceous groups

In addition to the four orders known to have survived from the Jurassic, the following, more or less at infraclass level, are known in the Cretaceous.

Theria *incertae sedis* or 'Theria of metatherian-eutherian grade.' No convenient term is available for Early to Middle Cretaceous Theria, including *Endotherium* and some Albian specimens from North America, that have definitely advanced beyond Pantotheria in molar structure and are related among themselves but cannot at present be confidently placed in Metatheria, Eutheria, or a possible, now undefinable common post-pantothere ancestry of the two.

Metatheria, order Marsupialia, Polyprotodonta (considered an order by Kirsch, 1968), or 'Marsupicarnivora' (of Ride, 1964). Late Cretaceous forms recognizable as marsupials; all now known are Didelphoidea.

Eutheria or Placentalia. Late Cretaceous forms recognizable as placentals. Distribution among several orders has been suggested, as will be mentioned subsequently.

Even at this level of hypothesizing, which I consider the lowest tolerable for general discussion of Mesozoic mammals as a whole, the indicated hypothesized groupings do not merit or receive complete confidence. For example, Kermack (1967) would remove the Amphilestinae, as Amphilestidae, from the Triconodonta *sensu stricto*, and he considers it possible that they are more nearly related to the Pantotheria.

AFFINITIES BETWEEN KNOWN TRIASSIC FAMILIES AND JURASSIC ORDERS

A next step is to seek affinities between the known Triassic families and the Jurassic orders, involving an increase in numbers of hypotheses and a marked decrease in confidence.

Least secure of all is the ordinal position of the Haramiyidae, known from isolated teeth, only, and not yet of agreed reference to the Mammalia (by any definition). An old idea (originated by O. C. Marsh) that they might be related or ancestral to multituberculates rested on so little evidence that it hardly warranted any confidence at all. However, the recent discovery of Kimmeridgian multituberculates, older and more primitive than any previously known, shows that some of them did have cheek teeth vaguely similar to haramiyids, and the hypothesis of relationship becomes respectable even if still inspiring incomplete confidence (Hahn, 1969).

Quite reasonable confidence is warranted in reference of the Sinoconodontidae to the Triconodonta (Patterson & Olson, 1961; Crompton & Jenkins, 1968). I do not think that this confidence is appreciably lessened by the possibility that some specimens originally referred to *Sinoconodon* may belong to *Morganucodon* (Rigney, 1963; Kermack, 1967) or, on the other hand, by the suggestion that *Sinoconodon* may be more closely related to *Morganucodon* than to *Triconodon* (Hopson & Crompton, 1969).

The morganucodonts are the best-known of all mammals earlier than late Cretaceous, although not all information on them is yet fully published. Up to a certain point an increase in information may lead to a decrease in consensus and confidence. The first students of these animals (Parrington, 1941, 1947; Kühne, 1949; Kermack, 1955; Peyer, 1956) expressed no doubts that they were triconodonts. However, with the exception of Peyer (who did not return to this subject), they all modified this view to some extent. Kühne (1950), following a suggestion from P. M. Butler, soon suggested that *Morganucodon* might be a pantothere, by which he meant to indicate relationship to *Docodon*, then considered a pantothere. Patterson (1956) sharply separated *Docodon* from the pantotheres and placed *Morganucodon* with it in an order Docodonta (originally named by Kretzoi for the late Jurassic forms, only). Kermack (e.g. 1967, but also earlier) soon accepted and subsequently has generally followed that view. Parrington did not later concern himself with formal classification. Continuing to call these teeth triconodont or triconodont-like, he also considered them ancestral to pantotheres (1967), with no implication as to docodont relationships. Kühne (1968) came to consider the morganucodonts as docodonts, distinct from pantotheres. Hopson & Crompton (1969) have argued, in a way that is quite persuasive as the limited evidence now stands, that the morganucodonts were triconodonts but that they represent an early group of Triconodonta from which the Docodonta were derived. Vandebroek (1964) has clung to the idea that the morganucodonts are triconodonts and have nothing to do either with the docodonts (which he continues to classify as pantotheres) or with the pantotheres *sensu stricto* (his 'Dryolestoidea').

Kuehneotherium, which had the unusual history of becoming widely known before it had a name, was perhaps first described from a single tooth, called 'Duchy 33,' referred to the Symmetrodonta. Unfortunately this is another instance, like *Eozostrodon-Morganucodon*, in which there is a question as to whether a prior name is synonymous with a later one. 'Duchy 33' was named *Kuhneon duchyense* by Kretzoi (1960), who had never studied this group, in a paper known to few non-Hungarians. The name *Kuhneon* almost surely is a prior synonym of *Kuehneotherium*, which I will nevertheless continue to use. Kermack *et al.* (1968) believe the Kuehneotheriidae to be ancestral both to the Jurassic Symmetrodonta and to the Jurassic Pantotheria (their 'Eupantotheria'). No subsequent dissent is known to me, although these relationships would be impossible according to Vandebroek (1964), who did not in that study accept this form of relationship between Symmetrodonta and Pantotheria. Following the consensus, it would be equally valid to place the family Kuehneotheriidae in the Pantotheria or the Symmetrodonta. Kermack *et al.* (1968) prefer the former alternative, Crompton & Jenkins (1967) the latter. No essential difference of opinion as to affinities is involved.

Having already expressed my limited confidence in such assignments, I have a slight and perhaps temporary preference for placing:

Haramiyidae, with a large question mark, in the Multituberculata.
Sinoconodontidae in the Triconodonta.
Morganucodontidae also in the Triconodonta.
Kuehneotheriidae in the Symmetrodonta.

By Late Jurassic times, the situation is clarified to some extent. Multituberculata, Triconodonta, Docodonta, Symmetrodonta, and Pantotheria were then sharply distinct groups, and it is highly improbable that any except the last, as represented at this time and level, were ancestral or closely related to any later groups except those clearly referable to the same orders. Ordinal relationships are discussed in a later section.

AFFINITIES OF CRETACEOUS THERIA

The quite inadequate evidence prevents confident assignment of known Early to Middle Cretaceous Theria to any particular order. *Endotherium* has generally been classified in the Insectivora (*sensu lato*), and indeed Saban (1958) placed it in the Tertiary family Pantolestidae. Slaughter at first (1965) followed Patterson (1956) in considering the Albian Theria from Texas as indeterminate between Metatheria and Eutheria, but he assigned some later specimens definitely to the Eutheria (Slaughter, 1968*a*), order and family not specified, and others to the Metatheria (Slaughter, 1968*b*), order Marsupialia, family Didelphidae. He has discussed the criteria involved in this symposium.

In the Late Cretaceous the situation is clearer. Adequate specimens, now abundant although very limited geographically (only one locality in Mongolia and several in western North America) can be confidently sorted into Allotheria (Multituberculata), Metatheria (Didelphoidea only), and Eutheria. (See especially Clemens, 1964, 1966; Kielan-Jaworowska, 1968; Lillegraven, 1969; Sloan & Van Valen, 1965.) The well-known dentitions and some still poorly known (or poorly published) other parts are remarkably alike as to general pattern in all these eutherians. The details, however, are highly diverse. On the basis of these variations, some genera have been assigned to the Insectivora (*Gypsonictops*, so placed in all the publications known to me), some to the 'Deltatheridia' (*Cimolestes*, referred here by Lillegraven, 1969), some to the Primates (*Purgatorius*, a mainly Paleocene genus, to which Van Valen & Sloan, 1965, have referred a single lower molar from the Cretaceous), and some to the Condylarthra (*Protungulatum* Sloan & Van Valen, 1965).

Those ordinal references may ultimately be substantiated but they are possibly misleading and almost certainly premature. All these genera are so similar that if we were dealing only with them it would be unjustified to place them in more than one order; indeed one could defend referring all to a single family of a purely phenetic classification or one ignoring later forms. The ordinal references depend on inferences as to phylogenetic affinities with lineages that later became so distinct as to require ordinal separation. The attempt to make classification more nearly vertical is justifiable, but the evidence here involved is shaky and the strategy questionable. For example, *Procerberus* is considered as a prototypal insectivore and ancestral leptictid by Van Valen (1967), but is placed in the 'Order Deltatheridia,' family Palaeoryctidae by Estes & Berberian (1970; with unpublished information from R. E. Sloan & W. A. Clemens), while Lillegraven (1969) also considers it palaeoryctid but as the ancestor of the order Taeniodonta, hence referable to the latter in a consistently vertical classification. From the published figure, (Van Valen & Sloan, 1965, fig. 1A), I question whether the single tooth that bears the whole burden of proof for the existence of Cretaceous primates belongs to the genus *Purgatorius* or in any case is a primate. Lillegraven (1969, fig. 40) shows *Cimolestes* as ancestral to both Creodonta and Fissipeda, which in a vertical classification would suggest reference of *Cimolestes* to the Carnivora in the classic or Matthewian sense and jetisoning the possibly artificial and probably unnecessary 'Order Deltatheridia.' Reasons for doubts and queries could be multiplied.

Much Late Cretaceous material is still undescribed, and this and future discoveries will doubtless clarify what is now a situation even more confused than may appear from some recent publications. In the meantime it would seem to me more consistent with the actual state of knowledge to consider all Late Cretaceous Eutheria as Insectivora *sensu lato* (or *latissimo*, if preferred), a varied group near the basal radiation of most or all eutherian orders. One could then speculate at will as to possible derivation of this order or that from one or another primitive stock of the Insectivora in that sense.

ORDINAL RELATIONSHIPS

The groupings so far discussed involve a considerable degree of hypothesizing, although presented as a tolerable minimum in that respect. They also represent a low level of simplicity in the sense that the union of these groups into some single conceptual framework would seem simpler than their consideration as so many separate entities. Simplest in that sense would be a single phylogenetic scheme including all the groups. That can be achieved, but at the expense of going so far beyond the firm evidence as to sacrifice credibility.

For a start, it is highly probable that all the eutherian orders diverged from a single stock and that the divergence was under way but not far advanced in the Late Cretaceous. It is also virtually certain that the Metatheria and Eutheria arose from a common ancestry. It is debatable, and is being debated, whether that unknown or, at least, not surely identified common ancestry was more like the Metatheria or the Eutheria, or quite different from either. It is a reasonable hypothesis that the metatherian-eutherian ancestry evolved from the Pantotheria, in a broad sense. It does not follow that any actually known species, genus, or family of pantotheres is ancestral to later Theria. Among the groups that are known, the Amphitheriidae (including the Peramurinae) may be closest to the ancestral stock, but it would be a miracle if the few specimens known, from only two ages in one small area, included a real ancestor. The single tooth named *Aegialodon* (Kermack, Lees, & Mussett, 1965) may also be near but hardly in the direct line of ancestry, and its significance may have to be reconsidered now that we know that lower molars with equally basined and tricuspid talonids occurred in pantotheres as early as the Kimmeridgian and do not necessarily indicate the existence of a 'true' protocone (Krusat, 1969, and Clemens, 1971).

A relationship between pantotheres and symmetrodonts has long been recognized. As far as I know, Vandebroek (1964) is the only recent student who has denied this and that was before adequate publication on *Kuehneotherium*. The Kuehneotheriidae have confirmed the existence and clarified the nature of the relationship. Kermack *et al.* (1968) have shown that this group is morphologically suitable as an (approximate) ancestry for both later symmetrodonts and later pantotheres. They prefer to consider it a pantothere (Order Eupantotheria, Suborder Amphitheria in their terminology), and so indicate that symmetrodonts arose from primitive pantotheres. *Kuehneotherium* resembles Late Jurassic symmetrodonts more than it does middle or Late Jurassic pantotheres (Crompton & Jenkins, 1967). Therefore although I agree completely with Kermack, Kermack and Mussett as to the relationships of the Kuehneotheriidae, I prefer to express them somewhat differently, referring that family to the Symmetrodonta and considering that the Pantotheria (in my sense of the word) arose from primitive symmetrodonts. The generally more conservative descendants also became specialized in their own way, and we have within the Symmetrodonta an adaptable, primitive Triassic group, the Kuehneotheriidae, and less adaptable, specialized terminal members in the Jurassic and Cretaceous which had become quite distinct from the Pantotheria.

One can go so far in simplification by phylogenetic unification with reasonable confidence, but increasing doubts arise with attempts to bring into this phylogeny the remaining orders: Triconodonta, Docodonta, Multituberculata, and Monotremata. Among these, the nontherian orders, a hypothetical connection with the Theria can hardly be imagined except by way of the Triconodonta, especially if the Morganucodontidae are referred to that order. Hopson & Crompton (1969) have shown that such a connection is indeed imaginable. Yet considerable imagination is required, and Kermack's studies (1967 and others there cited) have suggested that the most fundamental division in the Mammalia may be just here, between (in my terms) the Triassic Triconodonta and Symmetrodonta. Indeed Hopson and Crompton agree, although hypothesizing that the two groups *at this level* are nevertheless closely related. The main point of difference is that the Morganucodontidae, later Triconodonta, Multituberculata, and Monotremata have a comparatively small alisphenoid and an anterior lamina of the petrosal (or periotic) through which cranial nerve V_3 passes, while in late Theria (at least) the alisphenoid is large, V_3 passes through it, and the anterior lamina of the petrosal is lacking.

That difference seems fairly fundamental, and yet its interpretation is not entirely clear, if only because the condition is unknown in the Docodonta (*sensu stricto*, excluding Morganucodontidae), Symmetrodonta, Pantotheria, or any other Theria earlier than the late Cretaceous, that is, at least 100 million years after the distinction is hypothesized to have arisen. Further questions arise from the fact that neither condition is known in the premammalian therapsids (again see Hopson & Crompton, 1969), despite which one should not cavalierly discard the suggestion of MacIntyre (1967) that the morganucodontid (or 'nontherian') condition may be primitive for mammals. We really do not know that it did not occur in early Theria.

While there is some persistent doubt about a triconodont-symmetrodont (or pantothere) phylogenetic union, a fairly close connection between Triconodonta and Docodonta now seems to merit a limited but positive degree of confidence. Opinions vary from the view that *Morganucodon* is in fact a docodont with no special relationship to triconodonts (e.g. Kermack, 1967) to the opinion that *Morganucodon* (that is, *Eozostrodon*) is flatly a triconodont with only some coincidental resemblance to docodonts (Parrington, 1967). The hypothesis I prefer, without strong conviction, is that the Docodonta evolved into a dead end by rapid molar specialization of a peculiar sort from primitive triconodonts resembling the morganucodonts. (I long since abandoned my early opinion that the docodonts were pantotheres, as Vandebroek still maintains.)

The extreme simplist view of Mesozoic mammalian phylogeny would require that multituberculates be derived from early (presumably Triassic) triconodonts or, what I believe comes to about the same thing, morganucodonts, and that has indeed been suggested in a reasonably tentative way, especially and lately by Hopson & Crompton (1969). One could imagine multituberculate teeth evolving from those of morganucodonts or other triconodonts, but one could imagine almost anything, and in fact there is no evidence whatever that this occurred. Carrying the multituberculates back into Triassic haramiyids, itself highly dubious, does not help the case, because haramiyid teeth are likewise not at all like those of any triconodonts. In fact they are rather more like the teeth of some therapsids. The only clear evidence for multituberculate-triconodont relationships is that both had an anterior lamina of the petrosal pierced by the foramen for V_3, but that arrangement is not so complex as to rule out independent derivation from the therapsid condition, and as far as the still rather scanty published data go other features of cranial anatomy are quite different in the two.

Finally the derivation of the Monotremata must be considered. This order has no useful fossil record. It resembles some, at least, of the early 'nontheria' in having a petrosal lamina and reduced alisphenoid. Hopson & Crompton (1969) conclude that no known feature excludes the Morganucodontidae from the ancestry of the Monotremata, but they recognize that this is a negative statement and that the only positive resemblance is that in the vestigial molars of *Ornithorhynchus* and in the functional molars of some morganucodonts there are prominent superoexternal and inferointernal cingula. In this symposium Kielan-Jaworowska expressed the opinion that the Monotremata are closely related to or were derived from the Multituberculata. This relationship was suggested long ago, perhaps first by E. D. Cope in 1888, but the evidence available up to now has been (in my opinion) quite unconvincing. As the new evidence is not yet known to me in detail I must suspend judgment.

MONOPHYLY, POLYPHYLY, AND THE DEFINITION OF MAMMALIA

Extended arguments as to whether the Mammalia are monophyletic or polyphyletic are in part, but only in part, a *Scheinproblem* or a quibbling over words. No student of the subject doubts that at some time and somewhere there was one low taxon of animals, perhaps a single species or a few species capable of hybridization, that was ancestral to all the Mammalia. Therefore the Mammalia are monophyletic in that obvious and trivial sense. As descendants of that ancestral unit multiplied and diversified through time, the resulting class would be nominally monophyletic if its name were applied to the ancestral unit itself and nominally polyphyletic if a line were drawn across the evolving lineages at a later time, a matter necessarily of practical interest but of no significance as regards the phylogenetic pattern. Nevertheless there are significant problems and questions.

Has the ancestral unit of monophyletic development of the Mammalia been found and identified ? Obviously not.

Can it be ? Certainly one should never consider any discovery impossible, but this one is extremely improbable. The group we want, a species or even a small genus, was probably limited geographically and ecologically. The chance of preservation and discovery must be slight. The chance of correct and precise recognition would be slighter still, for that would require distinguishing this one group from others, collateral relatives and close ancestors and descendants, closely similar to it.

Is it possible to define a monophyletic group Mammalia including the unknown one ancestral unit and all its descendants but no other species ? I submit that this also is obviously impossible and that this concept of monophyly as applied to taxonomy and nomenclature is simply quixotic.

Do descendants of the common ancestry of all the animals we call Mammalia also include animals we do not call Mammalia ? This is a meaningful question and a difficult one, perhaps impossible to answer at present with complete confidence but a reasonable opinion is possible. My opinion is that the probable answer, on the basis of animals now known, is 'no!' In that case the Mammalia of present usage are phylogenetically a monophyletic group, however we might have to deal with them in the practice of classification.

Have some typically or, in current usage, diagnostically mammalian features arisen independently within separate lineages referred to the Mammalia ? This is also meaningful, and difficult but open to investigation. The answer depends on what particular characters we consider typical or diagnostic. For many commonly so considered the answer is clearly 'yes!', for example the origin of a dentary-squamosal joint or the loss of accessory (non-dentary) bones in the mandible. If, then, we do take such a character as diagnostic, the resulting group is polyphyletic, but only nominally so. That was true of my former definition, which I no longer accept in its too precise form (Simpson, 1959, 1960).

Did some of the groups we call mammals arise separately from ancestors we call reptiles or would call 'reptiles' if we knew them ? This is what is usually meant when the question of monophyly versus polyphyly is discussed. It is what I meant, although I was not very clear about it, when I endorsed the concept of polyphyly of the Mammalia more than 40 years ago, and I believe it is what Hopson and Crompton now mean when they seek a simplist, unified, in *that* sense monophyletic, overall phylogeny starting from a proximate common ancestor of the Morganucodontidae and Kuehneotheriidae. In this sense I am still unconvinced. I think the simplist hypothesis may be simplistic. There are four major groups that have not yet been shown to have had a common ancestor outside of the Therapsida or nominal Reptilia:

Monotremata
Multituberculata
Triconodonta-Docodonta
Theria *sensu lato*

Much depends on how the taxon Mammalia is defined. It is interesting and significant that in this symposium no one attempted a formal definition of the class, in the usual sense of 'definition,' and the problem was no more than mentioned. Nevertheless there was no dissent as to what is meant by the word 'mammal' or the taxon named Mammalia. A definition was given, in effect, by simple enumeration of the members of the set or taxon, which are the four groups just named with the contents previously indicated, no more and no less. A morphological definition could be derived from the characters of all these groups, but it can no longer usefully be given in simple typological terms, such as that a mammal is a vertebrate with only one bone in the lower jaw, with a dentary-squamosal joint, with a synapsid temporal region, or with a differentiated diphyodont dentition.

CLASSIFICATION

Translation of still highly hypothetical concepts of affinities into an evolutionary classification depends for its broadest features on how far one is willing to go with the simplest hypotheses. If one is willing to accept only such ordinal affinities as seem reasonably established by evidence so far attained and made public, say those with a subjective confidence or probability rating greater than 50%, I believe that one would have to recognize the four groups just enumerated as co-ordinate subclasses. That has been a usual arrangement for a long time, at least since my memoirs of the 1920's, with variations as to nomenclature and some contents of the major groups. It was for a time supported by the Kermacks and their associates (e.g. Kermack & Mussett, 1958) and is retained in essence in the useful and authoritative recent compilation by Butler *et al.* (1967).

The following arrangement, although not quite like any previous one, is an updating of that concept, with the nomenclature that I now prefer although no other student of the subject may agree with every word of it.

Class Mammalia
  Subclass Prototheria
      Order Monotremata
  Subclass Allotheria
      Order Multituberculata (many subtaxa)
        ?Multituberculata *incertae sedis:* Haramiyidae
  Subclass Eotheria
      Order Triconodonta (Sinoconodontidae, Morganucodontidae, Amphilestidae, Triconodontidae)
      Order Docodonta (Docodontidae only)
  Subclass Theria
    Infraclass Patriotheria
      Order Symmetrodonta (Kuehneotheriidae, Amphidontidae, Spalacotheriidae)
      Order Pantotheria (Amphitheriidae, Paurodontidae, Dryolestidae, Aegialodontidae)
    Infraclass Metatheria
      Order Marsupialia, or orders various (many subtaxa)
    Infraclass Eutheria
      Orders and subtaxa many

The concept of a basic dichotomy developed by the Kermacks and their associates (e.g. Kermack, 1967) and adopted also by Crompton and others (e.g. Hopson & Crompton, 1969) need involve only the demotion of the first three subclasses of the preceding arrangement to infraclasses and their union into a single subclass. Kermack (1967) has pointed out that this subclass would require a name and that Prototheria (in a greatly extended sense) would have priority, but he prefers to await confirmation of this major grouping and then to have a new name applied to it. If this dichotomous arrangement were to be adopted, I would not object to applying the name Prototheria to the whole group now tentatively designated as simply 'nontherian mammals.'

ODONTOLOGY

For many years discussions of Mesozoic mammals were largely concerned with the homologies of the cusps, and to some extent also of the other topographic features of their cheek teeth. Ten years ago I reviewed that subject in a somewhat skeptical way (Simpson, 1961). It is interesting that hardly any mention of it was made in the present symposium, and it seemed that most of the participants were taking it for granted that the essential problems had been solved—this in spite of the presence of Vandebroek, whose views have been widely different from those apparently being taken for granted by other participants.

One part of the problem may be considered as essentially solved, that of symmetrodont-pantothere molar cusp homologies. Further comparisons, especially of *Kuehneotherium*, leave some but only a little question that Patterson's views (1956, diagrammatically summarized in Simpson, 1961: 80) were probably correct. There is also at least a reasonable hypothesis as to homologies among Docodonta-Triconodonta-Symmetrodonta (especially Hopson & Crompton, 1969), although these seem to me worthy of much less confidence.

The main problem all along has been that although the lower molars are not markedly different, all known upper molars of pantotheres are quite unlike those of the tribosphenic dentitions of primitive Metatheria and Eutheria. The almost generally accepted solution now is that the prominent lingual cusp of pantotheres is homologous with the tribosphenic paracone, or in one version, is an amphicone, ancestral to both paracone and metacone of tribosphenic molars. The tribosphenic protocone would then be a neomorph. However, the extremely scanty known Lower to Middle Cretaceous Theria do not demonstrate this transition, and the belief that it was indicated by basining of the talonid and proliferation on it from one to three cusps is now apparently no longer valid. I therefore believe that this problem is not yet solved and will require further discoveries for a solution.

A more recent and, I think, on the whole presently more interesting development in odontology involves a more functional attitude toward dental morphology. Interest in genetic-phylogenetic studies focused attention on the unworn coronal pattern. However, in many (not all) mammals appreciable change occurs through wear on that pattern or its complete obliteration in a normal lifetime. The functional condition is therefore not just the unworn one, but the usually progressively changing condition through the normal life span. Even the genetic point of view requires modification, because natural selection will certainly favour dental morphology that is effective not only in completely unused condition but also, indeed particularly, in stages of use.

With appreciation of those facts, new methods of study have come to the fore, and especially two. One is increased and more detailed attention to wear on teeth and especially the resulting facets and striae. These were not ignored even in the 19th century, but the beginning of more careful technical study can be credited to Butler (e.g. 1952, 1961; Butler & Mills, 1959). Increasingly detailed studies, now with special reference to Mesozoic mammals, are due especially to Crompton and his associates (e.g., Crompton & Jenkins, 1967; Hopson & Crompton, 1969; Jenkins, 1969). There have also been increasing studies of dentitions actually in use, not only by occlusion patterns or empirical juxtaposition of inert dentitions, especially by still and motion pictures (Crompton & Hiiemäe, 1969*a*,*b*, 1970). Direct observations can be made only on living animals, but the results are applicable also to tribosphenic Cretaceous Theria. Indirect application to nontribosphenic mammals is more dubious, although a good start was made by Crompton & Hiiemäe (1970) and by Crompton in this symposium.

Another point, known from pre-scientific times but lately coming into fuller appreciation, is that dentitions are multipurpose organs, used not only for seizing and comminuting food (themselves two quite different functions), but also for grooming, fighting (both offensive and defensive), display, sex play, excavation, gnawing and chopping, and indeed many other particular functions. Every (e.g. 1960) has usefully attracted continued attention to this fact and to the longitudinal (lifetime) study of dentition by emphasizing one aspect which he calls 'thegosis': the sharpening of some teeth by wear in connection with their aggressive use. In this symposium Kühne has applied this special concept to some Mesozoic mammals. The concept was greeted with enthusiasm by some participants, by others welcomed but with considerable doubt as to whether the analysis of wear as bimodal is either clear-cut or sufficient.

ENVOI

I have touched on most of the themes of the symposium and a few others. The contributions by others speak for themselves, and while I trust that I have not done badly by them I have tried to review much of the subject in my own way rather than theirs. It is obviously true that I have not as yet even mentioned some pertinent aspects of the study of mammalian origins and early history. For example the background contribution here by Hopson on tooth replacement in some cynodonts is highly relevant, and attention might have been drawn to increasing knowledge of replacement in various Mesozoic mammals, for example in multituberculates by Szalay (1965). Some other studies on particular anatomical features may also at least be exemplified by citations: Barghusen & Hopson (1970) on the dentary-squamosal joint; Jenkins (1970) on the 'prototherian' (his quotes) level of postcranial anatomy; McKenna (1961) on the shoulder girdle of multituberculates; Moss & Kermack (1963) on Triassic mammal enamel.

Many years ago (in July 1926, to be exact) I published a somewhat callow attempt at ecological study of a fauna including Mesozoic mammals. Only quite recently has there been a significant return to that approach, with greatly improved knowledge of both Mesozoic mammals and ecology, notably by Estes & Berberian (1970) and by Clemens in this symposium.

Finally, attention should be particularly called to other recent general summaries of this subject, some already cited and some not: Hopson (1967, 1969); Kermack (1967); Parrington, 1967; Piveteau (1961); Vandebroek, 1964.

REFERENCES

BARGHUSEN, H. R. & HOPSON, J. A., 1970. Dentary-squamosal joint and the origin of mammals. *Science, N.Y.*, **168**: 573–575.

BUTLER, P. M., 1952. The milk-molars of Perissodactyla, with remarks on molar occlusion. *Proc. zool. Soc. Lond.*, **121**: 777–817.

BUTLER, P. M., 1961. Relationships between upper and lower molar patterns. International colloquium on the evolution of lower and non-specialized mammals. *Kon. Vlaamse Acad. Wetensch. Lett. Sch. Kunsten België*, Part 1, pp. 115–126. Brussels.

BUTLER, P. M., CLEMENS, W. A., GRAHAM, S. F., HOOIJER, D. A., KERMACK, K. A., PATTERSON, B., RIDE, W. D. L., RUSSELL, D. E., SAVAGE, R. J. G., SIMONS, E. L., TARLOW, L. B. H., THALER, L. & WHITWORTH, T., 1967. Mammalia. In W. G. Harland *et al.* (Eds), *The fossil record*, pp. 763–787. London: Geol. Soc. [It appears that the section on Mesozoic Mammalia is the work of Clemens, Kermack, Patterson, Ride, Russell & Tarlow.]

BUTLER, P. M. & MILLS, J. R. E., 1959. A contribution to the odontology of *Oreopithecus*. *Bull. Br. Mus. nat. hist.*, (*Geol.*), **4**: 1–25.

CASAMIQUELA, R. M., 1961. Sobre la presencia de un mamífero en el primer elenco (icnológico) de vertebrados del Jurásico de la Patagonia. *Physis*, **22**: 225–233.

CLEMENS, W. A., 1963*a*. Wealden mammalian fossils. *Palaeontology*, **6**: 55–69.

CLEMENS, W. A., 1963*b*. Late Jurassic mammalian fossils in the Sedgwick Museum, Cambridge. *Palaeontology*, **6**: 373–377.

CLEMENS, W. A., 1964. Fossil mammals of the type Lance formation, Wyoming, Part I, introduction and Multituberculata. *Univ. Calif. Publs geol. Sci.*, **48**: 1–105.

CLEMENS, W. A., 1966. Fossil mammals of the type Lance formation. Wyoming. Part II, Marsupialia. *Univ. Calif. Publs geol. Sci.*, **62**: 1–122.

CLEMENS, W. A., 1971. Mammalian evolution in the Cretaceous. In D. M. Kermack & K. A. Kermack (Eds), *Early Mammals. Zool. J. Linn. Soc.*, **50**, Suppl. 1: 165–180.

CROMPTON, A. W., 1964. A preliminary description of a new mammal from the upper Triassic of South Africa. *Proc. zool. Soc. Lond.*, **142**: 441–452.

CROMPTON, A. W. & HIIEMÄE, K., 1969*a*. Functional occlusion in tribosphenic molars. *Nature, Lond.*, **222**: 678–679.

CROMPTON, A. W. & HIIEMÄE, K., 1969*b*. How mammalian molar teeth work. *Discovery, Lond.*, **5**: 23–34.

CROMPTON, A. W. & HIIEMÄE, K., 1970. Molar occlusion and mandibular movements during occlusion in the American opossum, *Didelphis marsupialis* L. *Zool. J. Linn. Soc.*, **49**: 21–47.

CROMPTON, A. W. & JENKINS, F. A., JR., 1967. American Jurassic symmetrodonts and Rhaetic 'pantotheres.' *Science, N.Y.*, **155**: 1006–1009.

CROMPTON, A. W. & JENKINS, F. A., JR., 1968. Molar occlusion in late Triassic mammals. *Biol. Rev.*, **43**: 427–458.

ESTES, R. & BERBERIAN, P., 1970. Paleoecology of a later Cretaceous community from Montana. *Breviora*, **343**: 1–35.

EVERY, R. G., 1960. The significance of extreme mandibular movements. *Lancet*, 2 July 1960: 37–39.

GRAMBAST, L., MARTINEZ, M., MATTAUER, M. & THALER, L., 1967. *Perutherium altiplanense* nov. gen., nov. sp., premier mammifère mésozoique d'Amérique du Sud. *C. r. hebd. Séanc. Acad. Sci., Paris*, **264**: 707–710.

HAHN, G., 1969. Beiträge zue Fauna der Grube Guimarota nr. 3. Die Multituberculata. *Palaeontographica*, **133**, A-1-100.

HENKEL, S., 1966. Methoden zur Prospektion und Gewinnung kleiner Wirbeltier-Fossilien. *Neuer. Jb. Geol. Paläont. Mh.*, **3**: 178–184.

HENKEL, S. & KREBS, B., 1969. Zwei Saügetier-Unterkiefer aus der Unteren Kreide von Uña (Prov. Cuenca, Spanien), *Neues Jb. Geol. Paläont. Mh.*, **8**: 449–463.

HOPSON, J. A., 1967. Mammal-like reptiles and the origin of mammals. *Discovery*, **2**: 25–33.

HOPSON, J. A., 1969. The origin and adaptive radiation of mammal-like reptiles and non-Therian mammals. *Ann. N. Y. Acad. Sci.*, **167**: 199–216.

HOPSON, J. A. & CROMPTON, H. W., 1969. Origin of mammals. *Evolutionary Biology*, **3**: 15–72.

JENKINS, F. A., JR., 1969. Occlusion in *Docodon* (Mammalia, Docodonta). *Postilla*, **139**: 1–24.

JENKINS, F. A., JR., 1970. Cynodont postcranial anatomy and the 'prototherian' level of mammalian organization. *Evolution, Lancaster, Pa.*, **24**: 230–252.

KERMACK, D. M., KERMACK, K. A. & MUSSETT, F., 1968. The Welsh pantothere *Kuehneotherium praecursoris*. *J. Linn. Soc. (Zool.)*, **47**: 407–423.

KERMACK, K. A., 1965. Reconstructing one of our earliest ancestors: a triconodont. *Illustrated London News*, **227**: 1065.

KERMACK, K. A., 1967. The interrelations of early mammals. *J. Linn. Soc. (Zool.)*, **47**: 241–249.

KERMACK, K. A., LEES, P. M. & MUSSETT, F., 1965. *Aegialodon dawsoni*, a new trituberculo-sectorial tooth from the lower Wealden. *Proc. R. Soc.*, (Ser. B), **162**: 535–554.

KERMACK, K. A. & MUSSETT, F., 1958. The jaw articulation of the Docodonta and the classification of Mesozoic mammals. *Proc. R. Soc.*, (Ser. B), **148**: 204–215.

KERMACK, K. A. & MUSSETT, F., 1959. The first mammals. *Discovery, Lond.*, **20**: 144–151.

KIELAN-JAWOROWSKA, Z., 1968. Preliminary data on the upper Cretaceous eutherian mammals from Bayn Dzak, Gobi Desert. *Palaeont. pol.*, **19**: 171–191.

KIRSCH, J. A. W., 1968. Prodromus of the comparative serology of Marsupialia. *Nature, Lond.*, **217**: 418–420.

KRETZOI, M., 1960. Zur Benennung des ältesten Symmetrodonten. *Vertebr. hung.*, **2**: 307–309.

KRUSAT. G., 1969. Ein Pantotheria-Molar mit dreispitzigem Talonid aus dem Kimmeridge von Portugal. *Paläont. Z.*, **43**: 52–56.

KÜHNE, W. G., 1947. The geology of the fissure-filling 'Holwell 2': the age-determination of the mammalian teeth therein; and a report on the technique employed when collecting the teeth of *Eozostrodon* and Microcleptidae. *Proc. zool. Soc. Lond.*, **116**: 729–733.

KÜHNE, W. G., 1949. On a triconodont tooth of a new pattern from a fissure-filling in South Glamorgan. *Proc. zool. Soc. Lond.*, **119**: 345–350.

KÜHNE, W. G., 1950. A symmetrodont tooth from the Rhaeto-Lias. *Nature, Lond.*, **166**: 696.

KÜHNE, W. G., 1968. Kimeridge [sic!] mammals and their bearing on the phylogeny of the Mammalia. In E. J. Drake (Ed.), *Evolution and environment*, pp. 109–123. New Haven: Yale University Press.

KÜHNE, W. & CRUSAFONT PAIRÓ, J., 1969. Mamíferos del Wealdiense de Uña, cerca de Cuenca. *Acta geol. Hispan.*, **3**: 133–134.

LEDOUX, J., HARTENBERGER, J., MICHAUX, J., SUDRE, J. & THALER, L., 1966. Découverte d'un mammifère dans le Cretacé supérieur á dinosaures de Champ-Garimond prés de Fons (Gard). *C. r. hebd. séanc. Acad, Sci., Paris*, **362**: 1925–1928.

LEES, P. M., 1964. A flotation method of obtaining mammal teeth from Mesozoic bone-beds. *Curator*, **7**: 300–306.

LILLEGRAVEN, J. A., 1969. Latest Cretaceous mammals of upper part of Edmonton formation of Alberta, Canada, and review of marsupial-placental dichotomy in mammalian evolution. *Paleont. Contr. Univ. Kans.*, **50**: 1–122.

MACINTYRE, G., 1967. Foramen pseudovale and quasi-mammals. *Evolution, Lancaster, Pa.*, **21**: 834–841.

MCKENNA, M. C., 1961. On the shoulder girdle of the mammalian class Allotheria. *Am. Mus. Novit.*, **2066**: 1–27.

MILLS, J. R. E., 1964. The dentitions of *Peramus* and *Amphitherium*. *Proc. Linn. Soc. Lond.*, **175**: 117–133.

MOSS, M. L. & KERMACK, K. A., 1967. Enamel structure in two Triassic mammals. *J. dent. Res.*, **46**: 745–747.

PARRINGTON, F. R., 1941. On two mammalian teeth from the lower Rhaetic of Somerset. *Ann. Mag. nat. Hist.*, **8** (11): 140–144.

PARRINGTON, F. R., 1947. On a collection of Rhaetic mammalian teeth. *Proc. zool. Soc. Lond.*, **116**: 707–728.

PARRINGTON, F. R., 1967. The origins of mammals. *Advmt Sci., Lond.*, **24**: 165–173.

PATTERSON, B., 1956. Early Cretaceous mammals and the evolution of mammalian molar teeth. *Fieldiana, Geol.*, **13**: 1–105.

PATTERSON, B. & OLSON, E. C., 1961. A triconodontid mammal from the Triassic of Yunnan. International colloquium on the evolution of lower and non-specialized mammals. *Kon. Vlaamse Acad. Wetensch., Lett. Sch. Kunsten België*, Part 1, pp. 129–191. Brussels.

PEYER, B., 1956. Über Zähne von Haramyiden, von Triconodonten und von wahrscheinlich synapsiden Reptilien aus dem Rhät von Hallau. *Schweiz. palaeont. Abh.*, **72**: 1–72.

PIVETEAU, J., 1961. [Sections on Mesozoic mammals.] In J. Piveteau (Ed.), *Traité de paleóntologie*, **6** (1): 352–353, 523–531, 536–584. Paris: Masson.

RIDE, W. D. L., 1964. A review of Australian fossil marsupials. *J. Proc. R. Soc. West. Aust.*, **47**: 97–131.

RIGNEY, H. W., 1963. A specimen of *Morganucodon* from Yunnan. *Nature, Lond.*, **197**: 1122.

ROMER, A. S., 1969. Cynodont reptile with incipient mammalian jaw articulation. *Science, N.Y.*, **166**: 881–882.

SABAN, R., 1958. Insectivora. In J. Piveteau (Ed.), *Traité de paléontologie*, **6** (2): 822–909. Paris: Masson.

SHIKAMA, T., 1947. *Teilhardosaurus* and *Endotherium*, new Jurassic Reptilia and Mammalia from the Husin coal-field, South Manchuria. *Proc. Japan Acad.*, **23**: 76–84.

SIMPSON, G. G., 1959. Mesozoic mammals and the polyphyletic origin of mammals. *Evolution, Lancaster, Pa.*, **13**: 405–414.

SIMPSON, G. G., 1960. Diagnosis of the classes Reptilia and Mammalia. *Evolution, Lancaster, Pa.*, **14**: 388–392.

SIMPSON, G. G., 1961. Evolution of Mesozoic mammals. International colloquium on the evolution of lower and non-specialized mammals. *Kon. Vlaamse Acad. Wetensch. Lett. Sch. Kunsten België*, Part 1, pp. 57–95. Brussels.

SLAUGHTER, B. H., 1965. A therian from the lower Cretaceous (Albian) of Texas. *Postilla*, **93**: 1–18.

SLAUGHTER, B. H., 1968*a*. Earliest known eutherian mammals and the evolution of premolar occlusion. *Tex. J. Sci.*, **20**: 2–12.

SLAUGHTER, B. H., 1968*b*. Earliest known marsupials. *Science, N.Y.*, **162**: 254–255.

SLOAN, R. E. & VAN VALEN, L., 1965. Cretaceous mammals from Montana. *Science, N.Y.*, **148**: 220–227.

SZALAY, F. S., 1965. First evidence of tooth replacement in the subclass Allotheria (Mammalia). *Am. Mus. Novit.*, **2226**: 1–12.

VANDEBROEK, G., 1964. Recherches sur l'origine des mammifères. *Annls Soc. r. zool. Belg.*, **94**: 117–160.

VAN VALEN, L., 1967. New Paleocene insectivores and insectivore classification. *Bull. Am. Mus. nat., Hist.*, **135**: 217–284.

VAN VALEN, L. & SLOAN, R. E., 1965. The earliest primates. *Science, N.Y.*, **150**: 743–745.

YABE, H. & SHIKAMA, T., 1938. A new Jurassic Mammalia from South Manchuria. *Proc. imp. Acad. Tokyo*, **14**: 353–359.

Index